D0900291

SECOND EDITION

THE 8051 MICROCONTROLLER

I. Scott MacKenzie
University of Guelph
Guelph, Ontario

Prentice Hall, Upper Saddle River, New Jersey 07458

Library of Congress Cataloging-in- Publication Data
MacKenzie, I. Scott
The 8051 microcontroller / I. Scott MacKenzie—2nd ed.
p. cm.
Includes bibliographical references and index.
ISBN 0-02-373660-7
1. Intel 8051 (Computer)—Programming 2. Digital control systems.
I. Title. II. Title: Eight thousand fifty-one
QA76.8.127M23 1995
004.165—-dc20 94-8278
CIP

Cover Photo: **Lester Lefkowitz/Tony Stone Worldwide**
Editor: **Charles E. Stewart, Jr.**
Production Editor: **Stephen C. Robb**
Cover Designer: **Julia Z. Van Hook**
Production Manager: **Pamela D. Bennett**

Simon & Schuster Company / A Viacom Company
Upper Saddle River, New Jersey 07458

Printed in the United States of America
10 9 8 7

ISBN 0-02-373660-7

Prentice-Hall International (UK) Limited, London
Prentice-Hall of Australia Pty. Limited, Sydney
Prentice-Hall Canada Inc., Toronto
Prentice-Hall Hispanoamericana, S.A., Mexico
Prentice-Hall of India Private Limited, New Delhi
Prentice-Hall of Japan, Inc., Tokyo
Simon & Schuster Asia Pte. Ltd., Singapore
Editora Prentice-Hall do Brasil, Ltda., Rio de Janeiro

PREFACE

This book examines the hardware and software features of the MCS-51 family of microcontrollers. The intended audience is college or university students of electronics or computer technology, electrical or computer engineering, or practicing technicians or engineers interested in learning about microcontrollers

The means to effectively fulfill that audience's informational needs were tested and refined in the development of this book. In its prototype form, *The 8051 Microcontroller* has been the basis of a fifth semester course for college students in computer engineering. As detailed in Chapter 10, students build an 8051 single-board computer as part of this course. That computer, in turn, has been used as the target system for a final, sixth semester "project" course in which students design, implement, and document a "product" controlled by the 8051 microcontroller and incorporating original software and hardware.

Since the 8051—like all microcontrollers—contains a high degree of functionality, the book emphasizes architecture and programming rather than electrical details. The software topics are delivered in the context of Intel's assembler (ASM51) and linker/locator (RL51).

It is my view that courses on microprocessors or microcontrollers are inherently more difficult to deliver than courses in, for example, digital systems, because a linear sequence of topics is hard to devise. The very first program that is demonstrated to students brings with it significant assumptions, such as a knowledge of the CPU's programming model and addressing modes, the distinction between an address and the content of an address, and so on. For this reason, a course based on this book should not attempt to follow strictly the sequence presented. Chapter 1 is a good starting point, however. It serves as a general introduction to microcontrollers, with particular emphasis on the distinctions between microcontrollers and microprocessors.

Chapter 2 introduces the hardware architecture of the 8051 microcontroller, and its counterparts that form the MCS-51 family. Concise examples are presented using short sequences of instructions. Instructors should be prepared at this point to introduce, in parallel, topics from Chapters 3 and 7 and Appendices A and C to support the requisite software knowledge in these examples. Appendix A is particularly valuable, since it contains in a single figure the entire 8051 instruction set.

Chapter 3 introduces the instruction set, beginning with definitions of the 8051's addressing modes. The instruction set has convenient categories of instructions (data transfer, branch, etc.) which facilitate a step-wise presentation. Numerous brief examples demonstrate each addressing mode and each type of instruction.

Chapters 4, 5, and 6 progress through the 8051's on-chip features, beginning with the timers, advancing to the serial port (which requires a timer as a baud rate generator), and concluding with interrupts. The examples in these chapters are longer and more complex than those presented earlier. Instructors are wise not to rush into these chapters: it is essential that students gain solid understanding of the 8051's hardware architecture and instruction set before advancing to these topics.

Many of the topics in Chapter 7 will be covered, by necessity, in progressing through the first six chapters. Nevertheless, this chapter is perhaps the most important for developing in students the potential to undertake large-scale projects. Advanced topics such as assemble-time expression evaluation, modular programming, linking and locating, and macro programming will be a significant challenge for many students. At this point the importance of hands-on experience cannot be over-emphasized. Students should be encouraged to experiment by entering the examples in the chapter into the computer and observing the output and error messages provided by ASM51, RL51, and the object-to-hex conversion utility (OH).

Some advanced topics relating to programming methods, style, and the development environment are presented in Chapters 8 and 9. These chapters address larger, more conceptual topics important in professional development environments.

Chapter 10 presents several design examples incorporating selected hardware with supporting software. The software is fully annotated and is the real focus in these examples. The second edition includes two additional interfaces; a digital-to-analog output interface using an MC1408 8-bit DAC, and an analog-to-digital input interface using an ADC0804 8-bit ADC. One of the designs in Chapter 10 is the SBC-51—the 8051 single-board computer. The SBC-51 can form the basis of a course on the 8051 microcontroller. A short monitor program is included (see Appendix G) which is sufficient to get "up and running." A development environment also requires a host computer which doubles as a dumb terminal for controlling the SBC-51 after programs have been downloaded for execution.

Many dozens of students have wire-wrapped prototype versions of the SBC-51 during the years that I have taught 8051-based courses to computer engineering students. Shortly after the release of the first edition of this text, URDA, Inc. (Pittsburgh, Pennsylvania) began manufacturing and marketing a PC-board version of the SBC-51. This has proven to be a cost-effective solution to implementing a complete lecture-plus-lab package for teaching the 8051 microcontroller to technology students. Contact URDA at 1-800-338-0517 for more information.

Finally, each chapter contains questions further exploring the concepts presented. This new edition includes 128 end-of-chapter questions—almost double the number in the first edition. A solutions manual is available to instructors from the publisher.

The book makes extensive use of, and builds on, Intel's literature on the MCS-51 devices. In particular, Appendix C contains the definitions of all 8051 instructions and Appendix E contains the 8051 data sheet. Intel's cooperation is gratefully acknowledged. I also thank the following persons who reviewed the manuscript and offered invaluable comments, criticism, and suggestions: Antony Alumkal, Austin Community College; Omer Farook, Purdue University—Calumet; David Jones, Lenoir Community College; Roy Seigel, DeVry Institute; and Chandra Sekhar, Purdue University—Calumet.

I. Scott MacKenzie

CONTENTS

APPENDIXES

1

INTRODUCTION TO MICROCONTROLLERS

1.1 INTRODUCTION

Although computers have only been with us for a few decades, their impact has been profound, rivaling that of the telephone, automobile, or television. Their presence is felt by us all, whether computer programmers or recipients of monthly bills printed by a large computer system and delivered by mail. Our notion of computers usually categorizes them as "data processors," performing numeric operations with inexhaustible competence.

We confront computers of a vastly different breed in a more subtle context performing tasks in a quiet, efficient, and even humble manner, their presence often unnoticed. As a central component in many industrial and consumer products, we find computers at the supermarket inside cash registers and scales; at home in ovens, washing machines, alarm clocks, and thermostats; at play in toys, VCRs, stereo equipment, and musical instruments; at the office in typewriters and photocopiers; and in industrial equipment such as drill presses and phototypesetters. In these settings computers are performing "control" functions by interfacing with the "real world" to turn devices on and off and to monitor conditions. **Microcontrollers** (as opposed to microcomputers or microprocessors) are often found in applications such as these.

It's hard to imagine the present world of electronic tools and toys without the microprocessor. Yet this single-chip wonder has barely reached its twentieth birthday. In 1971 Intel Corporation introduced the 8080, the first successful microprocessor. Shortly thereafter, Motorola, RCA, and then MOS Technology and Zilog introduced similar devices: the 6800, 1801, 6502, and Z80, respectively. Alone these integrated circuits (ICs) were rather helpless (and they remain so); but as part of a single-board computer (SBC) they became the central component in useful products for learning about and designing with microprocessors. These SBCs, of which the *D2* by Motorola, *KIM-1* by MOS Technology, and *SDK-85* by Intel are the most memorable, quickly found their way into design labs at colleges, universities, and electronics companies.

A device similar to the microprocessor is the microcontroller. In 1976 Intel introduced the 8748, the first device in the MCS-48™ family of microcontrollers. Within a single integrated circuit containing over 17,000 transistors, the 8748 delivered a CPU,

1K byte of EPROM, 64 bytes of RAM, 27 I/O pins, and an 8-bit timer. This IC, and other MCS-48™ devices that followed, soon became an industry standard in control-oriented applications. Replacement of electromechanical components in products such as washing machines and traffic light controllers was a popular application initially, and remains so. Other products where microcontrollers can be found include automobiles, industrial equipment, consumer entertainment products, and computer peripherals. (Owners of an IBM *PC* need only look inside the keyboard for an example of a microcontroller in a minimum-component design.)

The power, size, and complexity of microcontrollers advanced an order of magnitude in 1980 with Intel's announcement of the 8051, the first device in the MCS-51™ family of microcontrollers. In comparison to the 8048, this device contains over 60,000 transistors, 4K bytes ROM, 128 bytes of RAM, 32 I/O lines, a serial port, and two 16-bit timers—a remarkable amount of circuity for a single IC (see Figure 1–1). New members have been added to the MCS-51™ family, and today variations exist virtually doubling these specifications. Siemens Corporation, a second source for MCS-51™ components, offers the SAB80515, an enhanced 8051 in a 68-pin package with six 8-bit I/O ports, 13 interrupt sources, and an 8-bit A/D converter with 8 input channels. The 8051 family is well established as one of the most versatile and powerful of the 8-bit microcontrollers, its position as a leading microcontroller entrenched for years to come.

This book is about the MCS-51™ family of microcontrollers. The following chapters introduce the hardware and software architecture of the MCS-51™ family, and demonstrate through numerous design examples how this family of devices can participate in electronic designs with a minimum of additional components.

In the following sections, through a brief introduction to computer architecture, we shall develop a working vocabulary of the many acronyms and buzz words that prevail

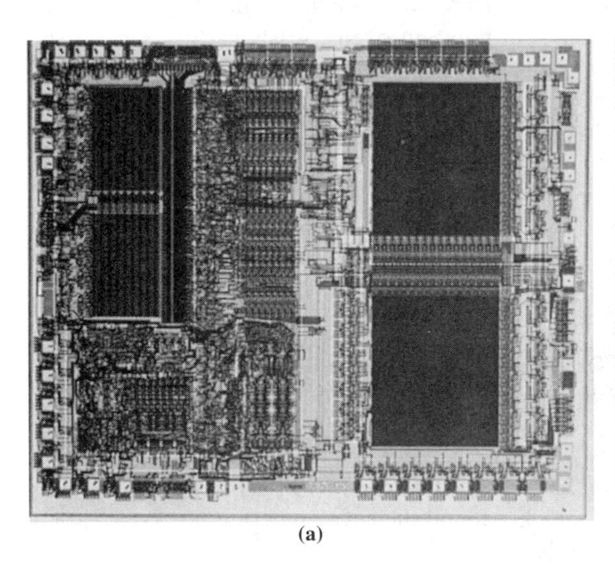

(a)

(b)

FIGURE 1–1
The 8051 microcontroller. (a) An 8051 die. (b) An 8751 EPROM. (Courtesy Intel Corp. Copyright 1991.)

(and often confound) in this field. Since many terms have vague and overlapping definitions subject to the prejudices of large corporations and the whims of various authors, our treatment is practical rather than academic. Each term is presented in its most common setting with a straightforward explanation.

1.2 TERMINOLOGY

To begin, a **computer** is defined by two key traits: (1) the ability to be programmed to operate on data without human intervention, and (2) the ability to store and retrieve data. More generally, a **computer system** also includes the **peripheral devices** for communicating with humans, as well as **programs** that process data. The equipment is **hardware,** the programs are **software.** Let's begin with computer hardware by examining Figure 1–2.

The absence of detail in the figure is deliberate, making it representative of all sizes of computers. As shown, a computer system contains a **central processing unit** (CPU) connected to **random access memory** (RAM) and **read-only memory** (ROM) via the **address bus, data bus,** and **control bus. Interface circuits** connect the system buses to **peripheral devices.** Let's discuss each of these in detail.

1.3 THE CENTRAL PROCESSING UNIT

The CPU, as the "brain" of the computer system, administers all activity in the system and performs all operations on data. Most of the CPU's mystique is undeserved, since it is just a collection of logic circuits that continuously performs two operations: fetching

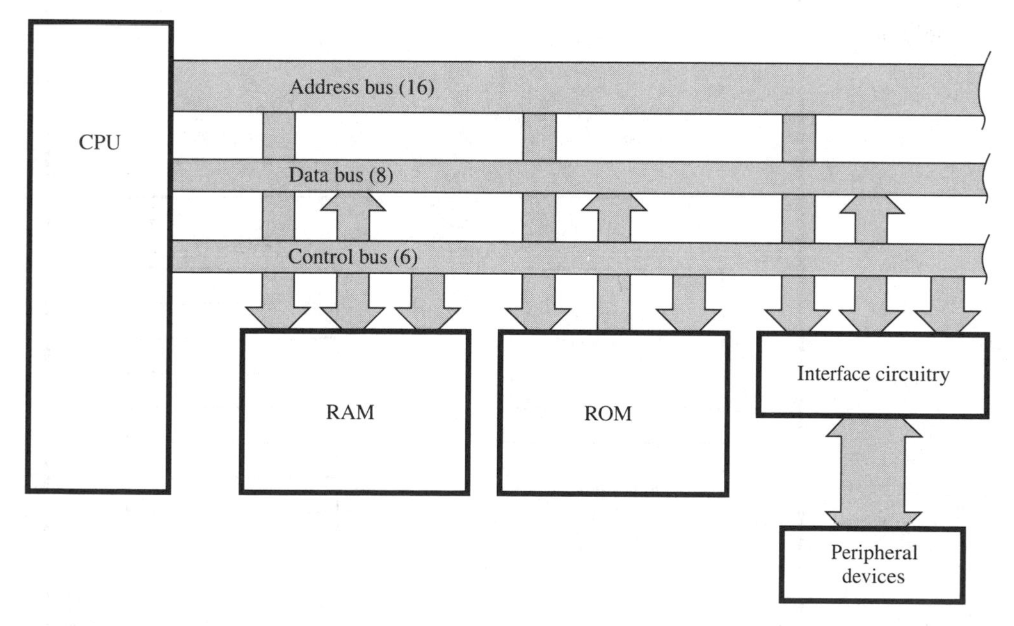

FIGURE 1–2
Block diagram of a microcomputer system

instructions and executing instructions. The CPU has the ability to understand and execute instructions based on a set of binary codes, each representing a simple operation. These instructions are usually arithmetic (add, subtract, multiply, divide), logic (AND, OR, NOT, etc.), data movement, or branch operations, and are represented by a set of binary codes called the **instruction set.**

Figure 1–3 is an extremely simplified view of the inside of a CPU. It shows a set of **registers** for the temporary storage of information, an **arithmetic and logic unit** (ALU) for performing operations on this information, an **instruction decode and control unit** that determines the operation to perform and sets in motion the necessary actions to perform it, and two additional registers. The **instruction register** (IR) holds the binary code for each instruction as it is executed, and the **program counter** (PC) holds the memory address of the next instruction to be executed.

Fetching an instruction from the system RAM is one of the most fundamental operations performed by the CPU. It involves the following steps: (a) the contents of the program counter are placed on the address bus, (b) a READ control signal is activated, (c) data (the instruction opcode) are read from RAM and placed on the data bus, (d) the opcode is latched into the CPU's internal instruction register, and (e) the program counter is incremented to prepare for the next fetch from memory. Figure 1–4 illustrates the flow of information for an instruction fetch.

The execution stage involves decoding (or deciphering) the opcode and generating control signals to gate internal registers in and out of the ALU and to signal the ALU to perform the specified operation. Due to the wide variety of possible operations, this explanation is somewhat limited in scope. It applies to a simple operation such as "incre-

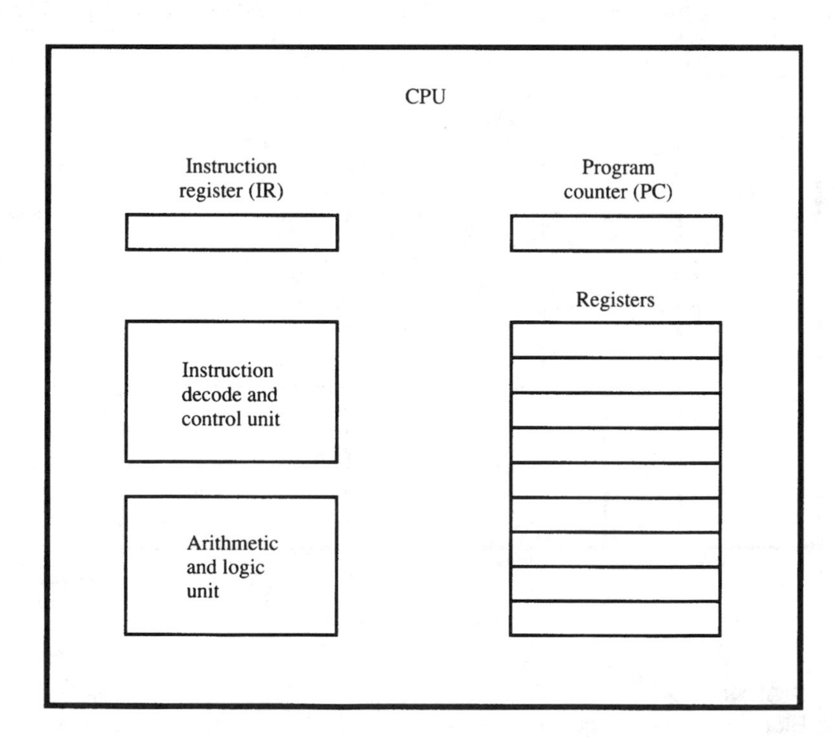

FIGURE 1–3
The central processing unit (CPU)

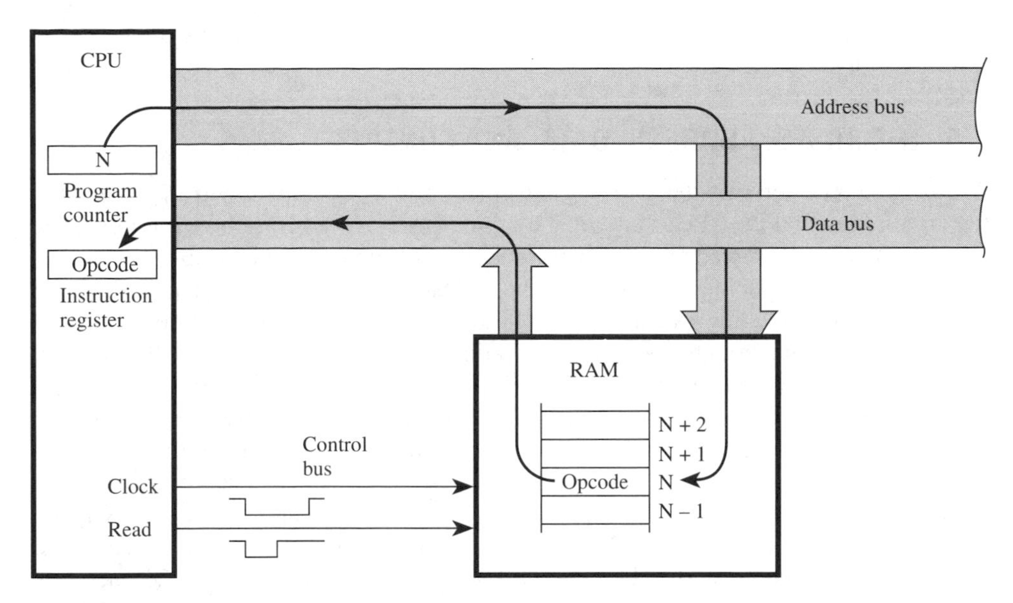

FIGURE 1–4
Bus activity for an opcode fetch cycle

ment register." More complex instructions require more steps, such as reading a second and third byte as data for the operation.

A series of instructions combined to perform a meaningful task is called a **program,** or **software,** and herein is the real mystique. The degree to which tasks are efficiently and correctly carried out is determined for the most part by the quality of software, not by the sophistication of the CPU. Programs, then, "drive" the CPU, and in doing so they occasionally go amiss, mimicking the frailties of their authors. Phrases such as "The computer made a mistake" are misguided. Although equipment breakdowns are inevitable, mistakes in results are usually a sign of poor programs or operator error.

1.4 SEMICONDUCTOR MEMORY: RAM AND ROM

Programs and data are stored in memory. The variations of computer memory are so vast, their accompanying terms so plentiful, and technology breakthroughs so frequent, that extensive and continual study is required to keep abreast of the latest developments. The memory devices directly accessible by the CPU consist of semiconductor ICs (integrated circuits) called RAM and ROM. There are two features that distinguish RAM and ROM: first, RAM is read/write memory while ROM is read-only memory; and second, RAM is volatile (the contents are lost when power is removed), while ROM is nonvolatile.

Most computer systems have a disk drive and a small amount of ROM, just enough to hold the short, frequently used software routines that perform input/output operations. User programs and data are stored on disk and are loaded into RAM for execution. With the continual drop in the per-byte cost of RAM, small computer systems often contain millions of bytes of RAM.

1.5 THE BUSES: ADDRESS, DATA, AND CONTROL

A **bus** is a collection of wires carrying information with a common purpose. Access to the circuitry around the CPU is provided by three buses: the **address bus, data bus,** and **control bus.** For each read or write operation, the CPU specifies the location of the data (or instruction) by placing an address on the address bus, and then activates a signal on the control bus indicating whether the operation is a read or write. Read operations retrieve a byte of data from memory at the location specified and place it on the data bus. The CPU reads the data and places it in one of its internal registers. For a write operation, the CPU outputs data on the data bus. Because of the control signal, memory recognizes the operation as a write cycle and stores the data in the location specified.

Most small computers have 16 or 20 address lines. Given n address lines, each with the possibility of being high (1) or low (0), 2^n locations can be accessed. A 16-bit address bus, therefore, can access $2^{16} = 65{,}536$ locations, and a 20-bit address can access $2^{20} = 1{,}048{,}576$ locations. The abbreviation K (for kilo) stands for $2^{10} = 1024$, therefore 16 bits can address $2^6 \times 2^{10} = 64$K locations, while 20 bits can address 1024K (or 1 Meg) locations.

The data bus carries information between the CPU and memory or between the CPU and I/O devices. Extensive research effort has been expended in determining the sort of activities that consume a computer's valuable execution time. Evidently computers spend up to two-thirds of their time simply moving data. Since the majority of move operations are between a CPU register and external RAM or ROM, the number of lines (the width) of the data bus is important for overall performance. This limitation-by-width is a bottleneck: There may be vast amounts of memory on the system, and the CPU may possess tremendous computational power, but access to the data—data movement between the memory and CPU via the data bus—is bottlenecked by the width of the data bus.

This trait is so important that it is common to add a prefix indicating the extent of this bottleneck. The phrase "16-bit computer" refers to a computer with 16 lines on its data bus. Most computers fit the 4-bit, 8-bit, 16-bit, or 32-bit classification, with overall computing power increasing as the width of the data bus increases.

Note that the data bus, as shown in Figure 1–2, is bidirectional and the address bus is unidirectional. Address information is always supplied by the CPU (as indicated by the arrow in Figure 1–2), yet data may travel in either direction depending on whether a read or write operation is intended.[1] Note also that the term "data" is used in a general sense: the "information" that travels on the data bus may be the instructions of a program, an address appended to an instruction, or the data used by the program.

The control bus is a hodgepodge of signals, each having a specific role in the orderly control of system activity. As a rule, control signals are timing signals supplied by the CPU to synchronize the movement of information on the address and data buses. Al-

[1]Address information is sometimes also provided by direct memory access (DMA) circuitry (in addition to the CPU).

though there are usually three signals, such as CLOCK, READ, and WRITE, for basic data movement between the CPU and memory, the names and operation of these signals are highly dependent on the specific CPU. The manufacturer's data sheets must be consulted for details.

1.6 INPUT/OUTPUT DEVICES

I/O devices, or "computer peripherals," provide the path for communication between the computer system and the "real world." Without these, computer systems would be rather introverted machines, of little use to the people who use them. Three classes of I/O devices are **mass storage, human interface,** and **control/monitor.**

1.6.1 Mass Storage Devices

Like semiconductor RAMs and ROMs, mass storage devices are players in the arena of memory technology—constantly growing, ever improving. As the name suggests, they hold large quantities of information (programs or data) that cannot fit into the computer's relatively small RAM or "main" memory. This information must be loaded into main memory before the CPU accesses it. Classified according to ease of access, mass storage devices are either **online** or **archival.** Online storage, usually on magnetic disk, is available to the CPU without human intervention upon the request of a program, and archival storage holds data that are rarely needed and require manual loading onto the system. Archival storage is usually on magnetic tapes or disks, although optical discs, such as CD-ROM or WORM technology, are now emerging and may alter the notion of archival storage due to their reliability, high capacity, and low cost.[2]

1.6.2 Human Interface Devices

The union of man and machine is realized by a multitude of human interface devices, the most common being the video display terminal (VDT) and printer. Although printers are strictly output devices that generate hardcopy output, VDTs are really two devices, since they contain a keyboard for input and a CRT (cathode-ray tube) for output. An entire field of engineering, called "ergonomics" or "human factors," has evolved from the necessity to design these peripheral devices with humans in mind, the goal being the safe, comfortable, and efficient mating of the characteristics of people with the machines they use. Indeed, there are more companies that manufacture this class of peripheral device than companies that manufacture computers. For most computer systems, there are at least three of these devices: a keyboard, CRT, and printer. Other human interface devices include the joystick, light pen, mouse, microphone, or loudspeaker.

1.6.3 Control/Monitor Devices

By way of control/monitor devices (and some meticulously designed interface electronics and software), computers can perform a myriad of control-oriented tasks, and per-

[2]"CD-ROM" stands for compact-disc read-only memory. "WORM" stands for write-once read-mostly. A CD-ROM contains 550 Mbytes of storage, enough to store the entire 32 volumes of the Encyclopedia Britannica.

form them unceasingly, without fatigue, far beyond the capabilities of humans. Applications such as temperature control of a building, home security, elevator control, home appliance control, and even welding parts of an automobile, are all made possible using these devices.

Control devices are outputs, or **actuators,** that can affect the world around them when supplied with a voltage or current (e.g., motors and relays). Monitoring devices are inputs, or **sensors,** that are stimulated by heat, light, pressure, motion, etc., and convert this to a voltage or current read by the computer (e.g., phototransistors, thermistors, and switches). The interface circuitry converts the voltage or current to binary data, or vice versa, and through software an orderly relationship between inputs and outputs is established. The hardware and software interfacing of these devices to microcontrollers is one of the main themes in this book.

1.7 PROGRAMS: BIG AND SMALL

The preceding discussion has focused on computer systems hardware with only a passing mention of the programs, or software, that make them work. The relative emphasis placed on hardware versus software has shifted dramatically in recent years. Whereas the early days of computing witnessed the materials, manufacturing, and maintenance costs of computer hardware far surpassing the software costs, today, with mass-produced LSI (large-scale integrated) chips, hardware costs are less dominant. It is the labor-intensive job of writing, documenting, maintaining, updating, and distributing software that constitutes the bulk of the expense in automating a process using computers.

Let's examine the different types of software. Figure 1–5 illustrates three levels of software between the user and the hardware of a computer system: the **application software,** the **operating system,** and the **input/output subroutines.**

FIGURE 1–5
Levels of software

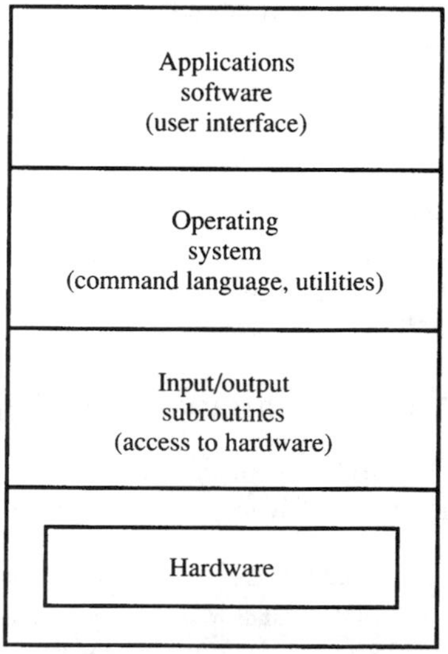

At the lowest level, the input/output subroutines directly manipulate the hardware of the system, reading characters from the keyboard, writing characters to the CRT, reading blocks of information from the disk, and so on. Since these subroutines are so intimately linked to the hardware, they are written by the hardware designers and are (usually) stored in ROM. (They are the BIOS—basic input/output system—on the IBM *PC*, for example.)

To provide close access to the system hardware for programmers, explicit entry and exit conditions are defined for the input/output subroutines. One only needs to initialize values in CPU registers and call the subroutine; the action is carried out with results returned in CPU registers or left in system RAM.

As well as a full complement of input/output subroutines, the ROM contains a start-up program that executes when the system is powered up or reset manually by the operator. The nonvolatile nature of ROM is essential here since this program must exist upon power-up. "Housekeeping" chores, such as checking for options, initializing memory, performing diagnostic checks, etc., are all performed by the start-up program. Last, but not least, a **bootstrap loader** routine reads the first track (a small program) from the disk into RAM and passes control to it. This program then loads the RAM-resident portion of the operating system (a large program) from the disk and passes control to it, thus completing the start-up of the system. There is a saying that "the system has pulled itself up by its own bootstraps."

The operating system is a large collection of programs that come with the computer system and provide the mechanism to access, manage, and effectively utilize the computer's resources. These abilities exist through the operating system's **command language** and **utility programs,** which in turn facilitate the development of applications software. If the applications software is well designed, the user interacts with the computer with little or no knowledge of the operating system. Providing an effective, meaningful, and safe user interface is one of the prime objectives in the design of applications software.

1.8 MICROS, MINIS, AND MAINFRAMES

Using this as a starting point, we classify computers by their size and power as microcomputers, minicomputers, or mainframe computers. A key trait of microcomputers is the size and packaging of the CPU: It is contained within a single integrated circuit—a **microprocessor.** On the other hand, minicomputers and mainframe computers, as well as being more complex in every architectural detail, have CPUs consisting of multiple ICs, ranging from several ICs (minicomputers) to several circuit boards of ICs (mainframes). This is necessary to achieve the high speeds and computational power of larger computers.

Typical microcomputers such as the IBM *PC,* Apple *Macintosh,* and Commodore *Amiga* incorporate a microprocessor as their CPU. The RAM, ROM, and interface circuits require many ICs, with the component count often increasing with computing power. Interface circuits vary considerably in complexity depending on the I/O devices. Driving the loudspeaker contained in most microcomputers, for example, requires only a couple of logic gates. The disk interface, however, usually involves many ICs, some in LSI packages.

Another feature separating micros from minis and mainframes is that microcomputers are single-user, single-task systems—they interact with one user, and they execute one program at a time. Minis and mainframes, on the other hand, are multiuser, multitasking systems—they can accommodate many users and programs simultaneously. Actually, the simultaneous execution of programs is an illusion resulting from "time slicing" CPU resources. (Multiprocessing systems, however, use multiple CPUs to execute tasks simultaneously.)

1.9 MICROPROCESSORS VS. MICROCONTROLLERS

It was pointed out above that microprocessors are single-chip CPUs used in microcomputers. How, then, do microcontrollers differ from microprocessors? This question can be addressed from three perspectives: **hardware architecture, applications,** and **instruction set features.**

1.9.1 Hardware Architecture

To highlight the difference between microcontrollers and microprocessors, Figure 1–2 is redrawn showing more detail (see Figure 1–6).

Whereas a microprocessor is a single-chip CPU, a microcontroller contains, in a single IC, a CPU and much of the remaining circuitry of a complete microcomputer system. The components within the dotted line in Figure 1–6 are an integral part of most microcontroller ICs. As well as the CPU, microcontrollers include RAM, ROM, a serial interface, a parallel interface, timer, and interrupt scheduling circuitry—all within the same IC. Of course, the amount of on-chip RAM does not approach that of even a modest microcomputer system; but, as we shall learn, this is not a limitation, since microcontrollers are intended for vastly different applications.

An important feature of microcontrollers is the built-in interrupt system. As control-oriented devices, microcontrollers are often called upon to respond to external stimuli (interrupts) in real time. They must perform fast context switching, suspending one process while executing another in response to an "event." The opening of a microwave oven's door is an example of an event that might cause an interrupt in a microcontroller-based product. Of course, most microprocessors can also implement powerful interrupt schemes, but external components are usually required. A microcontroller's on-chip circuitry includes all the interrupt handling circuitry necessary.

1.9.2 Applications

Microprocessors are most commonly used as the CPU in microcomputer systems. This is what they are designed for, and this is where their strengths lie. Microcontrollers, however, are found in small, minimum-component designs performing control-oriented activities. These designs were often implemented in the past using dozens or even hundreds of digital ICs. A microcontroller can aid in reducing the overall component count. All that is required is a microcontroller, a small number of support components, and a control program in ROM. Microcontrollers are suited to "control" of I/O devices in designs requiring a minimum component count, whereas microprocessors are suited to "processing" information in computer systems.

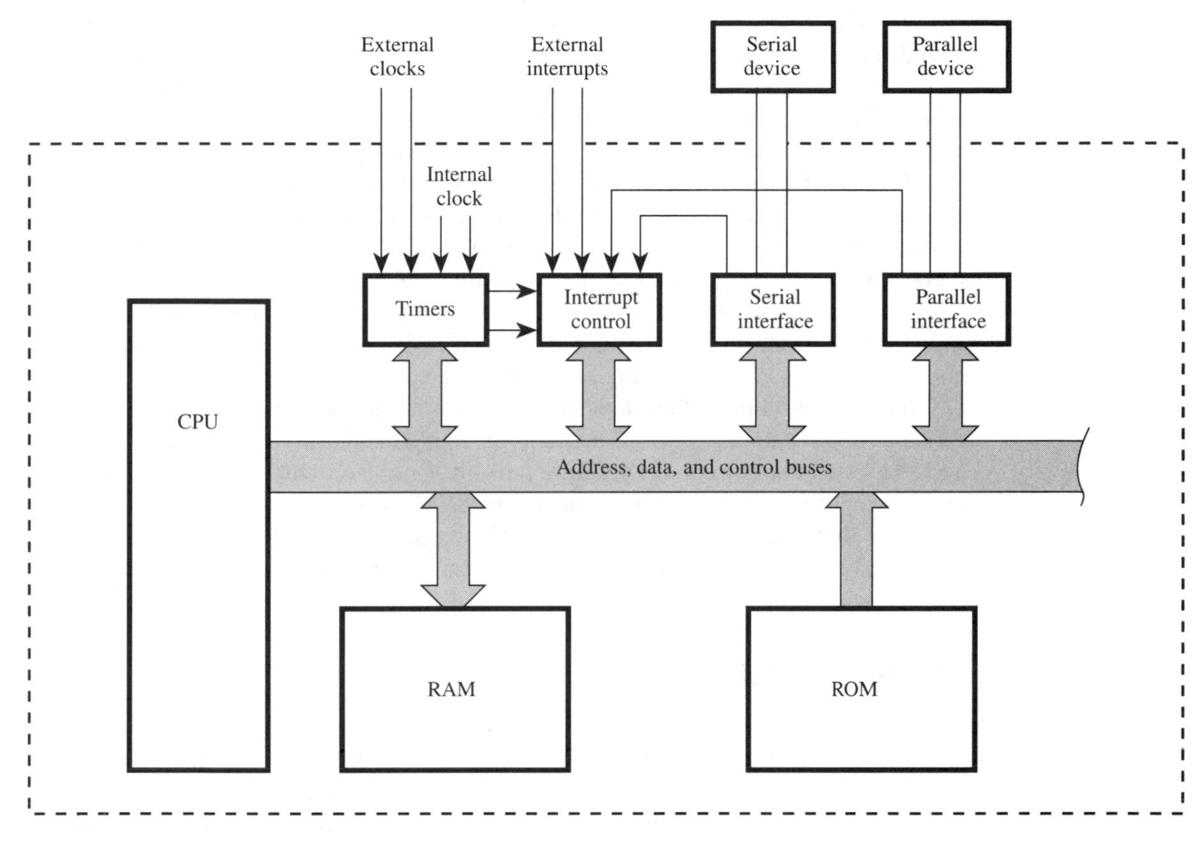

FIGURE 1–6
Detailed block diagram of a microcomputer system

1.9.3 Instruction Set Features

Due to the differences in applications, microcontrollers have somewhat different requirements for their instruction sets than microprocessors. Microprocessor instruction sets are "processing intensive," implying they have powerful addressing modes with instructions catering to operations on large volumes of data. Their instructions operate on nibbles, bytes, words, or even double words.[3] Addressing modes provide access to large arrays of data, using address pointers and offsets. Auto-increment and auto-decrement modes simplify stepping through arrays on byte, word, or double-word boundaries. Privileged instructions cannot execute within the user program. The list goes on.

Microcontrollers, on the other hand, have instruction sets catering to the control of inputs and outputs. The interface to many inputs and outputs uses a single bit. For example, a motor may be turned on and off by a solenoid energized by a 1-bit output port.

[3]The most common interpretation of these terms is 4 bits = 1 nibble, 8 bits = 1 byte, 16 bits = 1 word, and 32 bits = 1 double word.

Microcontrollers have instructions to set and clear individual bits and perform other bit-oriented operations such as logically ANDing, ORing, or EXORing bits, jumping if a bit is set or clear, and so on. This powerful feature is rarely present in microprocessors, which are usually designed to operate on bytes or larger units of data.

In the control and monitoring of devices (perhaps with a 1-bit interface), microcontrollers have built-in circuitry and instructions for input/output operations, event timing, and enabling and setting priority levels for interrupts caused by external stimuli. Microprocessors often require additional circuitry (serial interface ICs, interrupt controllers, timers, etc.) to perform similar operations. Nevertheless, the sheer processing capability of a microcontroller never approaches that of a microprocessor (all else being equal), since a great deal of the IC's "real estate" is consumed by the on-chip functions—at the expense of processing power, of course.

Since the on-chip real estate is at a premium in microcontrollers, the instructions must be extremely compact, with the majority implemented in a single byte. A design criterion is often that the control program must fit into the on-chip ROM, since the addition of even one external ROM adds too much cost to the final product. A tight encoding scheme for the instruction set is essential. This is rarely a feature of microprocessors; their powerful addressing modes bring with them a less-than-compact encoding of instructions.

1.10 NEW CONCEPTS

Microcontrollers, like other products considered in retrospect to be a breakthrough, have arrived out of two complementary forces: market need and new technology. The new technology is just that mentioned above: semiconductors with more transistors in less space, mass produced at a lower cost. The market need is the industrial and consumer appetite for more sophisticated tools and toys.[4] This encompasses a lot of territory. The most illustrative, perhaps, is the automobile dashboard. Witness the transformation of the car's "control center" over the past decade—made possible by the microcontroller and other technological developments. Once, drivers were content to know their speed; today they may find a display of fuel economy and estimated time of arrival. Once it was sufficient to know if a seatbelt was unfastened while starting the car; today, we are "told" which seatbelt is the culprit. If a door is ajar, we are again duly informed by the spoken word. (Perhaps the seatbelt is stuck in the door.)

This brings to mind a necessary comment. Microprocessors (and in this sense microcontrollers) have been dubbed "solutions looking for a problem." It seems they have proven so effective at reducing the complexity of circuitry in (consumer) products, that manufacturers are often too eager to include superfluous features simply because they are easy to design into the product. The result often lacks eloquence—a show-stopper initially, but an annoyance finally. The most stark example of this bells-and-whistles approach occurs in the recent appearance of products that talk. Whether automobiles, toys, or toasters, they are usually examples of tackiness and overdesign—1980s art deco, perhaps. Rest assured that once the dust has settled and the novelty has diminished, only the subtle and appropriate will remain.

---

[4]It is sometimes argued that "market need" is really "market want," spurred on by the self-propelled growth of technology.

Microcontrollers are specialized. They are not used in computers per se, but in industrial and consumer products. Users of such products are quite often unaware of the existence of microcontrollers: to them, the internal components are but an inconsequential detail of design. As examples, consider microwave ovens, programmable thermostats, electronic scales, and even cars. The electronics within each of these products typically incorporates a microcontroller interfacing to push buttons, switches, lights, and alarms on a front panel; yet user operation mimics that of the electromechanical predecessors, with the exception of some added features. The microcontroller is invisible to the user.

Unlike computer systems, which are defined by their ability to be programmed and then reprogrammed, microcontrollers are permanently programmed for one task. This comparison results in a stark architectural difference between the two. Computer systems have a high RAM-to-ROM ratio, with user programs executing in a relatively large RAM space and hardware interfacing routines executing in a small ROM space. Microcontrollers, on the other hand, have a high ROM-to-RAM ratio. The control program, perhaps relatively large, is stored in ROM, while RAM is used only for temporary storage. Since the control program is stored permanently in ROM, it has been dubbed **firmware.** In degrees of "firmness," it lies somewhere between software—the programs in RAM that are lost when power is removed—and hardware—the physical circuits. The difference between software and hardware is somewhat analogous to the difference between a page of paper (hardware) and words written on a page (software). Consider firmware as a standard form letter, designed and printed for a single purpose.

1.11 GAINS AND LOSSES: A DESIGN EXAMPLE

The tasks performed by microcontrollers are not new. What is new is that designs are implemented with fewer components than before. Designs previously requiring tens or even hundreds of ICs are implemented today with only a handful of components, including a microcontroller. The reduced component count, a direct result of the microcontroller's programmability and high degree of integration, usually translates into shorter development time, lower manufacturing cost, lower power consumption, and higher reliability. Logic operations that require several ICs can often be implemented within the microcontroller, with the addition of a control program.

One tradeoff is speed. Microcontroller-based solutions are never as fast as the discrete counterparts. Situations requiring extremely fast response to events (a minority of applications) are poorly handled by microcontrollers. As an example, consider in Figure 1–7 the somewhat trivial implementation of the NAND operation using an 8051 microcontroller.

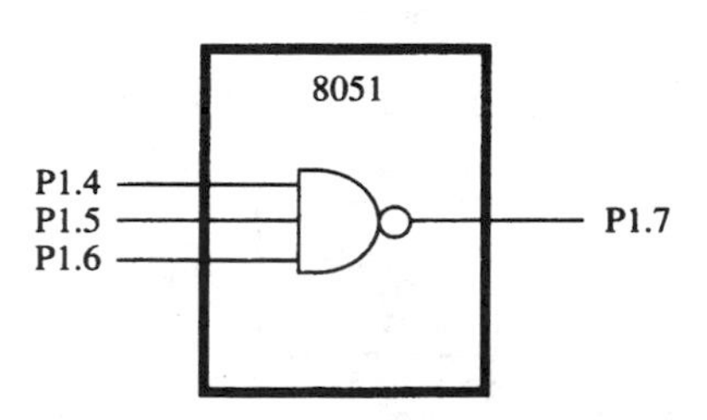

FIGURE 1–7
Microcontroller implementation of a simple logic operation

It is not at all obvious that a microcontroller could be used for such an operation, but it can. The software must perform the operations shown in the flowchart in Figure 1–8. The 8051 assembly language program for this logic operation is shown below.

```
LOOP:   MOV     C,P1.4      ;READ P1.4 BIT INTO CARRY FLAG
        ANL     C,P1.5      ;AND WITH P1.5
        ANL     C,P1.6      ;AND WITH P1.6
        CPL     C           ;CONVERT TO "NAND" RESULT
        MOV     P1.7,C      ;SEND TO P1.7 OUTPUT BIT
        SJMP    LOOP        ;REPEAT
```

If this program executes on an 8051 microcontroller, indeed the 3-input NAND function is realized. (It could be verified with a voltmeter or oscilloscope.) The propagation delay from an input transition to the correct output level is quite long, at least in comparison to the equivalent TTL (transistor-transistor logic) circuit. Depending on when the input changed relative to the program sensing the change, the delay is from 3 to 17 microseconds. (This assumes standard 8051 operation using a 12 MHz crystal.) The equivalent TTL propagation delay is on the order of 10 nanoseconds—about three orders of magnitude less. Obviously, there is no contest when comparing the speed of microcontrollers with TTL implementations of the same function.

In many applications, particularly those with human operation, whether the delays are measured in nanoseconds, microseconds, or milliseconds is inconsequential. (When

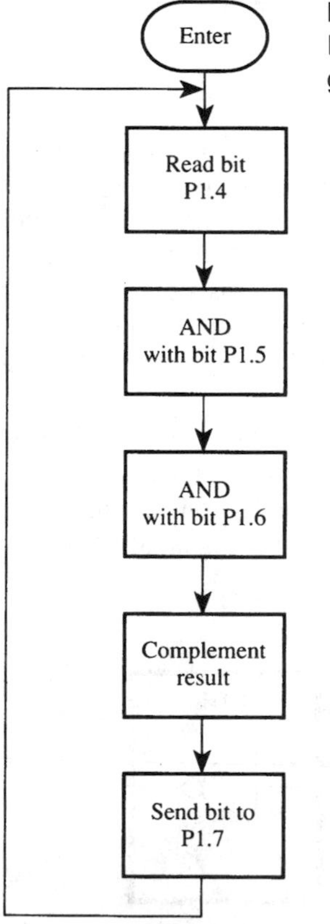

FIGURE 1–8
Flowchart for logic gate program

the oil pressure drops in your car, do you need to be informed within microseconds?) The logic gate example illustrates that microcontrollers can implement logic operations. Furthermore, as designs become complex, the advantages of the microcontroller-based design begin to take hold. The reduced component count has advantages, as mentioned earlier; but, also, the operations in the control program make it possible to introduce changes in design by modifying only the software. This has minimal impact on the manufacturing cycle.

This concludes our introduction to microcontrollers. In the next chapter, we begin our examination of the MCS-51™ family of devices.

PROBLEMS

1. What was the first widely used microprocessor? In what year was it introduced and by what company?
2. Two of the smaller microprocessor companies in the 1970s were MOS Technology and Zilog. Name the microprocessor that each of these companies introduced.
3. What year was the 8051 microcontroller introduced? What was the predecessor to the 8051 and in what year was it introduced?
4. Name the two types of semiconductor memory discussed in this chapter. Which type retains its contents when powered-off? What is the common term that describes this property?
5. Which register in a CPU always contains an address? What address is contained in this register?
6. During an opcode fetch, what is the information on the address and data buses? What is the direction of information flow on these buses during an opcode fetch?
7. How many bytes of data can be addressed by a computer system with an 18-bit address bus and an 8-bit data bus?
8. What is the usual meaning of "16-bits" in the phrase "16-bit computer"?
9. What is the difference between online storage and archival storage?
10. What type of technology is used for archival storage besides magnetic tape and disk?
11. With regard to computing systems, what is the goal of the field of engineering known as "human factors"?
12. Consider the following human interface devices: a joystick, a light pen, a mouse, a microphone, and a loudspeaker. Which are input devices? Which are output devices?
13. Of the three levels of software presented in this chapter, which is the lowest level? What is the purpose of this level of software?
14. What is the difference between an actuator and a sensor? Give an example of each.
15. What is firmware? Comparing a microcontroller-based system to a microprocessor-based system, which is more likely to rely on firmware? Why?
16. What is an important feature of a microcontroller's instruction set that distinguishes it from a microprocessor?
17. Name five products not mentioned in this chapter that are likely to use a microcontroller.

2

HARDWARE SUMMARY

2.1 MCS-51™ FAMILY OVERVIEW

The MCS-51™ is a family of microcontroller ICs developed, manufactured, and marketed by Intel Corporation. Other IC manufacturers, such as Siemens, Advanced Micro Devices, Fujitsu, and Philips are licensed "second source" suppliers of devices in the MCS-51™ family. Each microcontroller in the family boasts a complement of features suited to a particular design setting.

In this chapter the hardware architecture of the MCS-51™ family is introduced. Intel's data sheet for the entry-level devices (e.g., the 8051AH) is found in Appendix E. This appendix should be consulted for further details, for example, on electrical properties of these devices.

Many of the hardware features are illustrated with short sequences of instructions. Brief descriptions are provided with each example, but complete details of the instruction set are deferred to Chapter 3. See also Appendix A for a summary of the 8051 instruction set or Appendix C for definitions of each 8051 instruction.

The generic MCS-51™ IC is the 8051, the first device in the family offered commercially. Its features are summarized below.

- 4K bytes ROM (factory mask programmed)
- 128 bytes RAM
- Four 8-bit I/O (Input/Output) ports
- Two 16-bit timers
- Serial interface
- 64K external code memory space
- 64K external data memory space
- Boolean processor (operates on single bits)
- 210 bit-addressable locations
- 4 μs multiply/divide

TABLE 2–1
Comparison of MCS-51™ ICs

PART NUMBER	ON-CHIP CODE MEMORY	ON-CHIP DATA MEMORY	TIMERS
8051	4K ROM	128 bytes	2
8031	0K	128 bytes	2
8751	4K EPROM	128 bytes	2
8052	8K ROM	256 bytes	3
8032	0K	256 bytes	3
8752	8K EPROM	256 bytes	3

Other members of the MCS-51™ family offer different combinations of on-chip ROM or EPROM, on-chip RAM, or a third timer. Each of the MCS-51™ ICs is also offered in a low-power CMOS version (see Table 2–1).

The term "8051" loosely refers to the MCS-51™ family of microcontrollers. When discussion centers on an enhancement to the basic 8051 device, the specific part number

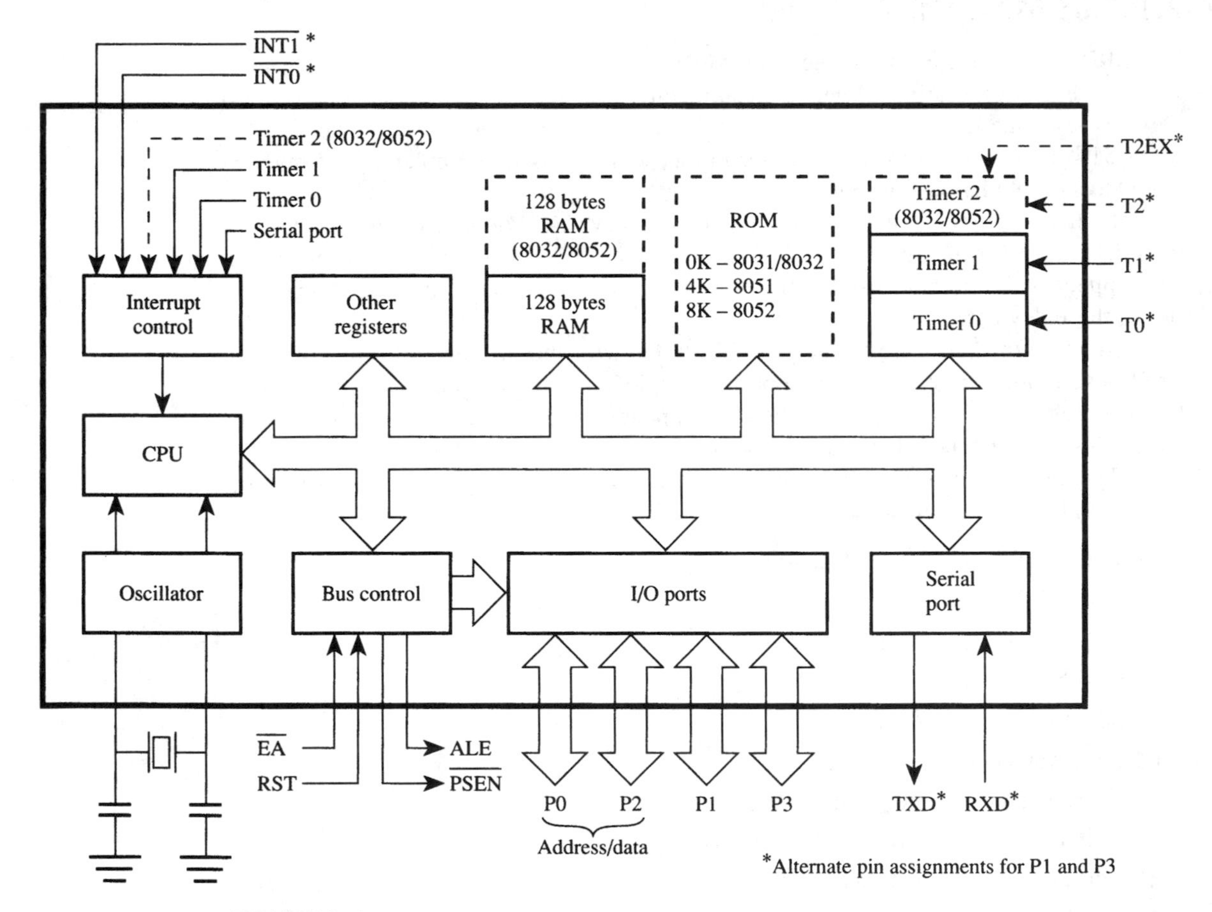

FIGURE 2–1
8051 block diagram

is used. The features mentioned above are shown in the block diagram in Figure 2–1. (See also Appendix D.)

2.2 ONCE AROUND THE PINS

This section introduces the 8051 hardware architecture from an external perspective—the pinouts (see Figure 2–2). A brief description of the function of each pin follows.

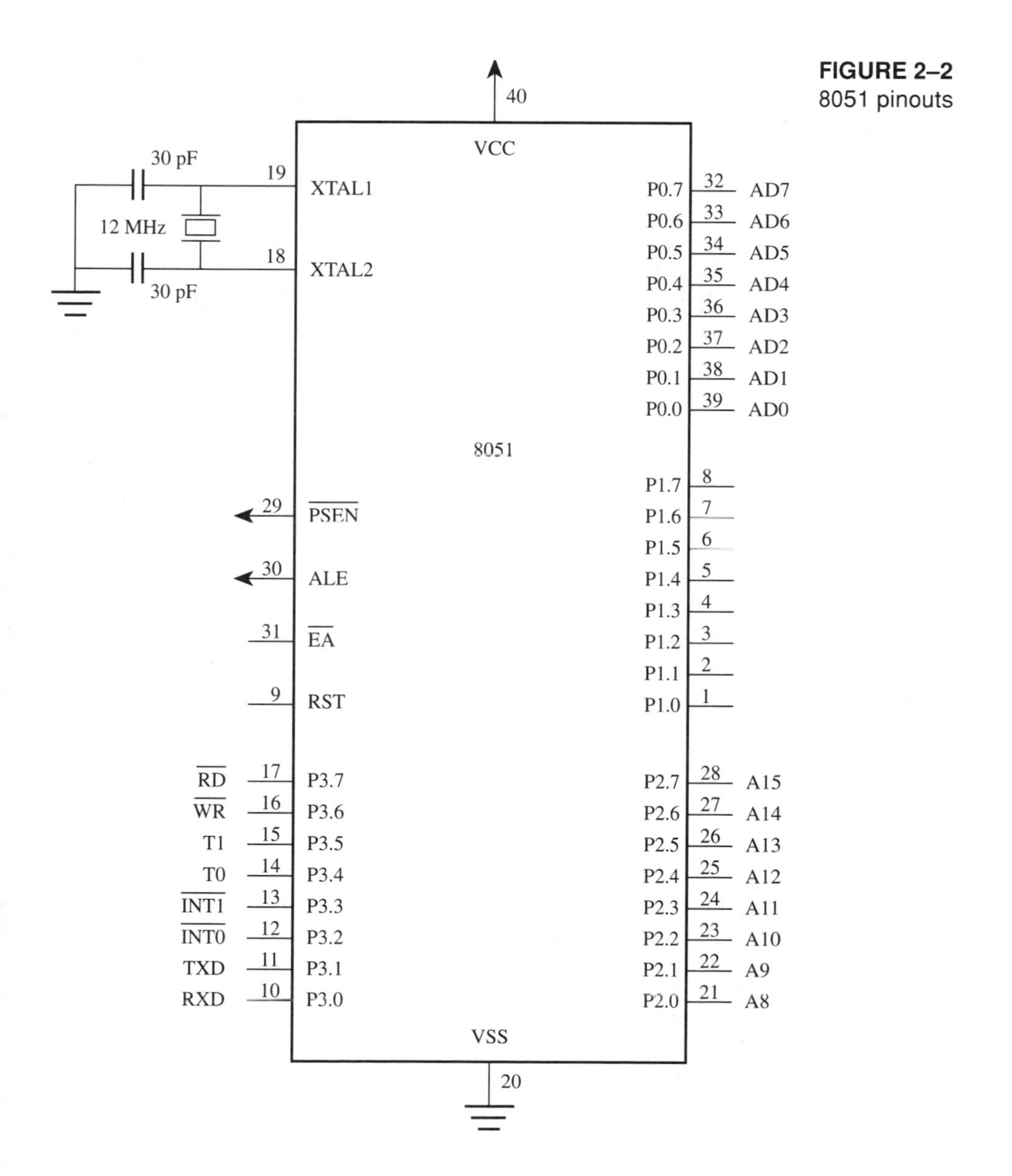

FIGURE 2–2
8051 pinouts

As evident in Figure 2–2, 32 of the 8051's 40 pins function as I/O port lines. However, 24 of these lines are dual-purpose (26 on the 8032/8052). Each can operate as I/O, or as a control line or part of the address or data bus.

Designs requiring a minimum of external memory or other external components use these ports for general purpose I/O. The eight lines in each port can be treated as a unit in interfacing to parallel devices such as printers, digital-to-analog converters, and so on. Or, each line can operate independently in interfacing to single-bit devices such as switches, LEDs, transistors, solenoids, motors, and loudspeakers.

2.2.1 Port 0

Port 0 is a dual-purpose port on pins 32–39 of the 8051 IC. In minimum-component designs, it is used as a general purpose I/O port. For larger designs with external memory, it becomes a multiplexed address and data bus. (See 2.6 External Memory.)

2.2.2 Port 1

Port 1 is a dedicated I/O port on pins 1–8. The pins, designated as P1.0, P1.1, P1.2, etc., are available for interfacing to external devices as required. No alternate functions are assigned for Port 1 pins; thus they are used solely for interfacing to external devices. Exceptions are the 8032/8052 ICs, which use P1.0 and P1.1 either as I/O lines or as external inputs to the third timer.

2.2.3 Port 2

Port 2 (pins 21–28) is a dual-purpose port serving as general purpose I/O, or as the high-byte of the address bus for designs with external code memory or more than 256 bytes of external data memory. (See 2.6 External Memory.)

2.2.4 Port 3

Port 3 is a dual-purpose port on pins 10–17. As well as general-purpose I/O, these pins are multifunctional, with each having an alternate purpose related to special features of the 8051. The alternate purpose of the Port 3 and Port 1 pins is summarized in Table 2–2.

TABLE 2–2
Alternate pin functions for port pins

BIT	NAME	BIT ADDRESS	ALTERNATE FUNCTION
P3.0	RXD	B0H	Receive data for serial port
P3.1	TXD	B1H	Transmit data for serial port
P3.2	$\overline{\text{INT0}}$	B2H	External interrupt 0
P3.3	$\overline{\text{INT1}}$	B3H	External interrupt 1
P3.4	T0	B4H	Timer/counter 0 external input
P3.5	T1	B5H	Timer/counter 1 external input
P3.6	$\overline{\text{WR}}$	B6H	External data memory write strobe
P3.7	$\overline{\text{RD}}$	B7H	External data memory read strobe
P1.0	T2	90H	Timer/counter 2 external input
P1.1	T2EX	91H	Timer/counter 2 capture/reload

2.2.5 $\overline{\text{PSEN}}$ (Program Store Enable)

The 8051 has four dedicated bus control signals. Program Store Enable ($\overline{\text{PSEN}}$) is an output signal on pin 29. It is a control signal that enables external program (code) memory. It usually connects to an EPROM's Output Enable ($\overline{\text{OE}}$) pin to permit reading of program bytes.

The $\overline{\text{PSEN}}$ signal pulses low during the fetch stage of an instruction. The binary codes of a program (opcodes) are read from EPROM, travel across the data bus, and are latched into the 8051's instruction register for decoding. When executing a program from internal ROM (8051/8052), $\overline{\text{PSEN}}$ remains in the inactive (high) state.

2.2.6 ALE (Address Latch Enable)

The ALE output signal on pin 30 will be familiar to anyone who has worked with Intel's 8085, 8088, or 8086 microprocessors. The 8051 similarly uses ALE for demultiplexing the address and data bus. When Port 0 is used in its alternate mode—as the data bus and the low-byte of the address bus—ALE is the signal that latches the address into an external register during the first half of a memory cycle. This done, the Port 0 lines are then available for data input or output during the second half of the memory cycle, when the data transfer takes place. (See 2.6 External Memory.)

The ALE signal pulses at a rate of 1/6th the on-chip oscillator frequency and can be used as a general-purpose clock for the rest of the system. If the 8051 is clocked from a 12 MHz crystal, the ALE signal oscillates at 2 MHz. The only exception is during the MOVX instruction, when one ALE pulse is missed. (See Figure 2–10.) This pin is also used for the programming input pulse for EPROM versions of the 8051.

2.2.7 $\overline{\text{EA}}$ (External Access)

The $\overline{\text{EA}}$ input signal on pin 31 is generally tied high (+5 V) or low (ground). If high, the 8051/8052 executes programs from internal ROM when executing in the lower 4K/8K of memory. If low, programs execute from external memory only (and $\overline{\text{PSEN}}$ pulses low accordingly). $\overline{\text{EA}}$ must be tied low for 8031/8032 ICs, since there is no on-chip program memory. If $\overline{\text{EA}}$ is tied low on an 8051/8052, internal ROM is disabled and programs execute from external EPROM. The EPROM versions of the 8051 also use the $\overline{\text{EA}}$ line for the +21 volt supply (V_{PP}) for programming the internal EPROM.

2.2.8 RST (Reset)

The RST input on pin 9 is the master reset for the 8051. When this signal is brought high for at least two machine cycles, the 8051 internal registers are loaded with appropriate values for an orderly system start-up. (See 2.8 Reset Operation.)

2.2.9 On-chip Oscillator Inputs

As shown in Figure 2–2, the 8051 features an on-chip oscillator that is typically driven by a crystal connected to pins 18 and 19. Stabilizing capacitors are also required as shown. The nominal crystal frequency is 12 MHz for most ICs in the MCS-51™ family, although the 80C31BH-1 can operate with crystal frequencies up to 16 MHz. The on-chip oscillator needn't be driven by a crystal. As shown in Figure 2–3, a TTL clock source can be connected to XTAL1 and XTAL2.

2.2.10 Power Connections

The 8051 operates from a single +5 volt supply. The V_{CC} connection is on pin 40, and the V_{SS} (ground) connection is on pin 20.

2.3 I/O PORT STRUCTURE

The internal circuitry for the port pins is shown in abbreviated form in Figure 2–4. Writing to a port pin loads data into a port latch that drives a field-effect transistor connected to the port pin. The drive capability is 4 low-power Schottky TTL loads for Ports 1, 2, and 3; and 8 LS loads for Port 0. (See Appendix E for more details.) Note that the pull-up resistor is absent on Port 0 (except when functioning as the external address/data bus). An external pull-up resistor may be needed, depending on the input characteristics of the device driven.

There is both a "read latch" and "read pin" capability. Instructions that require a read-modify-write operation (e.g., CPL P1.5) read the latch to avoid misinterpreting the voltage level in the event the pin is heavily loaded (e.g., when driving the base of a transistor). Instructions that input a port bit (e.g., MOV C,P1.5) read the pin. The port latch must contain a 1, in this case, otherwise the FET driver is ON and pulls the output low. A system reset sets all port latches, so port pins may be used as inputs without explicitly setting the port latches. If, however, a port latch is cleared (e.g., CLR P1.5), then it cannot function subsequently as an input unless the latch is set first (e.g., SETB P1.5).

Figure 2–4 does not show the circuitry for the alternate functions for Ports 0, 2, and 3. When the alternate function is in effect, the output drivers are switched to an internal address (Port 2), address/data (Port 0), or control (Port 3) signal, as appropriate.

2.4 MEMORY ORGANIZATION

Most microprocessors implement a shared memory space for data and programs. This is reasonable, since programs are usually stored on a disk and loaded into RAM for execution; thus both the data and programs reside in the system RAM. Microcontrollers, on the other hand, are rarely used as the CPU in "computer systems." Instead, they are employed as the central component in control-oriented designs. There is limited memory,

FIGURE 2–3
Driving the 8051 from a TTL oscillator

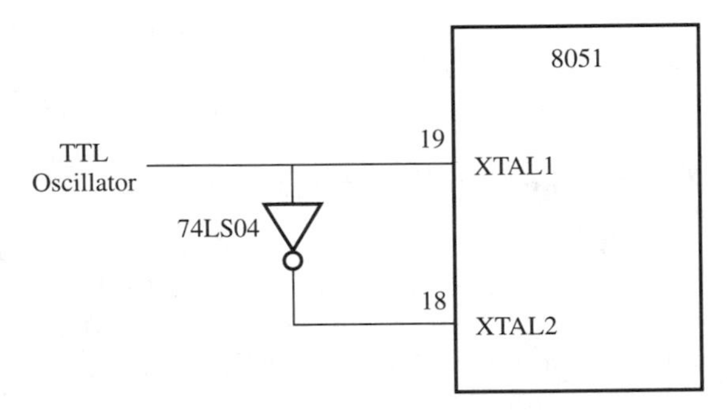

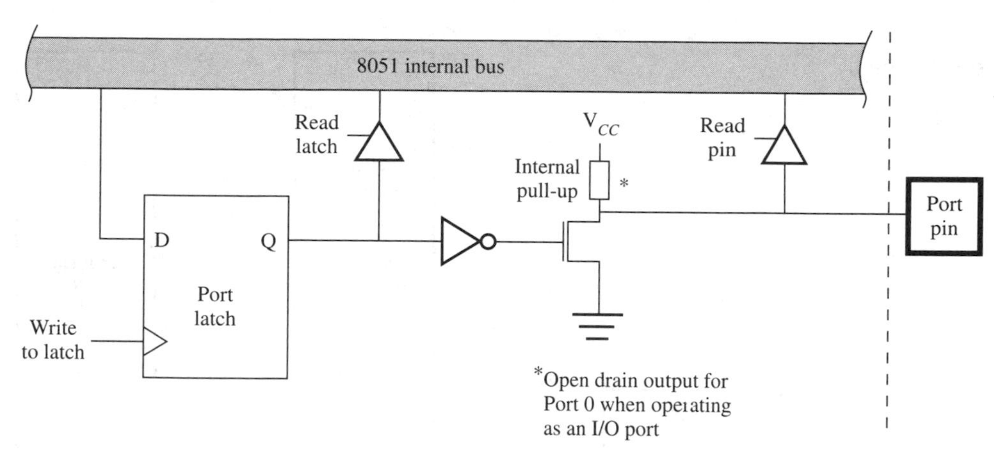

FIGURE 2–4
Circuitry for I/O ports

and there is no disk drive or disk operating system. The control program must reside in ROM.

For this reason, the 8051 implements a separate memory space for programs (code) and data. As shown in Table 2–1, both the code and data may be internal; however, both expand using external components to a maximum of 64K code memory and 64K data memory.

The internal memory consists of on-chip ROM (8051/8052 only) and on-chip data RAM. The on-chip RAM contains a rich arrangement of general-purpose storage, bit-addressable storage, register banks, and special function registers.

Two notable features are: (a) the registers and input/output ports are memory-mapped and accessible like any other memory location, and (b) the stack resides within the internal RAM, rather than in external RAM as typical of microprocessors.

Figure 2–5 summarizes the memory spaces for the ROM-less 8031 device without showing any detail of the on-chip data memory. (8032/8052 enhancements are summarized later.)

Figure 2–6 gives the details of the on-chip data memory. As shown, the internal memory space is divided between register banks (00H–1FH), bit-addressable RAM (20H–2FH), general-purpose RAM (30H–7FH), and special function registers (80H–FFH). Each of these sections of internal memory is discussed below.

2.4.1 General Purpose RAM

Although Figure 2–6 shows 80 bytes of general purpose RAM from addresses 30H to 7FH, the bottom 32 bytes from 00H to 2FH can be used similarly (although these locations have other purposes as discussed below).

Any location in the general-purpose RAM can be accessed freely using the direct or indirect addressing modes. For example, to read the contents of internal RAM address 5FH into the accumulator, the following instruction could be used:

```
MOV  A,5FH
```

FIGURE 2–5
Summary of the 8031 memory spaces

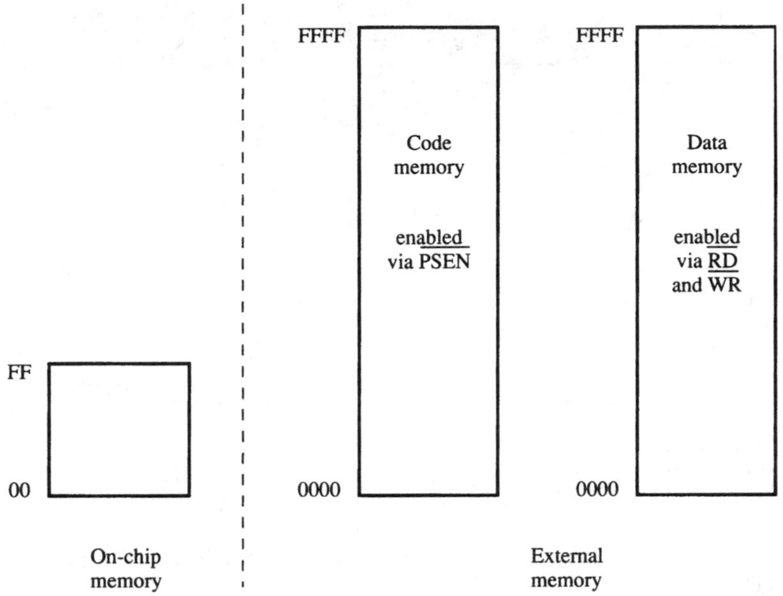

FF
00
On-chip memory

FFFF
Code memory
enabled via $\overline{PSEN}$
0000

FFFF
Data memory
enabled via $\overline{RD}$ and $\overline{WR}$
0000

External memory

This instruction moves a byte of data using direct addressing to specify the "source location" (i.e., address 5FH). The destination for the data is implicitly specified in the instruction opcode as the A accumulator. (Note: Addressing modes are discussed in detail in Chapter 3.)

Internal RAM can also be accessed using indirect addressing through R0 or R1. For example, the following two instructions perform the same operation as the single instruction above:

```
MOV  R0,#5FH
MOV  A,@R0
```

The first instruction uses immediate addressing to move the value 5FH into register R0, and the second instruction uses indirect addressing to move the data "pointed at by R0" into the accumulator.

2.4.2 Bit-addressable RAM

The 8051 contains 210 bit-addressable locations, of which 128 are at byte addresses 20H through 2FH, and the rest are in the special function registers (discussed below).

The idea of individually accessing bits through software is a powerful feature of most microcontrollers. Bits can be set, cleared, ANDed, ORed, etc., with a single instruction. Most microprocessors require a read-modify-write sequence of instructions to achieve the same effect. Furthermore, the 8051 I/O ports are bit-addressable, simplifying the software interface to single-bit inputs and outputs.

There are 128 general-purpose bit-addressable locations at byte addresses 20H through 2FH (8 bits/byte $\times$ 16 bytes = 128 bits). These addresses are accessed as bytes

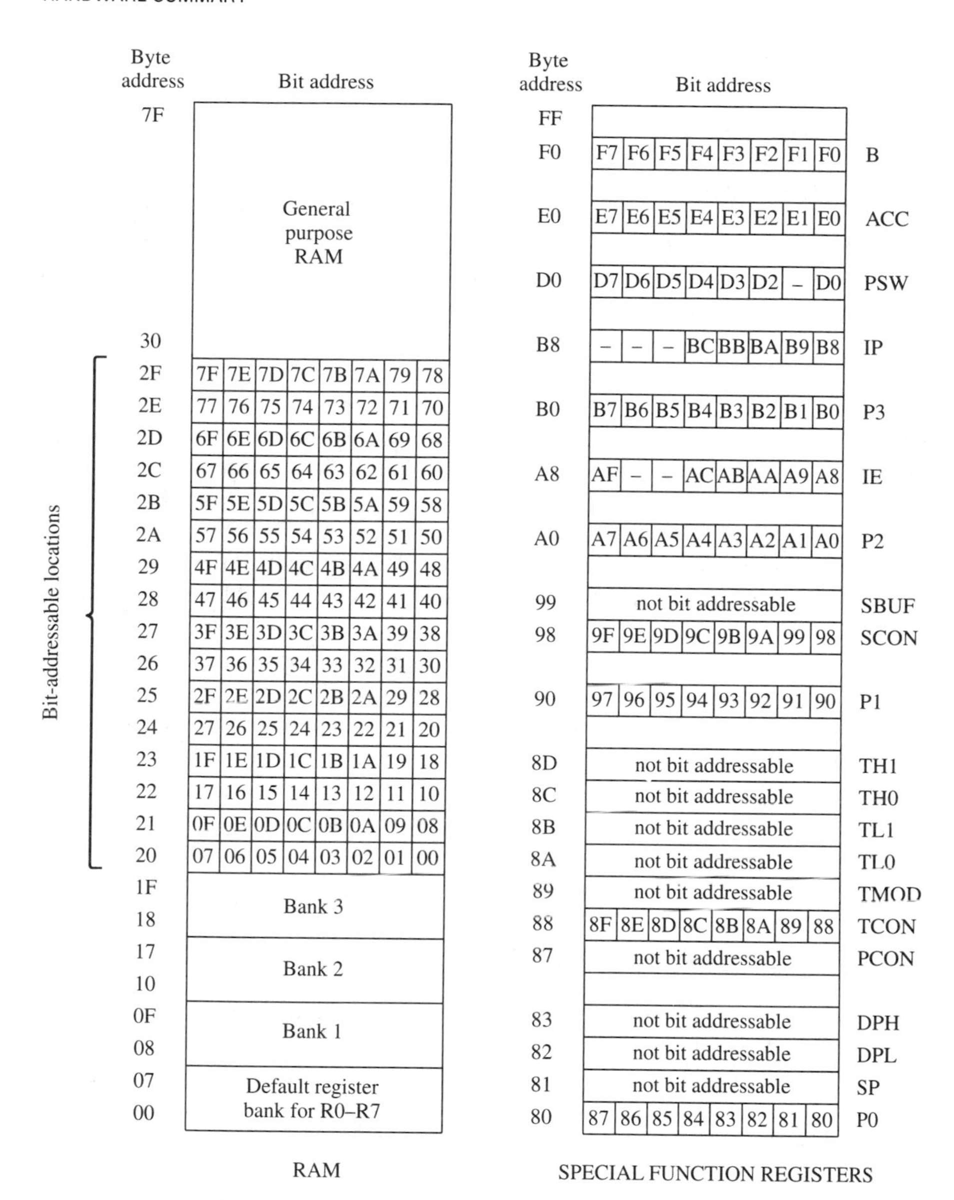

FIGURE 2–6
Summary of the 8051 on-chip data memory

or as bits, depending on the instruction. For example, to set bit 67H, the following instruction could be used:

```
SETB 67H
```

Referring to Figure 2–6, note that "bit address 67H" is the most-significant bit at "byte address 2CH." The instruction above has no effect on the other bits at this address. Most microprocessors would perform the same operation as follows:

```
MOV  A, 2CH              ;READ ENTIRE BYTE
ORL  A,#10000000B        ;SET MOST-SIGNIFICANT BIT
MOV  2CH,A               ;WRITE BACK ENTIRE BYTE
```

2.4.3 Register Banks

The bottom 32 locations of internal memory contain the register banks. The 8051 instruction set supports 8 registers, R0 through R7, and by default (after a system reset) these registers are at addresses 00H–07H. The following instruction, then, reads the contents of address 05H into the accumulator:

```
MOV  A,R5
```

This instruction is a 1-byte instruction using register addressing. Of course, the same operation could be performed in a 2-byte instruction using the direct address as byte 2:

```
MOV  A,05H
```

Instructions using registers R0 to R7 are shorter and faster than the equivalent instructions using direct addressing. Data values used frequently should use one of these registers.

The active register bank may be altered by changing the register bank select bits in the program status word (discussed below). Assuming, then, that register bank 3 is active, the following instruction writes the contents of the accumulator into location 18H:

```
MOV  R0,A
```

The idea of "register banks" permits fast and effective "context switching," whereby separate sections of software use a private set of registers independent of other sections of software.

2.5 SPECIAL FUNCTION REGISTERS

Internal registers on most microprocessors are accessed implicitly by the instruction set. For example, "INCA" on the 6809 microprocessor increments the contents of the A accumulator. The operation is specified implicitly within the instruction opcode. Similar access to registers is also used on the 8051 microcontroller. In fact, the 8051 instruction "INC A" performs the same operation.

The 8051 internal registers are configured as part of the on-chip RAM; therefore, each register also has an address.[1] This is reasonable for the 8051, since it has so many registers. As well as R0 to R7, there are 21 special function registers (SFRs) at the top of internal RAM, from addresses 80H to FFH. (See Figure 2–6 and Appendix D.) Note that most of the 128 addresses from 80H to FFH are not defined. Only 21 SFR addresses are defined (26 on the 8032/8052).

Although the accumulator (A) may be accessed implicitly as shown previously, most SFRs are accessed using direct addressing. Note in Figure 2–6 that some SFRs are

[1]The program counter and the instruction register are exceptions. Since these registers are rarely manipulated directly, nothing is gained by placing them in the on-chip RAM.

both bit-addressable and byte-addressable. Designers should be careful when accessing bits versus bytes. For example, the instruction:

```
SETB 0E0H
```

sets bit 0 in the accumulator, leaving the other bits unchanged. The trick is to recognize that E0H is both the byte address of the entire accumulator and the bit address of the least-significant bit in the accumulator. Since the SETB instruction operates on bits (not bytes), only the addressed bit is affected. Notice that the addressable bits within the SFRs have the five high-order address bits matching those of the SFR. For example, Port 1 is at byte address 90H or 10010000B. The bits within Port 1 have addresses 90H to 97H, or 10010xxxB.

The PSW is discussed in detail in the following section. The other SFRs are briefly introduced following the PSW, with detailed discussions deferred to later chapters.

2.5.1 Program Status Word

The program status word (PSW) at address D0H contains status bits as summarized in Table 2–3. Each of the PSW bits is examined below.

2.5.1.1 Carry Flag

The carry flag (CY) is dual-purpose. It is used in the traditional way for arithmetic operations: set if there is a carry out of bit 7 during an add, or set if there is a borrow into bit 7 during a subtract. For example, if the accumulator contains FFH, then the instruction

```
ADD  A,#1
```

leaves the accumulator equal to 00H and sets the carry flag in the PSW.

The carry flag is also the "Boolean accumulator," serving as a 1-bit register for Boolean instructions operating on bits. For example, the following instruction ANDs bit 25H with the carry flag and places the result back in the carry flag:

```
ANL  C,25H
```

TABLE 2–3
PSW (program status word) register summary

BIT	SYMBOL	ADDRESS	BIT DESCRIPTION
PSW.7	CY	D7H	Carry flag
PSW.6	AC	D6H	Auxiliary carry flag
PSW.5	F0	D5H	Flag 0
PSW.4	RS1	D4H	Register bank select 1
PSW.3	RS0	D3H	Register bank select 0 00 = bank 0; addresses 00H-07H 01 = bank 1; addresses 08H-0FH 10 = bank 2; addresses 10H-17H 11 = bank 3; addresses 18H-1FH
PSW.2	OV	D2H	Overflow flag
PSW.1	—	D1H	Reserved
PSW.0	P	D0H	Even parity flag

2.5.1.2 Auxiliary Carry Flag

When adding binary-coded-decimal (BCD) values, the auxiliary carry flag (AC) is set if a carry was generated out of bit 3 into bit 4 or if the result in the lower nibble is in the range 0AH–0FH. If the values added are BCD, then the add instruction must be followed by DA A (decimal adjust accumulator) to bring results greater than 9 back into range.

2.5.1.3 Flag 0

Flag 0 (F0) is a general-purpose flag bit available for user applications.

2.5.1.4 Register Bank Select Bits

The register bank select bits (RS0 and RS1) determine the active register bank. They are cleared after a system reset and are changed by software as needed. For example, the following three instructions enable register bank 3 and then move the contents of R7 (byte address 1FH) to the accumulator:

```
SETB RS1
SETB RS0
MOV  A,R7
```

When the above program is assembled, the correct bit addresses are substituted for the symbols "RS1" and "RS0." Thus, the instruction SETB RS1 is the same as SETB 0D4H.

2.5.1.5 Overflow Flag

The overflow flag (OV) is set after an addition or subtraction operation if there was an arithmetic overflow. When signed numbers are added or subtracted, software can examine this bit to determine if the result is in the proper range. When unsigned numbers are added, the OV bit can be ignored. Results greater than +127 or less than −128 will set the OV bit. For example, the following addition causes an overflow and sets the OV bit in the PSW:

```
Hex:    0F     Decimal:    15
       +7F                +127
        8E                 142
```

As a signed number, 8EH represents −116, which is clearly not the correct result of 142; therefore, the OV bit is set.

2.5.1.6 Parity Bit

The parity bit (P) is automatically set or cleared each machine cycle to establish even parity with the accumulator. The number of 1-bits in the accumulator plus the P bit is always even. If, for example, the accumulator contains 10101101B, P will contain 1 (establishing a total of 6 1-bits; i.e., an even number of 1s). The parity bit is most commonly used in conjunction with serial port routines to include a parity bit before transmission or to check for parity after reception.

2.5.2 B Register

The B register at address F0H is used along with the accumulator for multiply and divide operations. The MUL AB instruction multiplies the 8-bit unsigned values in A and B and leaves the 16-bit result in A (low-byte) and B (high-byte). The DIV AB instruction divides A by B leaving the integer result in A and the remainder in B. The B register can also be treated as a general-purpose scratch-pad register. It is bit-addressable through bit addresses F0H to F7H.

2.5.3 Stack Pointer

The stack pointer (SP) is an 8-bit register at address 81H. It contains the address of the data item currently on the top of the stack. Stack operations include "pushing" data on the stack and "popping" data off the stack. Pushing on the stack increments the SP before writing data, and popping from the stack reads data and then decrements the SP. The 8051 stack is kept in internal RAM and is limited to addresses accessible by indirect addressing. These are the first 128 bytes on the 8031/8051 or the full 256 bytes of on-chip RAM on the 8032/8052.

To reinitialize the SP with the stack beginning at 60H, the following instruction is used:

```
MOV  SP,#5FH
```

On the 8031/8051 this would limit the stack to 32 bytes, since the uppermost address of on-chip RAM is 7FH. The value 5FH is used, since the SP increments to 60H *before* the first push operation.

Designers may choose not to reinitialize the stack pointer and let it retain its default value upon system reset. The reset value of 07H maintains compatibility with the 8051's predecessor, the 8048, and results in the first stack write storing data in location 08H. If the application software does not reinitialize the SP, then register bank 1 (and perhaps 2 and 3) is not available, since this area of internal RAM is the stack.

The stack is accessed explicitly by the PUSH and POP instructions to temporarily store and retrieve data, or implicitly by the subroutine call (ACALL, LCALL) and return (RET, RETI) instructions to save and restore the program counter.

2.5.4 Data Pointer

The data pointer (DPTR), used to access external code or data memory, is a 16-bit register at addresses 82H (DPL, low-byte) and 83H (DPH, high-byte). The following three instructions write 55H into external RAM location 1000H:

```
MOV  A,#55H
MOV  DPTR,#1000H
MOVX @DPTR,A
```

The first instruction uses immediate addressing to load the data constant 55H into the accumulator. The second instruction also uses immediate addressing, this time to load the 16-bit address constant 1000H into the data pointer. The third instruction uses indirect addressing to move the value in A (55H) to the external RAM location whose address is in the DPTR (1000H).

2.5.5 Port Registers

The 8051 I/O ports consist of Port 0 at address 80H, Port 1 at address 90H, Port 2 at address A0H, and Port 3 at address B0H. Ports 0, 2, and 3 may not be available for I/O if external memory is used or if some of the 8051 special features are used (interrupts, serial port, etc.). Nevertheless, P1.2 to P1.7 are always available as general purpose I/O lines.

All ports are bit-addressable. This provides powerful interfacing possibilities. If a motor is connected through a solenoid and transistor driver to Port 1 bit 7, for example, it could be turned on and off using a single 8051 instruction:

```
SETB P1.7
```

might turn the motor on, and

```
CLR  P1.7
```

might turn it off.

The instructions above use the dot operator to address a bit within a bit-addressable byte location. The assembler performs the necessary conversion; thus, the following two instructions are the same:

```
CLR  P1.7
CLR  97H
```

The use of predefined assembler symbols (e.g., P1) is discussed in detail in Chapter 7.

As another example, consider the interface to a device with a status bit called BUSY, which is set when the device is busy and clear when it is ready. If BUSY connects to, say, Port 1 bit 5, the following loop could be used to wait for the device to become ready:

```
WAIT:        JB   P1.5,WAIT
```

This instruction means "if the bit P1.5 is set, jump to the label WAIT." In other words "jump back and check it again."

2.5.6 Timer Registers

The 8051 contains two 16-bit timer/counters for timing intervals or counting events. Timer 0 is at addresses 8AH (TL0, low-byte) and 8CH (TH0, high-byte), and Timer 1 is at addresses 8BH (TL1, low-byte) and 8DH (TH1, high-byte). Timer operation is set by the timer mode register (TMOD) at address 89H and the timer control register (TCON) at address 88H. Only TCON is bit-addressable. The timers are discussed in detail in Chapter 4.

2.5.7 Serial Port Registers

The 8051 contains an on-chip serial port for communicating with serial devices such as terminals or modems, or for interfaces with other ICs with a serial interface (A/D converters, shift registers, nonvolatile RAMs, etc.). One register, the serial data buffer (SBUF) at address 99H, holds both the transmit data and receive data. Writing to SBUF loads data for transmission; reading SBUF accesses received data. Various modes of operation are programmable through the bit-addressable serial port control register (SCON) at address 98H. Serial port operation is discussed in detail in Chapter 5.

2.5.8 Interrupt Registers

The 8051 has a 5-source, 2-priority level interrupt structure. Interrupts are disabled after a system reset and then enabled by writing to the interrupt enable register (IE) at address A8H. The priority level is set through the interrupt priority register (IP) at address B8H. Both registers are bit-addressable. Interrupts are discussed in detail in Chapter 6.

2.5.9 Power Control Register

The power control register (PCON) at address 87H contains miscellaneous control bits. These are summarized in Table 2–4.

The SMOD bit doubles the serial port baud rate when in Modes 1, 2, or 3. (See Chapter 5.) PCON bits 6, 5, and 4 are undefined. Bits 3 and 2 are general-purpose flag bits available for user applications.

The power control bits, power down (PD) and idle (IDL), were originally available in all MCS-51™ family ICs but are now implemented only in the CMOS versions. PCON is not bit-addressable.

2.5.9.1 Idle Mode

An instruction that sets the IDL bit will be the last instruction executed before entering idle mode. In idle mode the internal clock signal is gated off to the CPU, but not to the interrupt, timer, and serial port functions. The CPU status is preserved and all register contents are maintained. Port pins also retain their logic levels. ALE and $\overline{\text{PSEN}}$ are held high.

Idle mode is terminated by any enabled interrupt or by a system reset. Either condition clears the IDL bit.

2.5.9.2 Power Down Mode

An instruction that sets the PD bit will be the last instruction executed before entering power down mode. In power down mode, (1) the on-chip oscillator is stopped, (2) all

TABLE 2–4
PCON register summary

BIT	SYMBOL	DESCRIPTION
7	SMOD	Double-baud rate bit; when set, baud rate is doubled in serial port modes 1, 2, or 3
6	—	Undefined
5	—	Undefined
4	—	Undefined
3	GF1	General purpose flag bit 1
2	GF0	General purpose flag bit 0
1*	PD	Power down; set to activate power down mode; only exit is reset
0*	IDL	Idle mode; set to activate idle mode; only exit is an interrupt or system reset

*Only implemented in CMOS versions

functions are stopped, (3) all on-chip RAM contents are retained, (4) port pins retain their logic levels, and (5) ALE and $\overline{PSEN}$ are held low. The only exit is a system reset.

During power down mode, V_{CC} can be as low as 2V. Care should be taken not to lower V_{CC} until after power down mode is entered, and to restore V_{CC} to 5V at least 10 oscillator cycles before the RST pin goes low again (upon leaving power down mode).

2.6 EXTERNAL MEMORY

It is important that microcontrollers have expansion capabilities beyond the on-chip resources to avoid a potential design bottleneck. If any resources must be expanded (memory, I/O, etc.), then the capability must exist. The MCS-51™ architecture provides this in the form of a 64K external code memory space and a 64K external data memory space. Extra ROM and RAM can be added as needed. Peripheral interface ICs can also be added to expand the I/O capability. These become part of the external data memory space using memory-mapped I/O.

When external memory is used, Port 0 is unavailable as an I/O port. It becomes a multiplexed address (A0–A7) and data (D0–D7) bus, with ALE latching the low-byte of the address at the beginning of each external memory cycle. Port 2 is usually (but not always) employed for the high-byte of the address bus.

Before discussing the specific details of multiplexing the address and data buses, the general idea is presented in Figure 2–7. A nonmultiplexed arrangement uses 16 dedicated address lines and eight dedicated data lines, for a total of 24 pins. The multiplexed arrangement combines eight lines for the data bus and the low-byte of the address bus,

FIGURE 2–7
Multiplexing the address bus (low-byte) and data bus

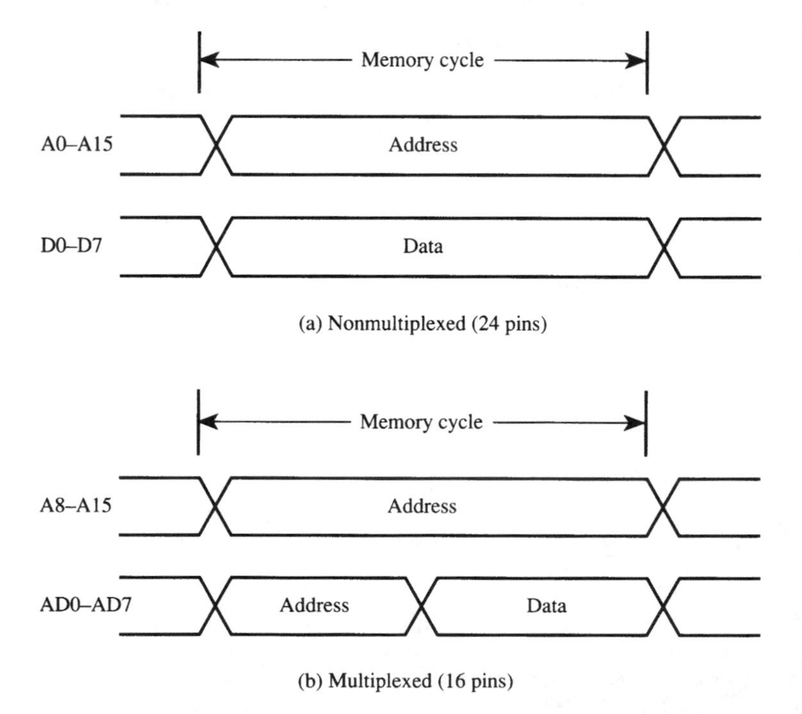

with another eight lines for the high-byte of the address bus—a total of 16 pins. The savings in pins allows other functions to be offered in a 40-pin DIP (dual inline package).

Here's how the multiplexed arrangement works: during the first half of each memory cycle, the low-byte of the address is provided on Port 0 and is latched using ALE. A 74HC373 (or equivalent) latch holds the low-byte of the address stable for the duration of the memory cycle. During the second half of the memory cycle, Port 0 is used as the data bus, and data are read or written depending on the operation.

2.6.1 Accessing External Code Memory

External code memory is read-only memory enabled by the $\overline{\text{PSEN}}$ signal. When an external EPROM is used, both Ports 0 and 2 are unavailable as general purpose I/O ports. The hardware connections for external EPROM memory are shown in Figure 2–8.

An 8051 machine cycle is 12 oscillator periods. If the on-chip oscillator is driven by a 12 MHz crystal, a machine cycle is 1 μs in duration. During a typical machine cycle, ALE pulses twice and 2 bytes are read from program memory. (If the current instruction is a 1-byte instruction, the second byte is discarded.) The timing for this operation, known as an opcode fetch, is shown in Figure 2–9.

2.6.2 Accessing External Data Memory

External data memory is read/write memory enabled by the $\overline{\text{RD}}$ and $\overline{\text{WR}}$—the alternate pin functions for P3.7 and P3.6. The only access to external data memory is with the

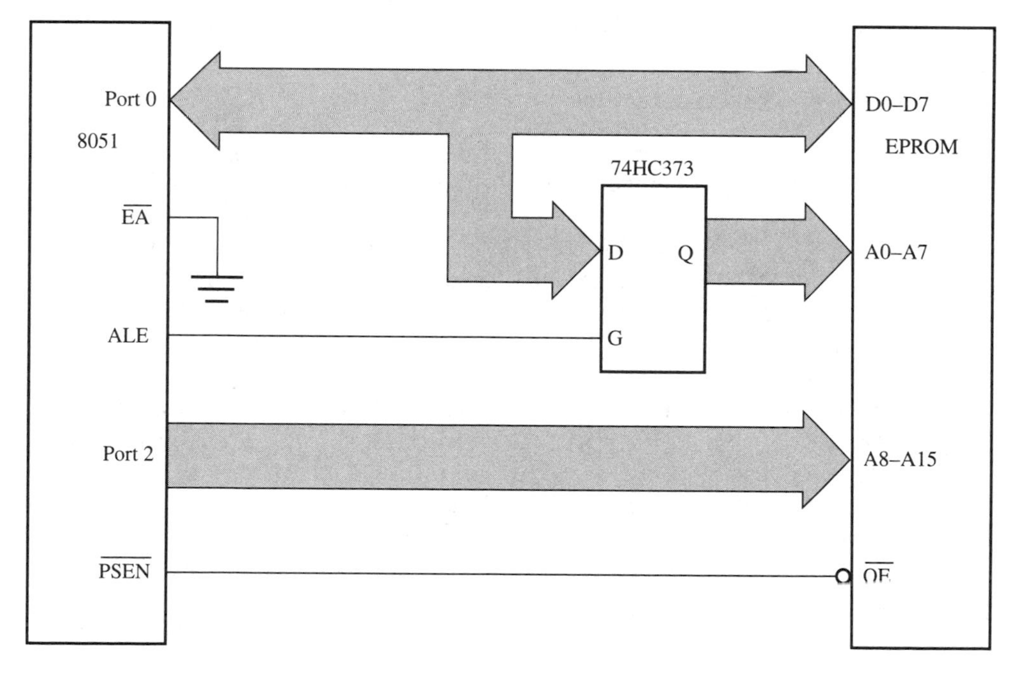

FIGURE 2–8
Accessing external code memory

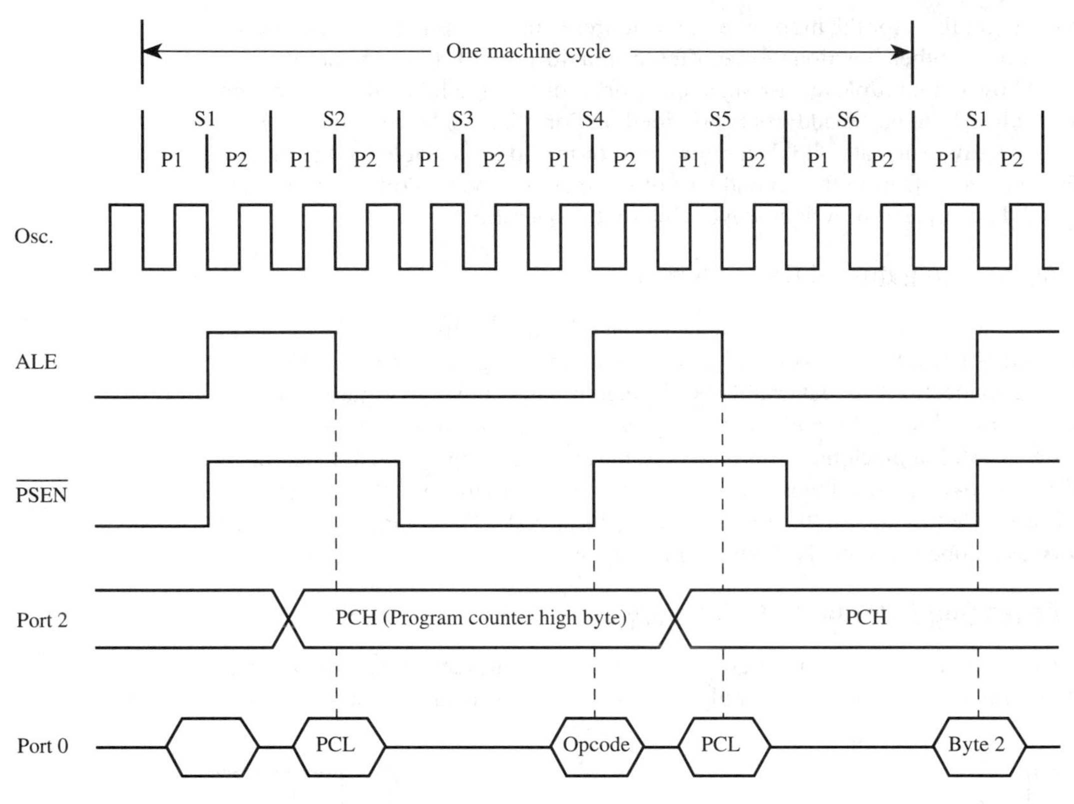

FIGURE 2–9
Read timing for external code memory

MOVX instruction, using either the 16-bit data pointer (DPTR), R0, or R1 as the address register.

RAMs may be interfaced to the 8051 the same way as EPROMs except the $\overline{\text{RD}}$ line connects to the RAM's output enable ($\overline{\text{OE}}$) line and $\overline{\text{WR}}$ connects to the RAM's write ($\overline{\text{W}}$) line. The connections for the address and data bus are the same as for EPROMs. Using Ports 0 and 2 as above, up to 64K bytes of external data RAM can be connected to the 8051.

A timing diagram for a read operation to external data memory is shown in Figure 2–10 for the MOVX A,@DPTR instruction. Notice that both an ALE pulse and a $\overline{\text{PSEN}}$ pulse are skipped in lieu of a pulse on the $\overline{\text{RD}}$ line to enable the RAM.[2]

The timing for a write cycle (MOVX @DPTR,A) is much the same except the $\overline{\text{WR}}$ line pulses low and data are outputted on Port 0. ($\overline{\text{RD}}$ remains high.)

Port 2 is relieved of its alternate function (of supplying the high-byte of the address) in minimum component systems, which use no external code memory and only a

[2]If MOVX instructions (and external RAM) are never used, then ALE pulses consistently at 1/6th the crystal frequency.

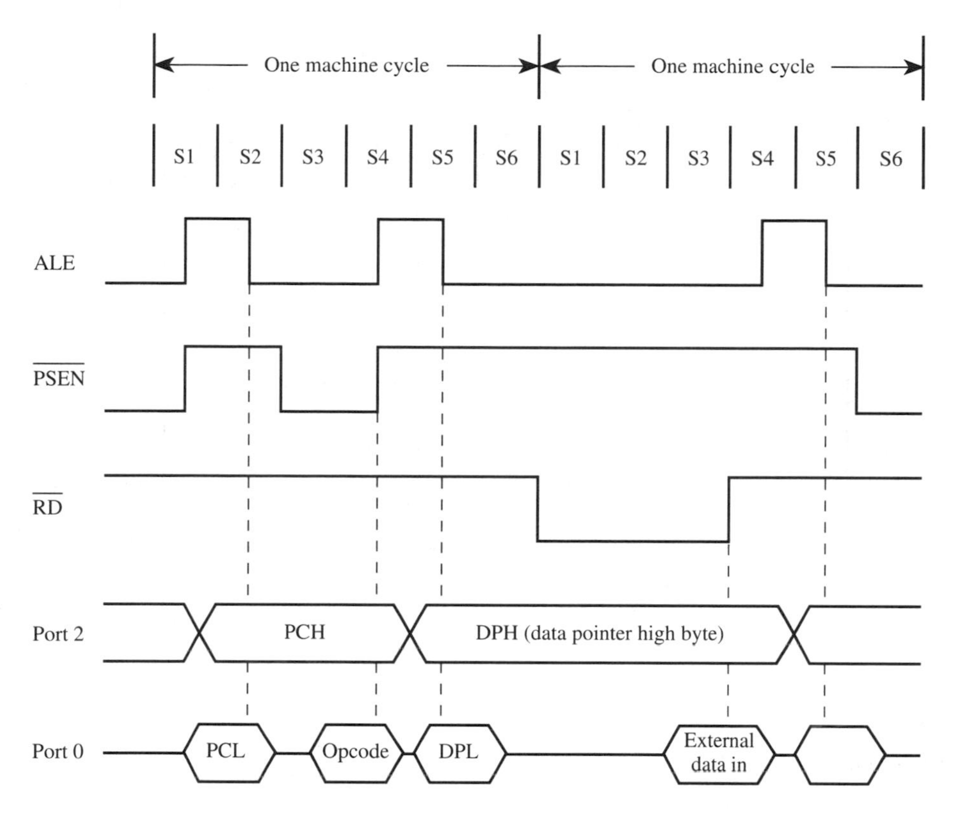

FIGURE 2–10
Timing for MOVX instruction

small amount of external data memory. Eight-bit addresses can access external data memory for small page-oriented memory configurations. If more than one 256-byte page of RAM is used, then a few bits from Port 2 (or some other port) can select a page. For example, a 1K byte RAM (i.e., four 256-byte pages) can be interfaced to the 8051 as shown in Figure 2–11.

Port 2 bits 0 and 1 must be initialized to select a page, and then a MOVX instruction is used to read or write data within that page. For example, assuming P2.0 = P2.1 = 0, the following instructions could be used to read the contents of external RAM address 0050H into the accumulator:

```
MOV  R0,#50H
MOVX A,@R0
```

In order to read the last address in this RAM, 03FFH, the two page select bits must be set. The following instruction sequence could be used:

```
SETB P2.0
SETB P2.1
MOV  R0,#0FFH
MOVX A,@R0
```

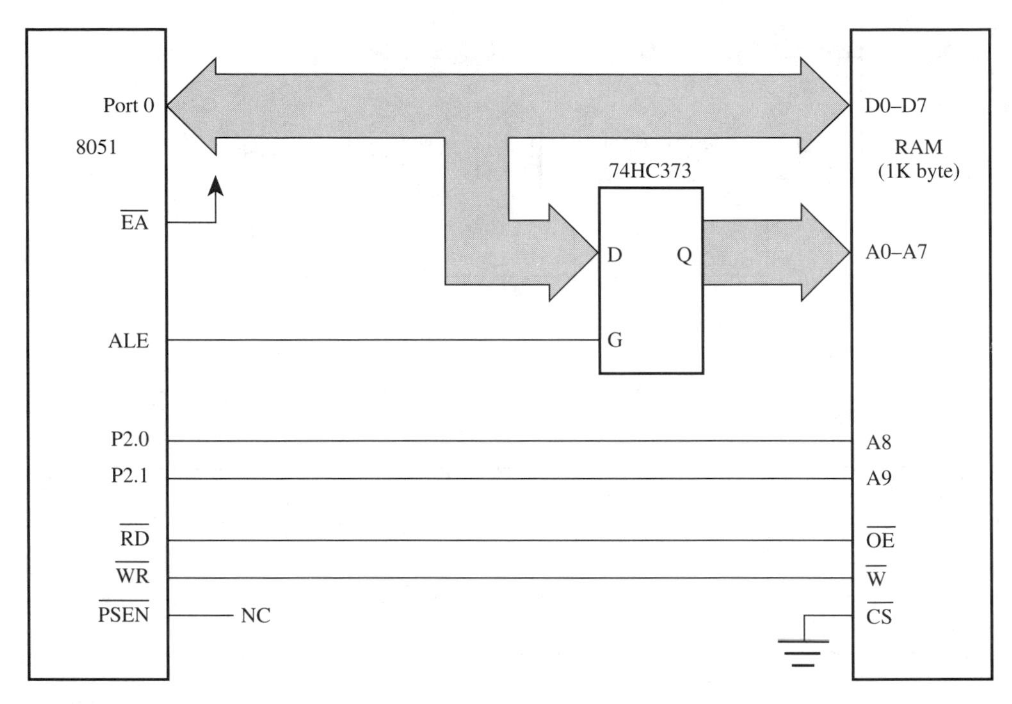

FIGURE 2–11
Interface to 1K RAM

A feature of this design is that Port 2 bits 2 to 7 are not needed as address bits, as they would be if the DPTR was the address register. P2.2 to P2.7 are available for I/O purposes.

2.6.3 Address Decoding

If multiple EPROMs and/or RAMs are interfaced to an 8051, address decoding is required. The decoding is similar to that required for most microprocessors. For example, if 8K byte EPROMs or RAMs are used, then the address bus must be decoded to select memory ICs on 8K boundaries: 0000H–1FFFH, 2000H–3FFFH, and so on.

Typically, a decoder IC such as the 74HC138 is used with its outputs connected to the chip select ($\overline{\text{CS}}$) inputs on the memory ICs. This is illustrated in Figure 2–12 for a system with multiple 2764 8K EPROMs and 6264 8K RAMs. Remember, due to the separate enable lines ($\overline{\text{PSEN}}$ for code memory, $\overline{\text{RD}}$ and $\overline{\text{WR}}$ for data memory), the 8051 can accommodate up to 64K *each* of EPROM and RAM.

2.6.4 Overlapping the External Code and Data Spaces

Since code memory is read-only, an awkward situation arises during the development of 8051 software. How is software "written into" a target system for debugging if it can only be executed from the "read-only" code space? A common trick is to overlap the external code and data memory spaces. Since $\overline{\text{PSEN}}$ is used to read code memory and $\overline{\text{RD}}$ is

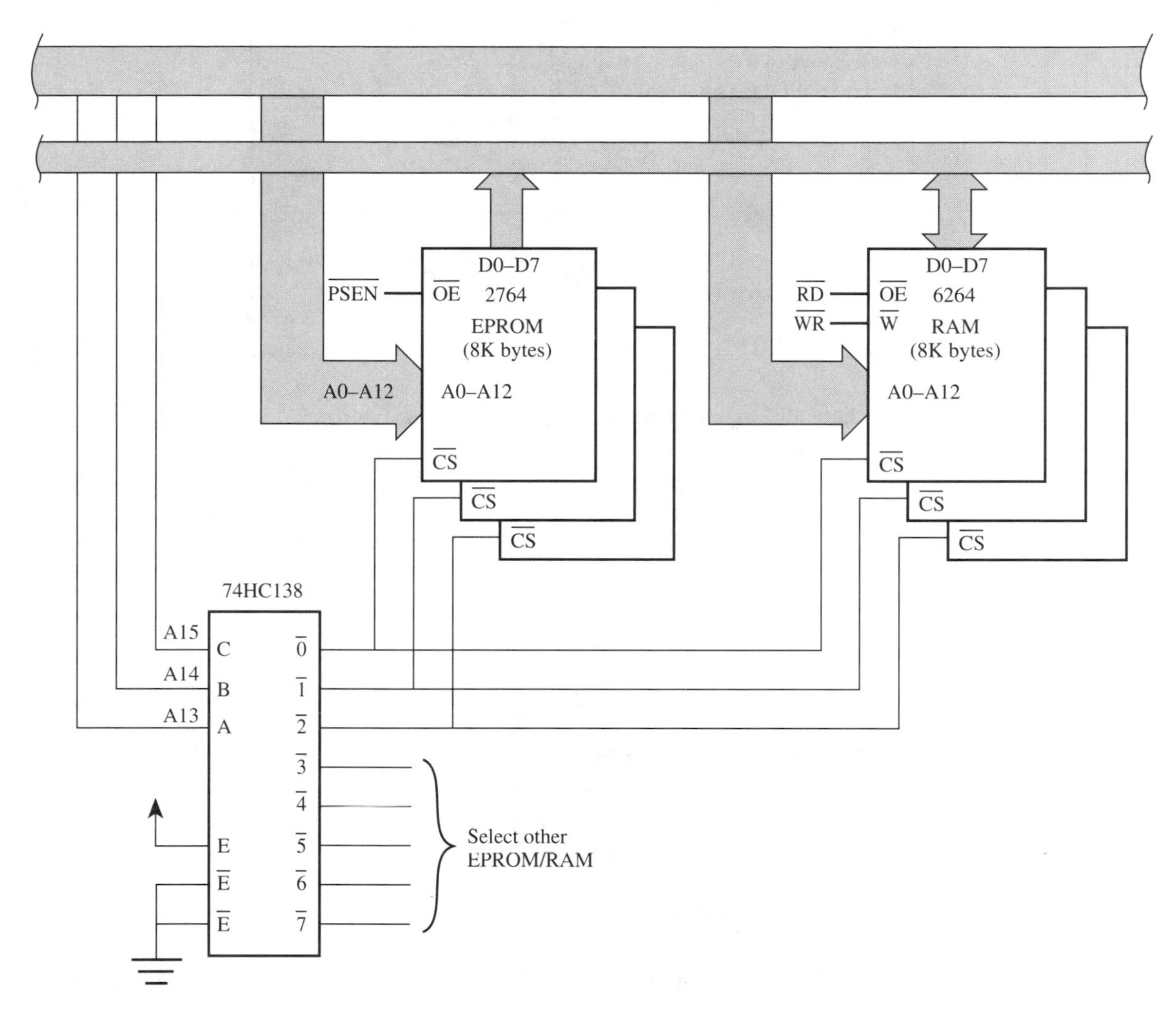

FIGURE 2–12
Address decoding

used to read data memory, a RAM can occupy code *and* data memory space by connecting its $\overline{\text{OE}}$ line to the logical AND (negative-input NOR) of $\overline{\text{PSEN}}$ and $\overline{\text{RD}}$. The circuit shown in Figure 2–13 allows the RAM IC to be written as data memory, and read as data *or* code memory. Thus a program can be loaded into the RAM (by writing to it as data memory) and executed (by accessing it as code memory).

2.7 8032/8052 ENHANCEMENTS

The 8032/8052 ICs (and the CMOS and/or EPROM versions) offer two enhancements to the 8031/8051 ICs. First, there is an additional 128 bytes of on-chip RAM from addresses 80H to FFH. So as not to conflict with the SFRs (which have the same ad-

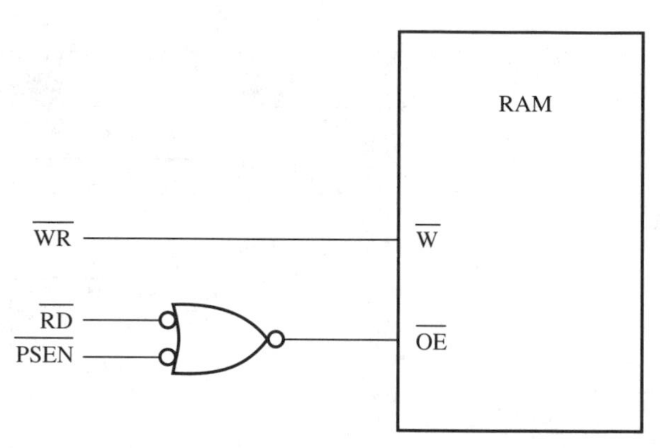

FIGURE 2–13
Overlapping the external code and data spaces

dresses), the additional 1/8K of RAM is only accessible using indirect addressing. An instruction such as

```
MOV  A,0F0H
```

moves the contents of the B register to the accumulator on all MCS-51™ ICs. The instruction sequence

```
MOV  R0,#0F0H
MOV  A,@R0
```

reads into the accumulator the contents of internal address F0H on the 8032/8052 ICs, but is undefined on the 8031/8051 ICs. The internal memory organization of the 8032/8052 ICs is summarized in Figure 2–14.

The second 8032/8052 enhancement is an additional 16-bit timer, Timer 2, which is programmed through five additional special function registers. These are summarized in Table 2–5. See Chapter 4 for more details.

2.8 RESET OPERATION

The 8051 is reset by holding RST high for at least two machine cycles and then returning it low. RST may be manually activated using a switch, or may be activated upon power-

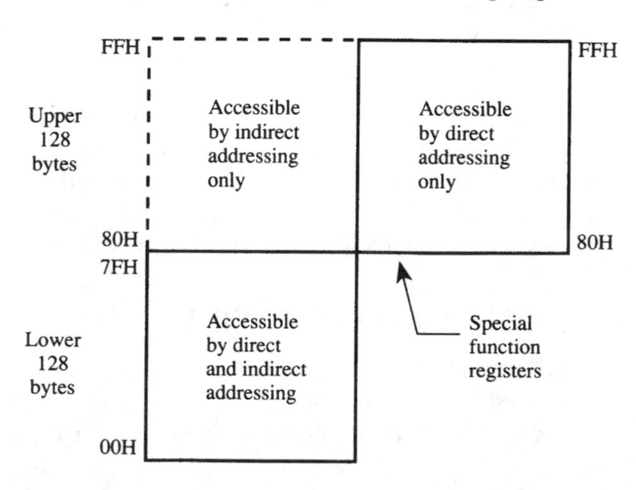

FIGURE 2–14
8032/52 memory spaces

TABLE 2–5
Timer 2 registers

REGISTER	ADDRESS	DESCRIPTION	BIT-ADDRESSABLE
T2CON	C8H	Control	Yes
RCAP2L	CAH	Low-byte capture	No
RCAP2H	CBH	High-byte capture	No
TL2	CCH	Timer 2 low-byte	No
TH2	CDH	Timer 2 high-byte	No

up using an R-C (resistor-capacitor) network. Figure 2–15 illustrates two circuits for implementing system reset.

The state of all the 8051 registers after a system reset is summarized in Table 2–6. The most important of these registers, perhaps, is the program counter, which is loaded with 0000H. When RST returns low, program execution always begins at the first location in code memory: address 0000H. The content of on-chip RAM is not affected by a reset operation.

2.9 SUMMARY

This chapter has summarized the 8051 hardware architecture. Before developing useful applications, though, we must understand the 8051 instruction set. The next chapter fo-

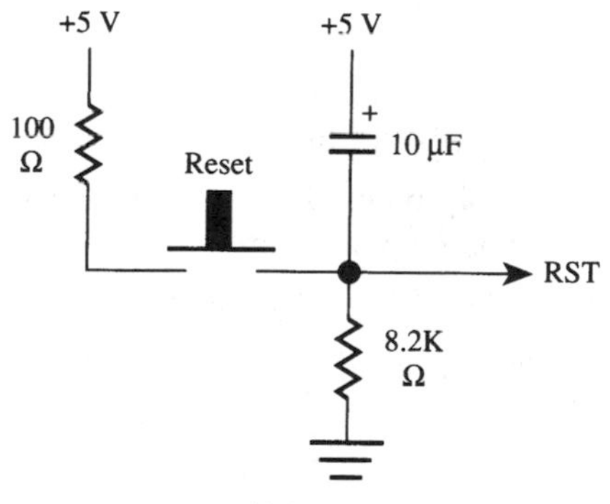

(a) Manual reset

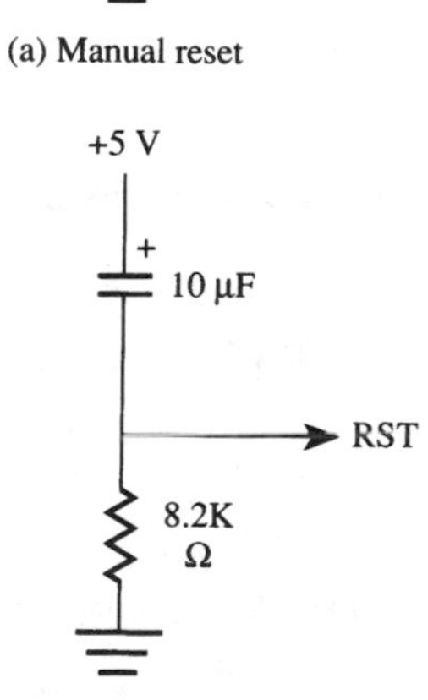

(b) Power-on reset

FIGURE 2–15
Two circuits for system reset. (a) Manual Reset (b) Power-on Reset

TABLE 2–6
Register values after system reset

REGISTER(S)	CONTENTS
Program counter	0000H
Accumulator	00H
B register	00H
PSW	00H
SP	07H
DPTR	0000H
Ports 0-3	FFH
IP (8031/8051)	XXX00000B
IP (8032/8052)	XX000000B
IE (8031/8051)	0XX00000B
IE (8032/8052)	0X000000B
Timer registers	00H
SCON	00H
SBUF	00H
PCON (HMOS)	0XXXXXXXB
PCON (CMOS)	0XXX0000B

cuses on the 8051 instructions and addressing modes. The discussions of the timer, serial port, and interrupt SFRs were deliberately sparse in this chapter, since dedicated chapters follow that examine these in detail.

PROBLEMS

1. Name four manufacturers of the 8051 microcontroller, besides Intel.
2. Which device in the MCS-51™ family would probably be used for a product that will be manufactured in large quantities with a large on-chip program?
3. What instruction could be used to set the least-significant bit at byte address 25H?
4. What instruction sequence could be used to place the logical OR of the bits at bit addresses 00H and 01H into bit address 02H?
5. What bit addresses are set to one as a result of the following instructions?

```
MOV  R0,#26H
MOV  @R0,#7AH
```

6. What 1-byte instruction has the same effect as the following 2-byte instruction?

```
MOV  0E0H,#55H
```

7. Illustrate an instruction sequence to store the value 0ABH in external RAM at address 9A00H.
8. How many special function registers are defined on the 8052?
9. What is the value of the 8051's stack pointer immediately after a system reset?
10. What instruction could be used to initialize the 8031 SP to create a 64-byte stack at the top of internal RAM?

11. A certain subroutine makes extensive use of registers R0–R7. Illustrate how this subroutine could switch the active register bank to bank 3 upon entry, and restore the previously active register bank upon exit.
12. The 80C31BH–1 can operate using a 16 MHz crystal connected to its XTAL1 and XTAL2 inputs. If MOVX instructions are not used, what is the frequency of the signal on ALE?
13. If an 8051 is operating from a 4 MHz crystal, what is the duration of a machine cycle?
14. If an 8051 is operating from a 10 MHz crystal, what is the frequency of the waveform on ALE? Assume the software is not accessing external RAM.
15. What is the duty cycle of ALE? Assume that software is not accessing external RAM. (Note: Duty cycle is defined as the proportion of time a pulse waveform is high.)
16. Section 2.8 states that the 8051 is reset if the RST pin is held high for a minimum of two machine cycles. (Note: As stated in the 8051's DC Characteristics in Appendix E, a "high" on RST is 2.5 volts minimum.)

    (a) If an 8051 is operating from an 8 MHz crystal, what is the minimum length of time for RST to be high to achieve a system reset?

    (b) Figure 2–15a shows an RC circuit for a manual reset. While the reset button is depressed, RST = 5 volts and the system is held in a reset state. How long after the reset button is released will the 8051 remain in a reset state?
17. How many low-power Schottky loads can be driven by the port line P1.7 on pin 8?
18. Name the 8051 control bus signals used to select external EPROMs and external RAMs.
19. What is the bit address of the most-significant bit at byte address 25H in the 8051's internal data memory?
20. What instruction sets the least-significant bit of the accumulator without affecting the other 7 bits?
21. Assuming the following instruction has just executed,

    ```
    MOV  A,#55H
    ```

    what is the state of the P bit in the program status word?
22. What instruction sequence could be used to copy the contents of R7 to external RAM location 100H?
23. Assume the first instruction executed following a system reset is a subroutine call. At what addresses in internal RAM is the program counter saved before branching to the subroutine?
24. What is the difference between the 8051's idle mode and power-down mode?
25. What instruction could be used to force the 8051 into power-down mode?
26. Illustrate how two 32K-byte static RAMs could be interfaced to the 8051 so that they occupy the full 64K external data space.

3

INSTRUCTION SET SUMMARY

3.1 INTRODUCTION

Just as sentences are made of words, programs are made of instructions. When programs are constructed from logical, well-thought-out sequences of instructions, fast, efficient, and even elegant programs result. Unique to each family of computers is its instruction set, a repertoire of primitive operations such as "add," "move," or "jump." This chapter introduces the MCS-51™ instruction set through an examination of addressing modes and examples from typical programming situations. Appendix A contains a summary chart of all the 8051 instructions. Appendix C provides a detailed description of each instruction. These appendices should be consulted for subsequent reference.

Programming techniques are not discussed, nor is the operation of the assembler program used to convert assembly language programs (mnemonics, labels, etc.) into machine language programs (binary codes). These topics are the subject of Chapter 7.

The MCS-51™ instruction set is optimized for 8-bit control applications. It provides a variety of fast, compact addressing modes for accessing the internal RAM to facilitate operations on small data structures. The instruction set offers extensive support for 1-bit variables, allowing direct bit manipulation in control and logic systems that require Boolean processing.

As typical of 8-bit processors, 8051 instructions have 8-bit opcodes. This provides a possibility of $2^8 = 256$ instructions. Of these, 255 are implemented and 1 is undefined. As well as the opcode, some instructions have one or two additional bytes for data or addresses. In all, there are 139 1-byte instructions, 92 2-byte instructions, and 24 3-byte instructions. The *Opcode Map* in Appendix B shows, for each opcode, the mnemonic, the number of bytes in the instruction, and the number of machine cycles to execute the instruction.

3.2 ADDRESSING MODES

When instructions operate on data, the question arises: "Where's the data?" The answer to this question lies in the 8051's "addressing modes." There are several possible addressing modes and there are several possible answers to the question, such as "in byte 2 of the instruction," "in register R4," "in direct address 35H," or perhaps "in external data memory at the address contained in the data pointer."

Addressing modes are an integral part of each computer's instruction set. They allow specifying the source or destination of data in different ways depending on the programming situation. In this section, we'll examine all the 8051 addressing modes and give examples of each. There are eight modes available:

- Register
- Direct
- Indirect
- Immediate
- Relative
- Absolute
- Long
- Indexed

3.2.1 Register Addressing

The 8051 programmer has access to 8 "working registers," numbered R0 through R7. Instructions using register addressing are encoded using the three least-significant bits of the instruction opcode to indicate 1 register within this logical address space. Thus, a function code and operand address can be combined to form a short (1-byte) instruction. (See Figure 3–1a.)

The 8051 assembly language indicates register addressing with the symbol *Rn* where *n* is from 0 to 7. For example, to add the contents of Register 7 to the accumulator, the following instruction is used

```
ADD  A,R7
```

and the opcode is 00101111B. The upper five bits, 00101, indicate the instruction, and the lower three bits, 111, the register. Convince yourself that this is the correct opcode by looking up this instruction in Appendix C.

There are four "banks" of working registers, but only one is active at a time. Physically, the register banks occupy the first 32 bytes of on-chip data RAM (addresses 00H–1FH) with PSW bits 4 and 3 determining the active bank. A hardware reset enables bank 0, but a different bank is selected by modifying PSW bits 4 and 3 accordingly. For example, the instruction

```
MOV  PSW,#00011000B
```

activates register bank 3 by setting the register bank select bits (RS1 and RS0) in PSW bit positions 4 and 3.

Some instructions are specific to a certain register, such as the accumulator, data pointer, etc., so address bits are not needed. The opcode itself indicates the register. These "register-specific" instructions refer to the accumulator as "A," the data pointer as "DPTR," the program counter as "PC," the carry flag as "C," and the accumulator-B register pair as "AB." For example,

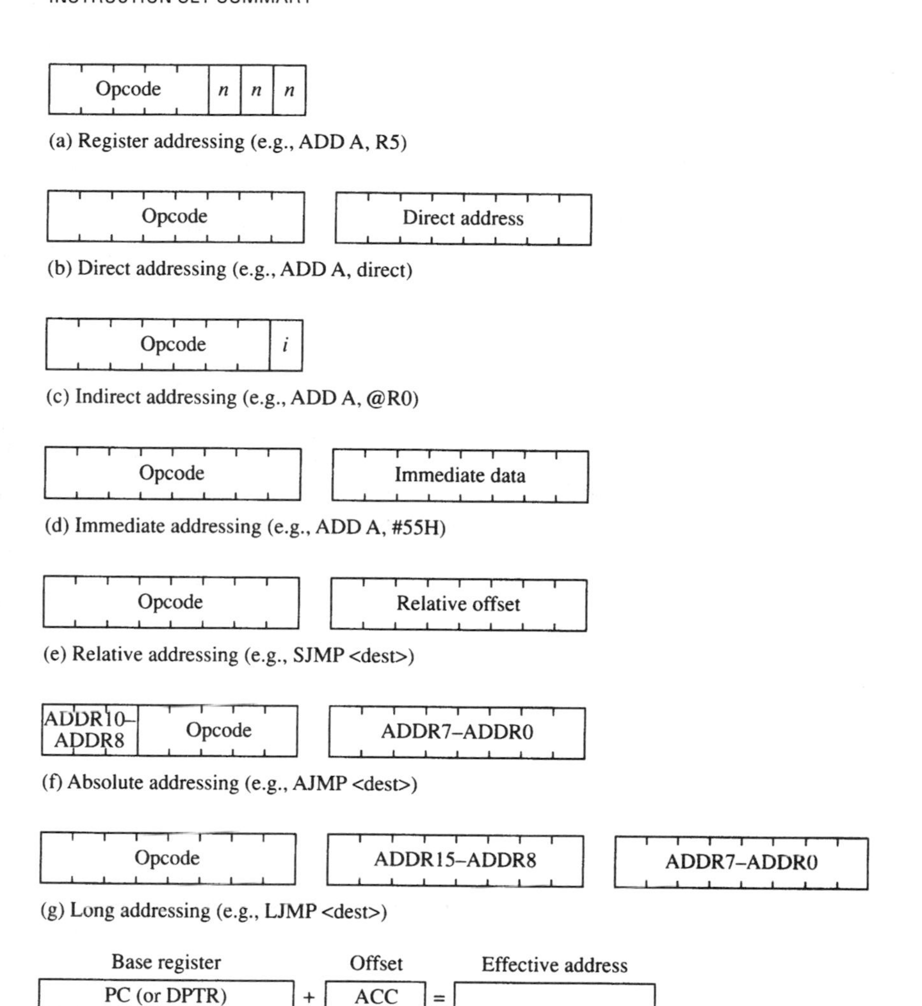

(h) Indexed addressing (e.g., MOVC A, @A + PC)

FIGURE 3–1
8051 Addressing modes. (a) Register addressing (b) Direct addressing (c) Indirect addressing (d) Immediate addressing (e) Relative addressing (f) Absolute addressing (g) Long addressing (h) Indexed addressing.

```
        INC  DPTR
```

is a 1-byte instruction that adds 1 to the 16-bit data pointer. Consult Appendix C to determine the opcode for this instruction.

3.2.2 Direct Addressing

Direct addressing can access any on-chip variable or hardware register. An additional byte is appended to the opcode specifying the location to be used. (See Figure 3–1b.)

Depending on the high-order bit of the direct address, one of two on-chip memory spaces is selected. When bit 7 = 0, the direct address is between 0 and 127 (00H–7FH) and the 128 low-order on-chip RAM locations are referenced. All I/O ports and special function, control, or status registers, however, are assigned addresses between 128 and 255 (80H–FFH). When the direct address byte is between these limits (bit 7 = 1), the corresponding special function register is accessed. For example, Ports 0 and 1 are assigned direct addresses 80H and 90H, respectively. It is not necessary to know the addresses of these registers; the assembler allows for and understands the mnemonic abbreviations ("P0" for Port 0, "TMOD" for timer mode register, etc.). As an example of direct addressing, the instruction

```
MOV  P1,A
```

transfers the contents of the accumulator to Port 1. The direct address of Port 1 (90H) is determined by the assembler and inserted as byte 2 of the instruction. The source of the data, the accumulator, is specified implicitly in the opcode. Using Appendix C as a reference, the complete encoding of this instruction is

```
10001001 - 1st byte (opcode)
10010000 - 2nd byte (address of P1)
```

3.2.3 Indirect Addressing

How is a variable identified if its address is determined, computed, or modified while a program is running? This situation arises when manipulating sequential memory locations, indexed entries within tables in RAM, multiple-precision numbers, or character strings. Register or direct addressing cannot be used, since they require operand addresses to be known at assemble-time.

The 8051 solution is indirect addressing. R0 and R1 may operate as "pointer" registers—their contents indicating an address in RAM where data are written or read. The least-significant bit of the instruction opcode determines which register (R0 or R1) is used as the pointer. (See Figure 3–1c.)

In 8051 assembly language, indirect addressing is represented by a commercial "at" sign (@) preceding R0 or R1. As an example, if R1 contains 40H and internal memory address 40H contains 55H, the instruction

```
MOV  A,@R1
```

moves 55H into the accumulator.

Indirect addressing is essential when stepping through sequential memory locations. For example, the following instruction sequence clears internal RAM from address 60H to 7FH:

```
            MOV  R0,#60H
LOOP:       MOV  @R0,#0
            INC  R0
            CJNE R0,#80H,LOOP
            (continue)
```

The first instruction initializes R0 with the starting address of the block of memory; the second instruction uses indirect addressing to move 00H to the location pointed at by R0; the third instruction increments the pointer (R0) to the next address; and the last instruction tests the pointer to see if the end of the block has been reached. The test uses 80H, rather than 7FH, because the increment occurs after the indirect move. This ensures the final location (7FH) is written to before terminating.

3.2.4 Immediate Addressing

When a source operand is a constant rather than a variable (i.e., the instruction uses a value known at assemble-time), then the constant can be incorporated into the instruction as a byte of "immediate" data. An additional instruction byte contains the value. (See Figure 3–1d.)

In assembly language, immediate operands are preceded by a number sign (#). The operand may be a numeric constant, a symbolic variable, or an arithmetic expression using constants, symbols, and operators. The assembler computes the value and substitutes the immediate data into the instruction. For example, the instruction

```
MOV  A,#12
```

loads the value 12 (0CH) into the accumulator. (It is assumed the constant "12" is in decimal notation, since it is not followed by "H.")

With one exception, all instructions using immediate addressing use an 8-bit data constant for the immediate data. When initializing the data pointer, a 16-bit constant is required. For example,

```
MOV  DPTR,#8000H
```

is a 3-byte instruction that loads the 16-bit constant 8000H into the data pointer.

3.2.5 Relative Addressing

Relative addressing is used only with certain jump instructions. A relative address (or offset) is an 8-bit signed value, which is added to the program counter to form the address of the next instruction executed. Since an 8-bit signed offset is used, the range for jumping is −128 to +127 locations. The relative offset is appended to the instruction as an additional byte. (See Figure 3–1e.)

Prior to the addition, the program counter is incremented to the address following the jump instruction; thus, the new address is relative to the next instruction, *not* the address of the jump instruction. (See Figure 3–2.)

Normally, this detail is of no concern to the programmer, since jump destinations are usually specified as labels and the assembler determines the relative offset accordingly. For example, if the label THERE represents an instruction at location 1040H, and the instruction

```
SJMP THERE
```

is in memory at locations 1000H and 1001H, the assembler will assign a relative offset of 3EH as byte 2 of the instruction (1002H +3EH = 1040H).

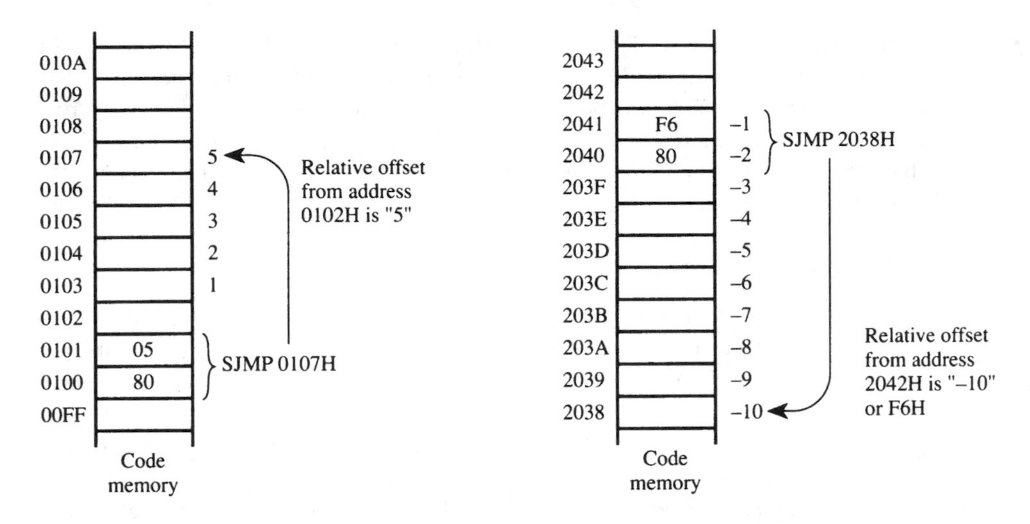

FIGURE 3–2
Calculating the offset for relative addressing. (a) Short jump ahead in memory. (b) Short jump back in memory.

Relative addressing offers the advantage of providing position-independent code (since "absolute" addresses are not used), but the disadvantage that the jump destinations are limited in range.

3.2.6 Absolute Addressing

Absolute addressing is used only with the ACALL and AJMP instructions. These 2-byte instructions allow branching within the current 2K page of code memory by providing the 11 least-significant bits of the destination address in the opcode (A10–A8) and byte 2 of the instruction (A7–A0). (See Figure 3–1f.)

The upper five bits of the destination address are the current upper five bits in the program counter, so the instruction following the branch instruction and the destination for the branch instruction must be within the same 2K page, since A15–A11 do not change. (See Figure 3–3.) For example, if the label THERE represents an instruction at address 0F46H, and the instruction

```
AJMP THERE
```

is in memory locations 0900H and 0901H, the assembler will encode the instruction as

```
11100001 - 1st byte (A10-A8 + opcode)
01000110 - 2nd byte (A7-A0)
```

The underlined bits are the low-order 11 bits of the destination address, 0F46H = 0000111101000110B. The upper 5 bits in the program counter will not change when this instruction executes. Note that both the AJMP instruction and the destination are within the 2K page bounded by 0800H and 0FFFH (see Figure 3–3), and therefore have the upper five address bits in common.

Absolute addressing offers the advantage of short (2-byte) instructions, but has the disadvantages of limiting the range for the destination and providing position-dependent code.

3.2.7 Long Addressing

Long addressing is used only with the LCALL and LJMP instructions. These 3-byte instructions include a full 16-bit destination address as bytes 2 and 3 of the instruction. (See Figure 3–1g.) The advantage is that the full 64K code space may be used, but the disadvantage is that the instructions are three bytes long and position-dependent. Position-dependence is a disadvantage because the program cannot execute at different addresses. If, for example, a program begins at 2000H and an instruction such as LJMP 2040H appears, then the program cannot be moved to, say, 4000H. The LJMP instruction would still jump to 2040H, which is not the correct location after the program has been moved.

3.2.8 Indexed Addressing

Indexed addressing uses a base register (either the program counter or the data pointer) and an offset (the accumulator) in forming the effective address for a JMP or MOVC instruction. (See Figure 3–1h.) Jump tables or look-up tables are easily created using indexed addressing. Examples are provided in Appendix C for the MOVC A, @A+<base-reg> and JMP @A+DPTR instructions.

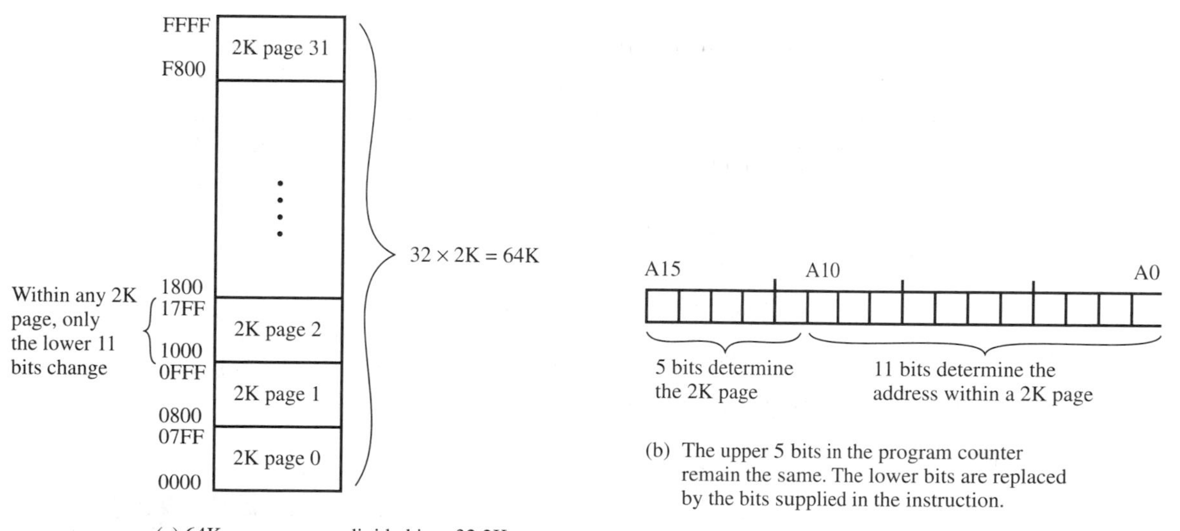

FIGURE 3–3
Instruction encoding for absolute addressing. (a) Memory map showing 2K pages (b) Within any 2K page, the upper 5 address bits are the same.

3.3 INSTRUCTION TYPES

The 8051 instructions are divided among five functional groups:

- □ Arithmetic
- □ Logical
- □ Data transfer
- □ Boolean variable
- □ Program branching

Appendix A provides a quick reference chart showing all the 8051 instructions by functional grouping. Once you are familiar with the instruction set, this chart should prove a handy and quick source of reference. We continue by examining instructions in each functional grouping from Appendix A.

3.3.1 Arithmetic Instructions

The arithmetic instructions are grouped together in Appendix A. Since four addressing modes are possible, the ADD A instruction can be written in different ways:

```
ADD  A,7FH                   (direct addressing)
ADD  A,@R0                   (indirect addressing)
ADD  A,R7                    (register addressing)
ADD  A,#35H                  (immediate addressing)
```

All arithmetic instructions execute in 1 machine cycle except the INC DPTR instruction (2 machine cycles) and the MUL AB and DIV AB instructions (4 machine cycles). (Note that one machine cycle takes 1 μs if the 8051 is operating from a 12 MHz clock.)

The 8051 provides powerful addressing of its internal memory space. Any location can be incremented or decremented using direct addressing without going through the accumulator. For example, if internal RAM location 7FH contains 40H, then the instruction

```
INC  7FH
```

increments this value, leaving 41H in location 7FH.

One of the INC instructions operates on the 16-bit data pointer. Since the data pointer generates 16-bit addresses for external memory, incrementing it in one operation is a useful feature. Unfortunately a decrement data pointer instruction is not provided and requires a sequence of instructions such as the following:

```
            DEC  DPL                 ;DECREMENT LOW-BYTE OF DPTR
            MOV  R7,DPL              ;MOVE TO R7
            CJNE R7,#0FFH,SKIP       ;IF UNDERFLOW TO FF
            DEC  DPH                 ;DECREMENT HIGH-BYTE TOO
SKIP:       (continue)
```

The high- and low-bytes of the DPTR must be decremented separately; however, the high-byte (DPH) is only decremented if the low-byte (DPL) underflows from 00H to FFH.

The MUL AB instruction multiplies the accumulator by the data in the B register and puts the 16-bit product into the concatenated B (high-byte) and accumulator (low-byte) registers. DIV AB divides the accumulator by the data in the B register, leaving the 8-bit quotient in the accumulator and the 8-bit remainder in the B register. For example, if A contains 25 (19H) and B contains 6 (06H), the instruction

```
DIV  AB
```

divides the contents of A by the contents of B. The A accumulator is left with the value 4 and the B accumulator is left with the value 1. ($25 \div 6 = 4$ with a remainder of 1.)

For BCD (binary-coded decimal) arithmetic, ADD and ADDC must be followed by a DA A (decimal adjust) operation to ensure the result is in range for BCD. Note that DA A will not convert a binary number to BCD; it produces a meaningful result only as the second step in the addition of 2 BCD bytes. For example, if A contains the BCD value 59 (59H), then the instruction sequence

```
ADD  A,#1
DA   A
```

first adds 1 to A, leaving the result 5AH, then adjusts the result to the correct BCD value of 60 (60H). ($59 + 1 = 60$.)

3.3.2 Logical Instructions

The 8051 logical instructions (see Appendix A) perform Boolean operations (AND, OR, Exclusive OR, and NOT) on bytes of data on a bit-by-bit basis. If the accumulator contains 00110101B, then the following AND logical instruction

```
ANL  A,#01010011B
```

leaves the accumulator holding 00010001B. This is illustrated below.

```
        01010011              (immediate data)
AND     00110101              (original value of A)
        00010001              (result in A)
```

Since the addressing modes for the logical instructions are the same as those for arithmetic instructions, the AND logical instruction can take several forms:

```
ANL  A,55H                    (direct addressing)
ANL  A,@R0                    (indirect addressing)
ANL  A,R6                     (register addressing)
ANL  A,#33H                   (immediate addressing)
```

All logical instructions using the accumulator as one of the operands execute in one machine cycle. The others take two machine cycles.

Logical operations can be performed on any byte in the internal data memory space without going through the accumulator. The "XRL direct,#data" instruction offers a quick and easy way to invert port bits, as in

```
XRL  P1,#0FFH
```

This instruction performs a read-modify-write operation. The eight bits at Port 1 are read; then each bit read is exclusive ORed with the corresponding bit in the immediate data. Since the eight bits of immediate data are all 1s, the effect is to complement each bit read (e.g., $A \oplus 1 = \overline{A}$). The result is written back to Port 1.

The rotate instructions (RL A and RR A) shift the accumulator one bit to the left or right. For a left rotation, the MSB rolls into the LSB position. For a right rotation, the LSB rolls into the MSB position. The RLC A and RRC A variations are 9-bit rotates using the accumulator and the carry flag in the PSW. If, for example, the carry flag contains 1 and A contains 00H, then the instruction

```
RRC  A
```

leaves the carry flag clear and A equal to 80H. The carry flag rotates into ACC.7 and ACC.0 rotates into the carry flag.

The SWAP A instruction exchanges the high and low nibbles within the accumulator. This is a useful operation in BCD manipulations. For example, if the accumulator contains a binary number that is known to be less than 100_{10}, it is quickly converted to BCD as follows:

```
MOV  B,#10
DIV  AB
SWAP A
ADD  A,B
```

Dividing the number by 10 in the first two instructions leaves the tens digit in the low nibble of the accumulator, and the ones digit in the B register. The SWAP and ADD instructions move the tens digit to the high nibble of the accumulator, and the ones digit to the low nibble.

3.3.3 Data Transfer Instructions

3.3.3.1 Internal RAM

The instructions that move data within the internal memory spaces (see Appendix A) execute in either one or two machine cycles. The instruction format

```
MOV     <destination>, <source>
```

allows data to be transferred between any two internal RAM or SFR locations without going through the accumulator. Remember, the upper 128 bytes of data RAM (8032/8052) are accessed only by indirect addressing, and the SFRs are accessed only by direct addressing.

A feature of the MCS-51™ architecture differing from most microprocessors is that the stack resides in on-chip RAM and grows upward in memory, toward higher memory addresses. The PUSH instruction first increments the stack pointer (SP), then copies the byte into the stack. PUSH and POP use direct addressing to identify the byte being saved or restored, but the stack itself is accessed by indirect addressing using the SP register. This means the stack can use the upper 128 bytes of internal memory on the 8032/8052.

The upper 128 bytes of internal memory are not implemented in the 8031/8051 devices. With these devices, if the SP is advanced above 7FH (127), the PUSHed bytes are lost and the POPed bytes are indeterminate.

Data transfer instructions include a 16-bit MOV to initialize the data pointer (DPTR) for look-up tables in program memory, or for 16-bit external data memory accesses.

The instruction format

```
XCH  A,<source>
```

causes the accumulator and the addressed byte to exchange data. An exchange "digit" instruction of the form

```
XCHD A,@Ri
```

is similar, but only the low-order nibbles are exchanged. For example, if A contains F3H, R1 contains 40H, and internal RAM address 40H contains 5BH, then the instruction

```
XCHD A,@R1
```

leaves A containing FBH and internal RAM location 40H containing 53H.

3.3.3.2 External RAM

The data transfer instructions that move data between internal and external memory use indirect addressing. The indirect address is specified using a 1-byte address (@Ri, where Ri is either R0 or R1 of the selected register bank), or a 2-byte address (@DPTR). The disadvantage in using 16-bit addresses is that all 8 bits of Port 2 are used as the high-byte of the address bus. This precludes the use of Port 2 as an I/O port. On the other hand, 8-bit addresses allow access to a few Kbytes of RAM, without sacrificing all of Port 2. (See Chapter 2, "Accessing External Data Memory.")

All data transfer instructions that operate on external memory execute in 2 machine cycles and use the accumulator as either the source or destination operand.

The read and write strobes to external RAM ($\overline{RD}$ and $\overline{WR}$) are activated only during the execution of a MOVX instruction. Normally, these signals are inactive (high), and if external data memory is not used, they are available as dedicated I/O lines.

3.3.3.3 Look-Up Tables

Two data transfer instructions are available for reading look-up tables in program memory. Since they access program memory, the look-up tables can only be read, not updated. The mnemonic is MOVC for "move constant." MOVC uses either the program counter or the data pointer as the base register and the accumulator as the offset.

The instruction

```
MOVC A,@A+DPTR
```

can accommodate a table of 256 entries, numbered 0 through 255. The number of the desired entry is loaded into the accumulator and the data pointer is initialized to the beginning of the table. The instruction

```
MOVC A,@A+PC
```

works the same way, except the program counter is used as the base address, and the table is accessed through a subroutine. First, the number of the desired entry is loaded into the accumulator, then the subroutine is called. The setup and call sequence would be coded as follows:

```
            MOV  A,ENTRY_NUMBER
            CALL LOOK_UP
            .
            .
            .
LOOK_UP:    INC  A
            MOVC A,@A+PC
            RET
TABLE:      DB   data,data,data,data, . . .
```

The table immediately follows the RET instruction in program memory. The INC instruction is needed because the PC points to the RET instruction when MOVC executes. Incrementing the accumulator will effectively bypass the RET instruction when the table look-up takes place.

3.3.4 Boolean Instructions

The 8051 processor contains a complete Boolean processor for single-bit operations. The internal RAM contains 128 addressable bits, and the SFR space supports up to 128 other addressable bits. All port lines are bit-addressable, and each can be treated as a separate single-bit port. The instructions that access these bits are not only conditional branches, but also a complete repertoire of move, set, clear, complement, OR, and AND instructions. Such bit operations—one of the most powerful features of the MCS-51™ family of microcontrollers—are not easily obtained in other architectures with byte-oriented operations.

The available Boolean instructions are shown in Appendix A. All bit accesses use direct addressing with bit addresses 00H–7FH in the lower 128 locations, and bit addresses 80H–FFH in the SFR space. Those in the lower 128 locations at byte addresses 20H–2FH are numbered sequentially from bit 0 of address 20H (bit 00H) to bit 7 of address 2FH (bit 7FH).

Bits may be set or cleared in a single instruction. Single-bit control is common for many I/O devices, including output to relays, motors, solenoids, status LEDs, buzzers, alarms, loudspeakers, or input from a variety of switches or status indicators. If an alarm is connected to Port 1 bit 7, for example, it might be turned on by setting the port bit,

```
SETB P1.7
```

and turned off by clearing the port bit

```
CLR  P1.7
```

The assembler will do the necessary conversion of the symbol "P1.7" into the correct bit address, 97H.

Note how easily an internal flag can be moved to a port pin:

```
        MOV  C,FLAG
        MOV  P1.0,C
```

In this example, FLAG is the name of any addressable bit in the lower 128 locations or the SFR space. An I/O line (the LSB of Port 1, in this case) is set or cleared depending on whether the flag bit is 1 or 0.

The carry bit in the program status word (PSW) is used as the single-bit accumulator of the Boolean processor. Bit instructions that refer to the carry bit as "C" assemble as carry-specific instructions (e.g., CLR C). The carry bit also has a direct address, since it resides in the PSW register, which is bit-addressable. Like other bit-addressable SFRs, the PSW bits have predefined mnemonics that the assembler will accept in lieu of the bit address. The carry flag mnemonic is "CY," which is defined as bit address 0D7H. Consider the following two instructions:

```
        CLR  C
        CLR  CY
```

Both have the same effect; however, the former is a 1-byte instruction, while the latter is a 2-byte instruction. In the latter case, the second byte is the direct address of the specified bit—the carry flag.

Note that the Boolean instructions include ANL (AND logical) and ORL (OR logical) operations, but not the XRL (exclusive OR logical) operation. An XRL operation is simple to implement. Suppose, for example, it is required to form the exclusive OR of two bits, BIT1 and BIT2, and leave the result in the carry flag. The instructions are shown below.

```
        MOV  C,BIT1
        JNB  BIT2,SKIP
        CPL  C
SKIP:   (continue)
```

First, BIT1 is moved to the carry flag. If BIT2 = 0, then C contains the correct result; that is, BIT1 $\oplus$ BIT2 = BIT1 if BIT2 = 0. If BIT2 = 1, C contains the complement of the correct result. Complementing C completes the operation.

3.3.4.1 Bit Testing

The code in the example above uses the JNB instruction, one of a series of bit-test instructions that jump if the addressed bit is set (JC, JB, JBC) or if the addressed bit is not set (JNC, JNB). In the above case, if BIT2 = 0 the CPL instruction is skipped. JBC (jump if bit set then clear bit) executes the jump if the addressed bit is set, and also clears the bit; thus, a flag can be tested and cleared in a single instruction.

All PSW bits are directly addressable, so the parity bit or the general purpose flags, for example, are also available for bit-test instructions.

3.3.5 Program Branching Instructions

As evident in Appendix A, there are numerous instructions to control the flow of programs, including those that call and return from subroutines or branch conditionally or unconditionally. These possibilities are enhanced further by the three addressing modes for the program branching instructions.

There are three variations of the JMP instruction: SJMP, LJMP, and AJMP (using relative, long, and absolute addressing, respectively). Intel's assembler (ASM51) allows the use of the generic JMP mnemonic if the programmer does not care which variation is encoded. Assemblers from other companies may not offer this feature. The generic JMP assembles to AJMP if the destination contains no forward reference and is within the same 2K page (as the instruction following the AJMP). Otherwise, it assembles to LJMP. The generic CALL instruction (see below) works the same way.

The SJMP instruction specifies the destination address as a relative offset, as shown in the earlier discussion on addressing modes. Since the instruction is two bytes long (an opcode plus a relative offset), the jump distance is limited to −128 to +127 bytes relative to the address following the SJMP.

The LJMP instruction specifies the destination address as a 16-bit constant. Since the instruction is three bytes long (an opcode plus two address bytes), the destination address can be anywhere in the 64K program memory space.

The AJMP instruction specifies the destination address as an 11-bit constant. As with SJMP, this instruction is two bytes long, but the encoding is different. The opcode contains 3 of the 11 address bits, and byte 2 holds the low-order eight bits of the destination address. When the instruction is executed, these 11 bits replace the low-order 11 bits in the PC, and the high-order five bits in the PC stay the same. The destination, therefore, must be within the same 2K block as the instruction following the AJMP. Since there is 64K of code memory space, there are 32 such blocks, each beginning at a 2K address boundary (0000H, 0800H, 1000H, 1800H, etc., up to F800H; see Figure 3–3).

In all cases the programmer specifies the destination address to the assembler in the usual way—as a label or as a 16-bit constant. The assembler will put the destination address into the correct format for the given instruction. If the format required by the instruction will not support the distance to the specified destination address, a "destination out of range" message is given.

3.3.5.1 Jump Tables

The JMP @A+DPTR instruction supports case-dependent jumps for jump tables. The destination address is computed at execution time as the sum of the 16-bit DPTR register and the accumulator. Typically, the DPTR is loaded with the address of a jump table, and the accumulator acts as an index. If, for example, five "cases" are desired, a value from 0 through 4 is loaded into the accumulator and a jump to the appropriate case is performed as follows:

```
MOV  DPTR,#JUMP_TABLE
MOV  A,INDEX_NUMBER
RL   A
JMP  @A+DPTR
```

The RL A instruction above converts the index number (0 through 4) to an even number in the range 0 through 8, because each entry in the jump table is a 2-byte address:

```
JUMP_TABLE: AJMP CASE0
            AJMP CASE1
            AJMP CASE2
            AJMP CASE3
```

3.3.5.2 Subroutines and Interrupts

There are two variations of the CALL instruction: ACALL and LCALL, using absolute and long addressing, respectively. As with JMP, the generic CALL mnemonic may be used with Intel's assembler if the programmer does not care which way the address is encoded. Either instruction pushes the contents of the program counter on the stack and loads the program counter with the address specified in the instruction. Note that the PC will contain the address of the instruction *following* the CALL instruction when it gets pushed on the stack. The PC is pushed on the stack low-byte first, high-byte second. The bytes are popped from the stack in the reverse order. For example, if an LCALL instruction is in code memory at locations 1000H–1002H and the SP contains 20H, then LCALL (a) pushes the return address (1003H) on the internal stack, placing 03H in 21H and 10H in 22H; (b) leaves the SP containing 22H; and (c) jumps to the subroutine by loading the PC with the address contained in bytes 2 and 3 of the instruction.

The LCALL and ACALL instructions have the same restrictions on the destination address as the LJMP and AJMP instructions just discussed.

Subroutines should end with a RET instruction, which returns execution to the instruction following the CALL. There is nothing magical about the way the RET instruction gets back to the main program. It simply "pops" the last two bytes off the stack and places them in the program counter. It is a cardinal rule of programming with subroutines that they should always be entered with a CALL instruction, and they should always be left with a RET instruction. Jumping in or out of a subroutine any other way usually fouls up the stack and causes the program to crash.

RETI is used to return from an interrupt service routine (ISR). The only difference between RET and RETI is that RETI signals the interrupt control system that the interrupt in progress is done. If there is no interrupt pending at the time RETI is executed, then RETI is functionally identical to RET. Interrupts and the RETI instruction are discussed in more detail in Chapter 6.

3.3.5.3 Conditional Jumps

The 8051 offers a variety of conditional jump instructions. All of these specify the destination address using relative addressing and so are limited to a jump distance of −128 to +127 bytes from the instruction following the conditional jump instruction. Note, however, that the user specifies the destination address the same way as with the other jumps, as a label or 16-bit constant. The assembler does the rest.

There is no 0-bit in the PSW. The JZ and JNZ instructions test the accumulator data for that condition.

The DJNZ instruction (decrement and jump if not zero) is for loop control. To execute a loop N times, load a counter byte with N and terminate the loop with a DJNZ to the beginning of the loop, as shown below for $N = 10$.

```
        MOV  R7,#10
LOOP:   (begin loop)
        .
        .
        .
        (end loop)
        DJNZ R7,LOOP
        (continue)
```

The CJNE instruction (compare and jump if not equal) is also used for loop control. Two bytes are specified in the operand field of the instruction and the jump is executed only if the two bytes are not equal. If, for example, a character has just been read into the accumulator from the serial port and it is desired to jump to an instruction identified by the label TERMINATE if the character is CONTROL-C (03H), then the following instructions could be used:

```
        CJNE A,#03H,SKIP
        SJMP TERMINATE
SKIP:   (continue)
```

Since the jump occurs only if A $\neq$ CONTROL-C, a skip is used to bypass the terminating jump instruction except when the desired code is read.

Another application of this instruction is in "greater than" or "less than" comparisons. The two bytes in the operand field are taken as unsigned integers. If the first is less than the second, the carry flag is set. If the first is greater than or equal to the second, the carry flag is cleared. For example, if it is desired to jump to BIG if the value in the accumulator is greater than or equal to 20H, the following instructions could be used:

```
        CJNE A,#20H,$+3
        JNC  BIG
```

The jump destination for CJNE is specified as "$+3." The dollar sign ($) is a special assembler symbol representing the address of the current instruction. Since CJNE is a 3-byte instruction, "$+3" is the address of the next instruction, JNC. In other words, the CJNE instruction follows through to the JNC instruction *regardless* of the result of the compare. The sole purpose of the compare is to set or clear the carry flag. The JNC instruction decides whether or not the jump takes place. This example is one instance in which the 8051 approach to a common programming situation is more awkward than with most microprocessors; however, as we shall see in Chapter 7, the use of macros allows powerful instruction sequences, such as the example above, to be constructed and executed using a single mnemonic.

PROBLEMS

1. What is the hexadecimal opcode for the following instruction?

```
        INC  DPTR
```

2. What is the hexadecimal opcode for the following instruction?

```
        DEC  R6
```

3. What instruction is represented by the opcode 5DH?
4. What instruction is represented by the opcode FFH?
5. List all the 8051's 3-byte instructions with an opcode ending in 5H.
6. Illustrate how the contents of internal address 50H could be transferred to the accumulator, using indirect addressing.
7. What opcode is undefined on the 8051?
8. The following is an 8051 instruction:

```
MOV        50H,#0FFH
```

   a) What is the opcode for this instruction?

   b) How many bytes long in this instruction?

   c) Explain the purpose of each byte of this instruction.

   d) How many machine cycles are required to execute this instruction?

   e) If an 8051 is operating from a 16 MHz crystal, how long does this instruction take to execute?

9. What is the relative offset for the instruction

```
SJMP AHEAD
```

   if the instruction is in locations 0400H and 0401H, and the label AHEAD represents the instruction at address 041FH?

10. What is the relative offset for the instruction

```
SJMP BACK
```

   if the instruction is in locations A050H and A051H, and the label BACK represents the instruction at address 9FE0H?

11. Assume the instruction

```
AJMP AHEAD
```

   is in code memory at addresses 2FF0H and 2FF1H, and the label AHEAD corresponds to an instruction at address 2F96H. What are the hexadecimal machine-language bytes for this instruction?

12. At a certain point in a program, it is desired to jump to the label EXIT if the accumulator equals the carriage return ASCII code. What instruction(s) would be used?
13. The instruction

```
SJMP BACK
```

   is in code memory at address 0100H and 0101H and the label BACK corresponds to an instruction at address 00AEH. What are the hexadecimal machine-language bytes for this instruction?

14. What does the following instruction do?

```
SETB 0D7H
```

What is a better way to perform the same operation? Why?

15. What is the difference between the following two instructions?

```
INC  A
INC  ACC
```

16. What are the machine-language bytes for the instruction

```
LJMP ONWARD
```

if the label ONWARD represents the instruction at address A0F6H?

17. Assume accumulator A contains 5AH. What is the result in accumulator A after the following instruction executes?

```
XRL  A,#0FFH
```

18. Assume the PSW contains 0C0H and accumulator A contains 50H just before the following instruction executes:

```
RLC  A
```

What is the content of accumulator A after this instruction executes?

19. What instruction sequence could be used to create a 5 μs low-going pulse on P1.7? Assume P1.7 is high initially and the 8051 is operating from a 12 MHz crystal.
20. Write a program to create an 83.3 kHz square wave on P1.0. (Assume 12 MHz operation.)
21. Write a program to generate a 4 μs active-high pulse on P1.7 every 200 μs.
22. Write programs to implement the logic operations shown in Figure 3–4.
23. For part (a) above, what is the worst-case propagation delay time from an input transition to an output transition?
24. What is the content of accumulator A after the following instruction sequence executes?

```
MOV  A,#7FH
MOV  50H,#29H
MOV  R0,#50H
XCHD A,@R0
```

25. What are the machine-language bytes for the following instruction?

```
SETB P2.6
```

26. What instruction sequence could be used to copy Flag 0 in the PSW to the port pin P1.5?
27. Under what circumstances will Intel's assembler (ASM51) convert a generic JMP instruction to LJMP?
28. The 8051 internal memory is initialized as follows immediately prior to the execution of a RET instruction:

Internal Address	Contents	SFRs	Contents
0B	9A	SP	0B
0A	78	PC	0200
09	56	A	55
08	34		
07	12		

What is the content of the PC after the RET instruction executes?

29. An 8051 subroutine is shown below:

```
SUB:     MOV  R0,#20H
LOOP:    MOV  @R0,#0
         INC  R0
         CJNE R0,#80H,LOOP
         RET
```

a) What does this subroutine do?

b) In how many machine cycles does each instruction execute?

c) How many bytes long is each instruction?

d) Convert the subroutine to machine language.

e) How long does this subroutine take to execute? (Assume 12 MHz operation.)

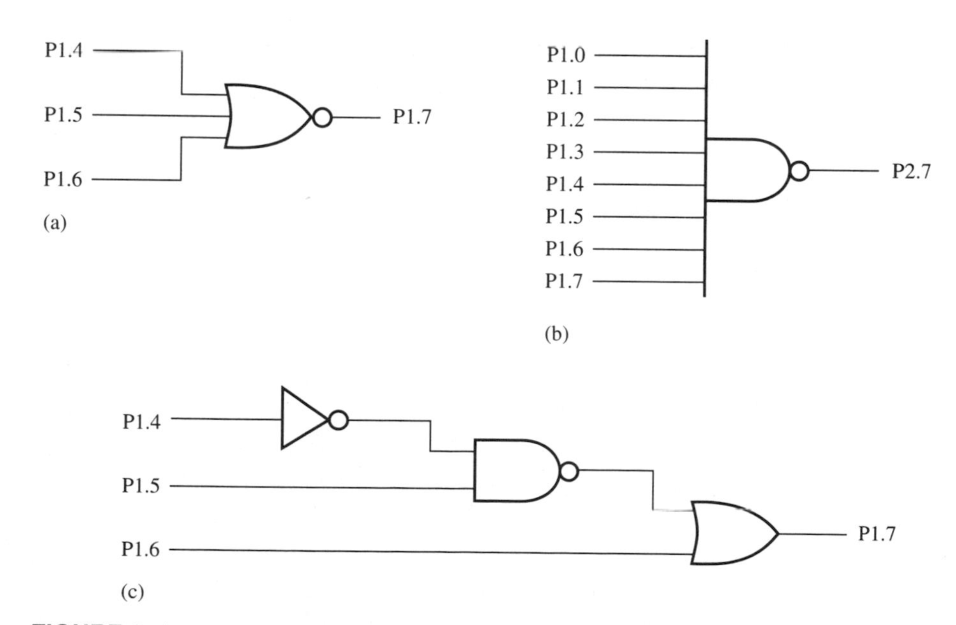

FIGURE 3–4
Logic gate programming problems. (a) 3-input NOR (b) 8-input NAND (c) 3-gate logic operation

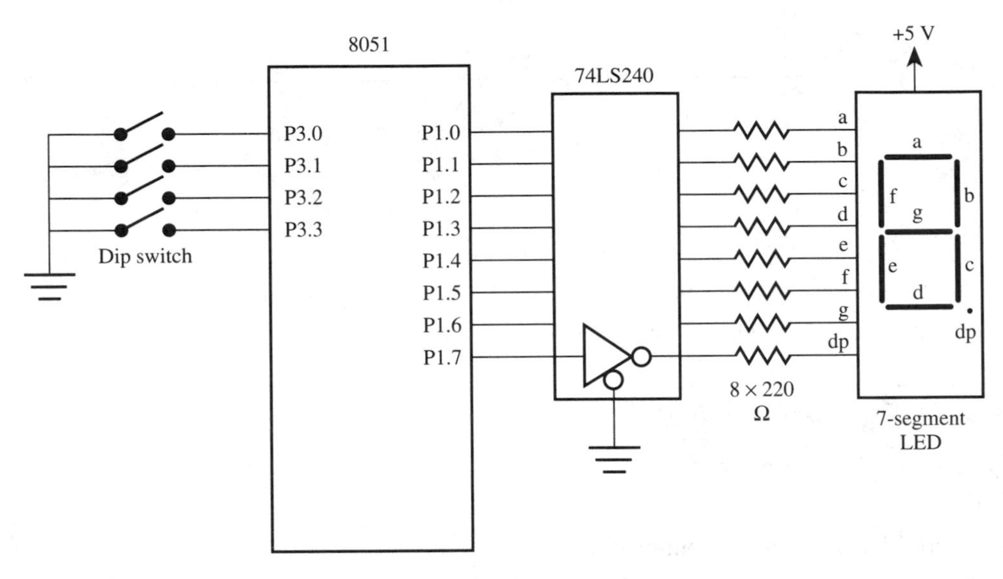

FIGURE 3–5
Interface to a DIP switch and 7-segment LED

30. A 4-bit DIP switch and a common-anode 7-segment LED are connected to an 8051 as shown in Figure 3–5. Write a program that continually reads a 4-bit code from the DIP switch and updates the LEDs to display the appropriate hexadecimal character. For example, if the code 1100B is read, the hexadecimal character "C" should appear; thus, segments *a* through *g* respectively should be ON, OFF, OFF, ON, ON, ON, and OFF. Note that setting an 8051 port pin to "1" turns the corresponding segment "ON." (See Figure 3–5.)

4

TIMER OPERATION

4.1 INTRODUCTION

In this chapter we examine the 8051's on-chip timers. We begin with a simplified view of timers as they are commonly used with microprocessors or microcontrollers.

A timer is a series of divide-by-two flip-flops that receive an input signal as a clocking source. The clock is applied to the first flip-flop, which divides the clock frequency by 2. The output of the first flip-flop clocks the second flip-flop, which also divides by 2, and so on. Since each successive stage divides by 2, a timer with n stages divides the input clock frequency by 2^n. The output of the last stage clocks a timer overflow flip-flop, or **flag,** which is tested by software or generates an interrupt. The binary value in the timer flip-flops can be thought of as a "count" of the number of clock pulses (or "events") since the timer was started. A 16-bit timer, for example, would count from 0000H to FFFFH. The overflow flag is set on the FFFFH-to-0000H overflow of the count.

The operation of a simple timer is illustrated in Figure 4–1 for a 3-bit timer. Each stage is shown as a type-D negative-edge-triggered flip-flop operating in divide-by-two mode (i.e., the $\overline{Q}$ output connects to the D input). The flag flip-flop is simply a type-D latch, set by the last stage in the timer. It is evident in the timing diagram in Figure 4–1b that the first stage (Q_0) toggles at 1/2 the clock frequency, the second stage at 1/4 the clock frequency, and so on. The count is shown in decimal, and is easily verified by examining the state of the three flip-flops. For example, the count "4" occurs when $Q_2 = 1$, $Q_1 = 0$, and $Q_0 = 0$ ($4_{10} = 100_2$).

Timers are used in virtually all control-oriented applications, and the 8051 timers are no exception. There are two 16-bit timers each with four modes of operation. A third 16-bit timer with three modes of operation is added on the 8052. The timers are used for (a) interval timing, (b) event counting, or (c) baud rate generation for the built-in serial port. Each is a 16-bit timer, therefore the 16th or last stage divides the input clock frequency by $2^{16} = 65,536$.

In interval timing applications, a timer is programmed to overflow at a regular interval and set the timer overflow flag. The flag is used to synchronize the program to perform an action such as checking the state of inputs or sending data to outputs. Other ap-

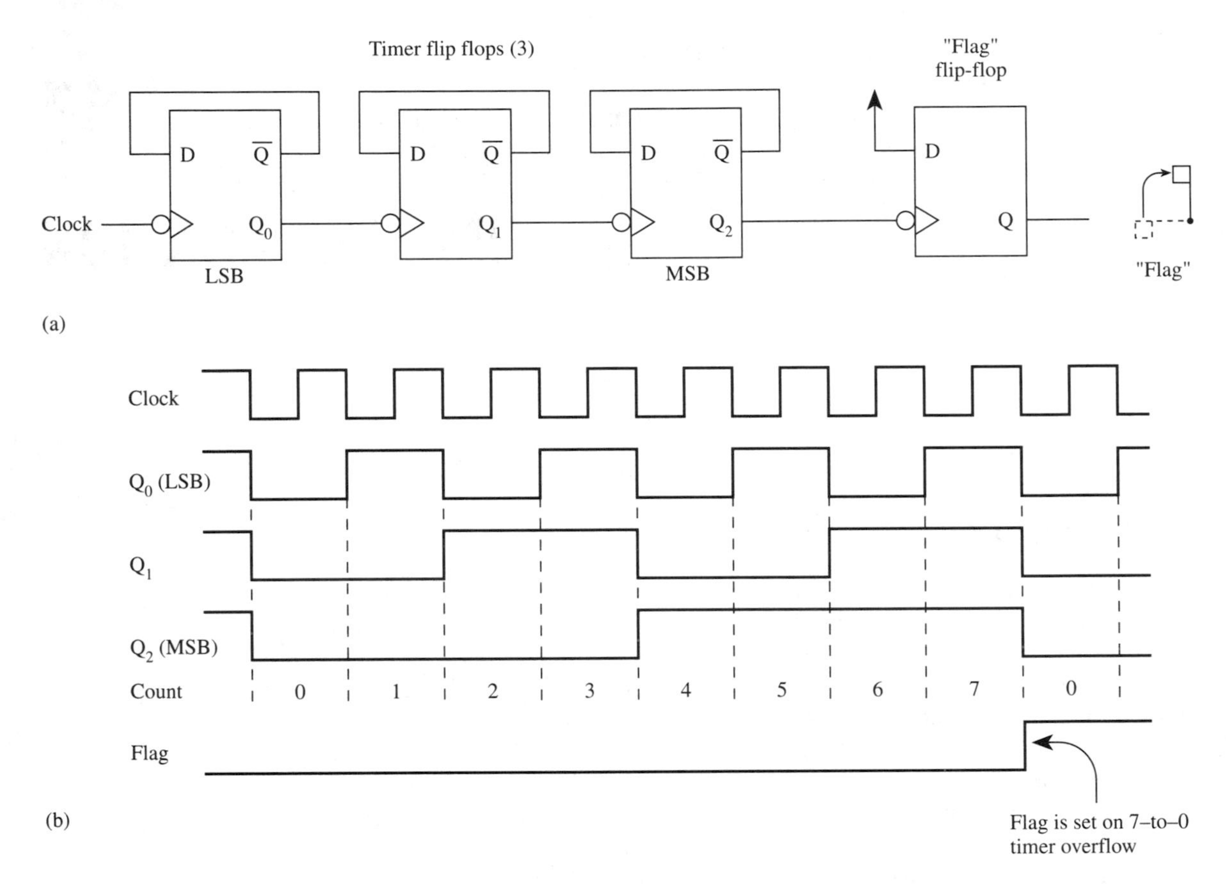

FIGURE 4–1
A 3-bit timer. (a) Schematic (b) Timing diagram.

plications can use the regular clocking of the timer to measure the elapsed time between two conditions (e.g., pulse width measurements).

Event counting is used to determine the number of occurrences of an event, rather than to measure the elapsed time between events. An "event" is any external stimulus that provides a 1-to-0 transition to a pin on the 8051 IC. The timers can also provide the baud rate clock for the 8051's internal serial port.

The 8051 timers are accessed using six special function registers. (See Table 4–1.) An additional 5 SFRs provide access to the third timer in the 8052.

4.2 TIMER MODE REGISTER (TMOD)

The TMOD register contains two groups of four bits that set the operating mode for Timer 0 and Timer 1. (See Table 4–2 and Table 4–3.)

TMOD is not bit-addressable, nor does it need to be. Generally, it is loaded once by software at the beginning of a program to initialize the timer mode. Thereafter, the timer can be stopped, started, and so on by accessing the other timer SFRs.

TABLE 4–1
Timer special function registers

TIMER SFR	PURPOSE	ADDRESS	BIT-ADDRESSABLE
TCON	Control	88H	Yes
TMOD	Mode	89H	No
TL0	Timer 0 low-byte	8AH	No
TL1	Timer 1 low-byte	8BH	No
TH0	Timer 0 high-byte	8CH	No
TH1	Timer 1 high-byte	8DH	No
T2CON*	Timer 2 control	C8H	Yes
RCAP2L*	Timer 2 low-byte capture	CAH	No
RCAP2H*	Timer 2 high-byte capture	CBH	No
TL2*	Timer 2 low-byte	CCH	No
TH2*	Timer 2 high-byte	CDH	No

*8032/8052 only

TABLE 4–2
TMOD (timer mode) register summary

BIT	NAME	TIMER	DESCRIPTION
7	GATE	1	Gate bit. When set, timer only runs while $\overline{INT1}$ is high
6	$C/\overline{T}$	1	Counter/timer select bit. 1 = event counter 0 = interval timer
5	M1	1	Mode bit 1 (see Table 4–3)
4	M0	1	Mode bit 0 (see Table 4–3)
3	GATE	0	Timer 0 gate bit
2	$C/\overline{T}$	0	Timer 0 counter/timer select bit
1	M1	0	Timer 0 M1 bit
0	M0	0	Timer 0 M0 bit

TABLE 4–3
Timer modes

M1	M0	MODE	DESCRIPTION
0	0	0	13-bit timer mode (8048 mode)
0	1	1	16-bit timer mode
1	0	2	8-bit auto-reload mode
1	1	3	Split timer mode: Timer 0: TL0 is an 8-bit timer controlled by timer 0 mode bits; TH0, the same except controlled by timer 1 mode bits Timer 1: stopped

4.3 TIMER CONTROL REGISTER (TCON)

The TCON register contains status and control bits for Timer 0 and Timer 1 (see Table 4–4). The upper four bits in TCON (TCON.4–TCON.7) are used to turn the timers on and off (TR0, TR1), or to signal a timer overflow (TF0, TF1). These bits are used extensively in the examples in this chapter.

The lower four bits in TCON (TCON.0–TCON.3) have nothing to do with the timers. They are used to detect and initiate external interrupts. Discussion of these bits is deferred until Chapter 6, when interrupts are discussed.

4.4 TIMER MODES AND THE OVERFLOW FLAG

Each timer is discussed below. Since there are two timers on the 8051, the notation "x" is used to imply either Timer 0 or Timer 1; thus, "THx" means either TH1 or TH0 depending on the timer.

The arrangement of timer registers TLx and THx and the timer overflow flags TFx is shown in Figure 4–2 for each mode.

4.4.1 13-Bit Timer Mode (Mode 0)

Mode 0 is a 13-bit timer mode that provides compatibility with the 8051's predecessor, the 8048. It is not generally used in new designs. (See Figure 4–2a.) The timer high-byte (THx) is cascaded with the five least-significant bits of the timer low-byte (TLx) to form a 13-bit timer. The upper three bits of TLx are not used.

TABLE 4–4
TCON (timer control) register summary

BIT	SYMBOL	BIT ADDRESS	DESCRIPTION
TCON.7	TF1	8FH	Timer 1 overflow flag. Set by hardware upon overflow; cleared by software, or by hardware when processor vectors to interrupt service routine
TCON.6	TR1	8EH	Timer 1 run control bit. Set/cleared by software to turn timer on/off
TCON.5	TF0	8DH	Timer 0 overflow flag
TCON.4	TR0	8CH	Timer 0 run control bit
TCON.3	IE1	8BH	External interrupt 1 edge flag. Set by hardware when a falling edge is detected on $\overline{\text{INT 1}}$; cleared by software, or by hardware when CPU vectors to interrupt service routine
TCON.2	IT1	8AH	External interrupt 1 type flag. Set/cleared by software for falling-edge/low-level activated external interrupt
TCON.1	IE0	89H	External interrupt 0 edge flag
TCON.0	IT0	88H	External interrupt 0 type flag

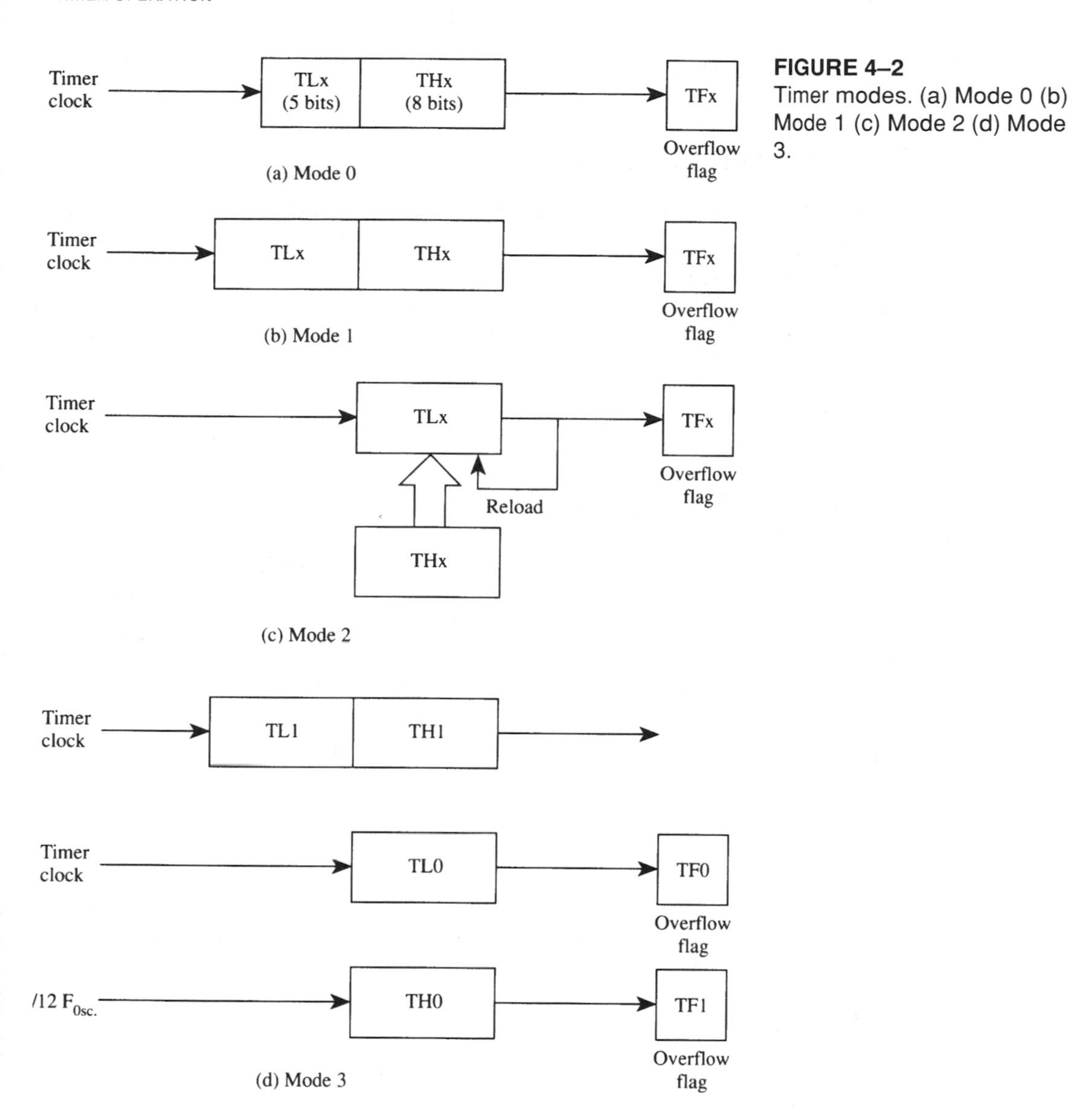

FIGURE 4–2
Timer modes. (a) Mode 0 (b) Mode 1 (c) Mode 2 (d) Mode 3.

4.4.2 16-Bit Timer Mode (Mode 1)

Mode 1 is a 16-bit timer mode and is the same as mode 0, except the timer is operating as a full 16-bit timer. The clock is applied to the combined high and low timer registers (TLx/THx). As clock pulses are received, the timer counts up: 0000H, 0001H, 0002H, etc. An overflow occurs on the FFFFH-to-0000H transition of the count and sets the timer overflow flag. The timer continues to count. The overflow flag is the TFx bit in TCON that is read or written by software. (See Figure 4–2b.)

The most-significant bit (MSB) of the value in the timer registers is THx bit 7, and the least-significant bit (LSB) is TLx bit 0. The LSB toggles at the input clock frequency divided by 2, while the MSB toggles at the input clock frequency divided by 65,536 (i.e., 2^{16}). The timer registers (TLx/THx) may be read or written at any time by software.

4.4.3 8-Bit Auto-Reload Mode (Mode 2)

Mode 2 is 8-bit auto-reload mode. The timer low-byte (TLx) operates as an 8-bit timer while the timer high-byte (THx) holds a reload value. When the count overflows from FFH to 00H, not only is the timer flag set, but the value in THx is loaded into TLx; counting continues from this value up to the next FFH-to-00H transition, and so on. This mode is convenient, since timer overflows occur at specific, periodic intervals once TMOD and THx are initialized. (See Figure 4–2c.)

4.4.4 Split Timer Mode (Mode 3)

Mode 3 is the split timer mode and is different for each timer. Timer 0 in mode 3 is split into two 8-bit timers. TL0 and TH0 act as separate timers with overflows setting the TF0 and TF1 bits respectively.

Timer 1 is stopped in mode 3, but can be started by switching it into one of the other modes. The only limitation is that the usual Timer 1 overflow flag, TF1, is not affected by Timer 1 overflows, since it is connected to TH0.

Mode 3 essentially provides an extra 8-bit timer: The 8051 appears to have a third timer. When Timer 0 is in mode 3, Timer 1 can be turned on and off by switching it out of and into its own mode 3. It can still be used by the serial port as a baud rate generator, or it can be used in any way not requiring interrupts (since it is no longer connected to TF1).

4.5 CLOCKING SOURCES

Figure 4–2 does not show how the timers are clocked. There are two possible clock sources, selected by writing to the counter/timer ($C/\overline{T}$) bit in TMOD when the timer is initialized. One clocking source is used for interval timing, the other for event counting.

4.5.1 Interval Timing

If $C/\overline{T} = 0$, continuous timer operation is selected and the timer is clocked from the on-chip oscillator. A divide-by-12 stage is added to reduce the clocking frequency to a value reasonable for most applications.

When continuous timer operation is selected, the timer is used for **interval timing.** The timer registers (TLx/THx) increment at a rate of 1/12th the frequency of the on-chip oscillator; thus, a 12 MHz crystal would yield a clock rate of 1 MHz. Timer overflows occur after a fixed number of clocks, depending on the initial value loaded into the timer registers, TLx/THx.

4.5.2 Event Counting

If $C/\overline{T} = 1$, the timer is clocked from an external source. In most applications, this external source supplies the timer with a pulse upon the occurrence of an "event"—the timer

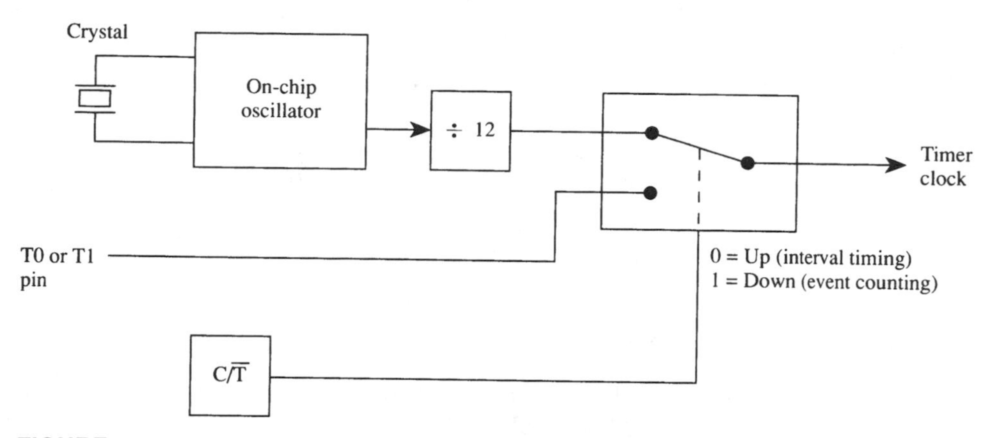

FIGURE 4–3
Clocking source

is **event counting.** The number of events is determined in software by reading the timer registers TLx/THx, since the 16-bit value in these registers increments for each event.

The external clock source comes by way of the alternate functions of the Port 3 pins. Port 3 bit 4 (P3.4) serves as the external clocking input for Timer 0 and is known as "T0" in this context. P3.5, or "T1," is the clocking input for Timer 1. (See Figure 4–3.)

In counter applications, the timer registers are incremented in response to a 1-to-0 transition at the external input, Tx. The external input is sampled during S5P2 of every machine cycle; thus, when the input shows a high in one cycle and a low in the next, the count is incremented. The new value appears in the timer registers during S3P1 of the cycle following the one in which the transition is detected. Since it takes two machine cycles (2 μs) to recognize a 1-to-0 transition, the maximum external frequency is 500 kHz (assuming 12 MHz operation).

4.6 STARTING, STOPPING, AND CONTROLLING THE TIMERS

Figure 4–2 illustrates the various configurations for the timer registers, TLx and THx, and the timer overflow flags, TFx. The two possibilities for clocking the timers are shown in Figure 4–3. We now demonstrate how to start, stop, and control the timers.

The simplest method for starting and stopping the timers is with the run control bit, TRx, in TCON. TRx is clear after a system reset; thus, the timers are disabled (stopped) by default. TRx is set by software to start the timers. (See Figure 4–4.)

Since TRx is in the bit-addressable register TCON, it is easy to start and stop the timers within a program. For example, Timer 0 is started by

```
SETB TR0
```

and stopped by

```
CLR  TR0
```

The assembler will perform the necessary symbolic conversion from "TR0" to the correct bit address. SETB TR0 is exactly the same as SETB 8CH.

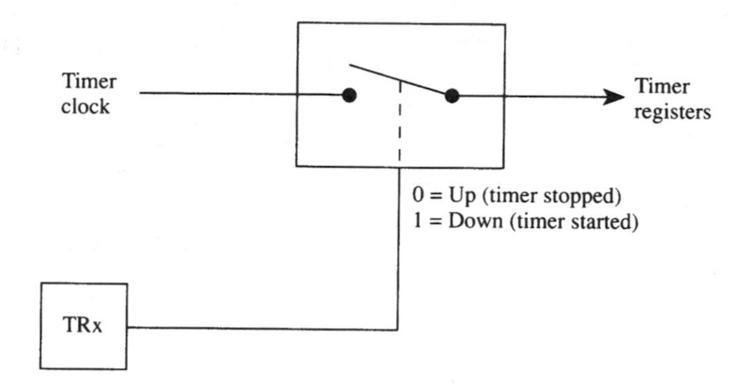

FIGURE 4–4
Starting and stopping the timers

Another method for controlling the timers is with the GATE bit in TMOD and the external input $\overline{\text{INTx}}$. Setting GATE = 1 allows the timer to be controlled by $\overline{\text{INTx}}$. This is useful for pulse width measurements as follows. Assume $\overline{\text{INT0}}$ is low but pulses high for a period of time to be measured. Initialize Timer 0 for mode 2, 16-bit timer mode, with TL0/TH0 = 0000H, GATE = 1, and TR0 = 1. When $\overline{\text{INT0}}$ goes high, the timer is "gated on" and is clocked at a rate of 1 MHz. When $\overline{\text{INT0}}$ goes low, the timer is "gated off" and the duration of the pulse in microseconds is the count in TL0/TH0. ($\overline{\text{INT0}}$ can be programmed to generate an interrupt when it returns low.)

To complete the picture, Figure 4–5 illustrates Timer 1 operating in mode 1 as a 16-bit timer. As well as the timer registers TL1/TH1 and the overflow flag TF1, the diagram shows the possibilities for the clocking source and for starting, stopping, and controlling the timer.

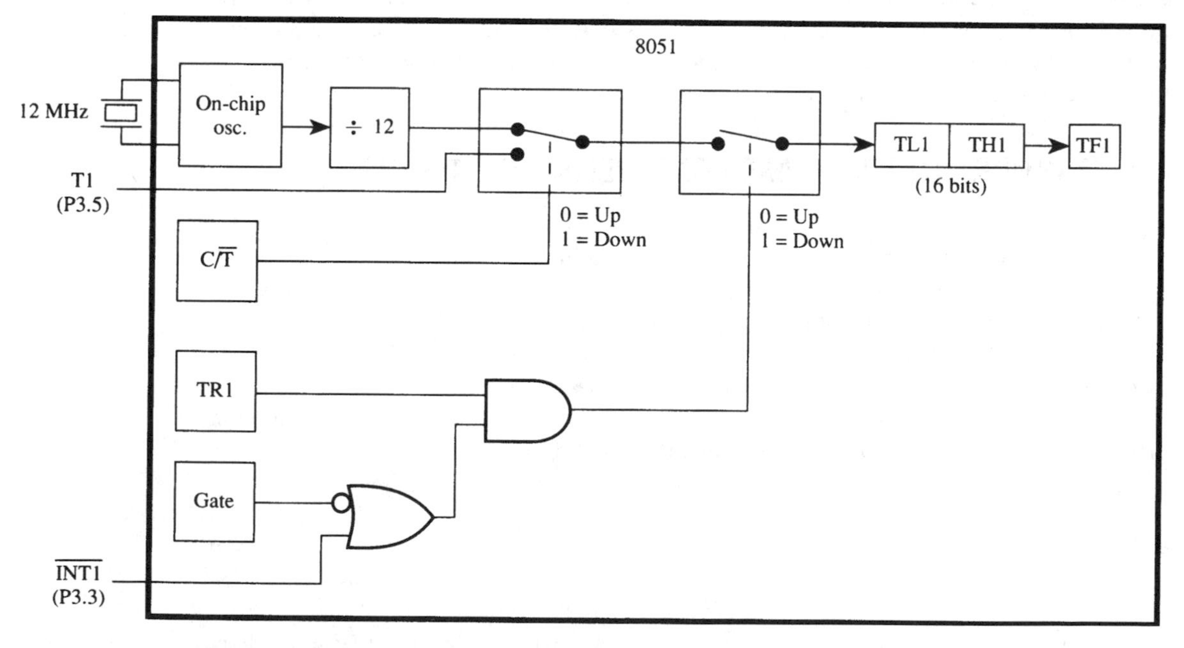

FIGURE 4–5
Timer 1 operating in mode 1

4.7 INITIALIZING AND ACCESSING TIMER REGISTERS

The timers are usually initialized once at the beginning of a program to set the correct operating mode. Thereafter, within the body of a program, the timers are started, stopped, flag bits tested and cleared, timer registers read or updated, and so on, as required in the application.

TMOD is the first register initialized, since it sets the mode of operation. For example, the following instruction initializes Timer 1 as a 16-bit timer (mode 1) clocked by the on-chip oscillator (interval timing):

```
        MOV  TMOD,#00010000B
```

The effect of this instruction is to set M1 = 0 and M0 = 1 for mode 1, leave $C/\overline{T} = 0$ and GATE = 0 for internal clocking, and clear the Timer 0 mode bits. (See Table 4–2.) Of course, the timer does not actually begin timing until its run control bit, TR1, is set.

If an initial count is necessary, the timer registers TL1/TH1 must also be initialized. Remembering that the timers count up and set the overflow flag on an FFFFH-to-0000H transition, a 100 μs interval could be timed by initializing TL1/TH1 to 100 counts less than 0000H. The correct value is −100 or FF9CH. The following instructions do the job:

```
        MOV  TL1,#9CH
        MOV  TH1,#0FFH
```

The timer is then started by setting the run control bit as follows:

```
        SETB TR1
```

The overflow flag is automatically set 100 μs later. Software can sit in a "wait loop" for 100 μs using a conditional branch instruction that returns to itself as long as the overflow flag is not set:

```
WAIT:   JNB  TF1,WAIT
```

When the timer overflows, it is necessary to stop the timer and clear the overflow flag in software:

```
        CLR  TR1
        CLR  TF1
```

4.7.1 Reading a Timer "On the Fly"

In some applications, it is necessary to read the value in the timer registers "on the fly." There is a potential problem that is simple to guard against in software. Since two timer registers must be read, a "phase error" may occur if the low-byte overflows into the high-byte between the two read operations. A value may be read that never existed. The solution is to read the high-byte first, then the low-byte, and then read the high-byte again. If the high-byte has changed, repeat the read operations. The instructions below read the contents of the timer registers TL1/TH1 into registers R6/R7, correctly dealing with this problem.

```
AGAIN:      MOV  A,TH1
            MOV  R6,TL1
            CJNE A,TH1,AGAIN
            MOV  R7,A
```

4.8 SHORT INTERVALS AND LONG INTERVALS

What is the range of intervals that can be timed? This issue is examined assuming the 8051 is operating from a 12 MHz crystal. The on-chip oscillator is divided by 12 and clocks the timers at a rate of 1 MHz.

The shortest possible interval is limited, not by the timer clock frequency, but by software. Presumably, something must occur at regular intervals, and it is the duration of instructions that limit this for very short intervals. The shortest instruction on the 8051 is one machine cycle or one microsecond. Table 4–5 summarizes the techniques for creating intervals of various lengths. (Operation from a 12 MHz crystal is assumed.)

Example 4–1: Pulse Wave Generation

Write a program that creates a periodic waveform on P1.0 with as high a frequency as possible. What are the frequency and duty cycle of the waveform?

Very short intervals (i.e., high frequencies) can be programmed without using the timers. Here's the program:

```
8100            5               ORG     8100H
8100 D290       6     LOOP:     SETB    P1.0   ;one machine cycle
8102 C290       7               CLR     P1.0   ;one machine cycle
8104 80FA       8               SJMP    LOOP   ;two machine cycles
                9               END
```

This program creates a pulse waveform on P1.0 with a period of 4 μs: high-time = 1 μs, low-time = 3μs. The frequency is 250 kHz and the duty cycle is 25%. (See Figure 4–6.)

It might appear at first that the instructions in Figure 4–6 are misplaced, but they're not. The SETB P1.0 instruction, for example, does not actually set the port bit until the end of the instruction, during S6P2.

The period of the output signal can be lengthened somewhat by inserting NOP (no operation) instructions into the loop. Each NOP adds 1 μs to the period of the output sig-

TABLE 4–5
Techniques for programming timed intervals (12 MHz operation)

MAXIMUM INTERVAL IN MICROSECONDS	TECHNIQUE
≈10	Software tuning
256	8-bit timer with auto-reload
65536	16-bit timer
No limit	16-bit timer plus software loops

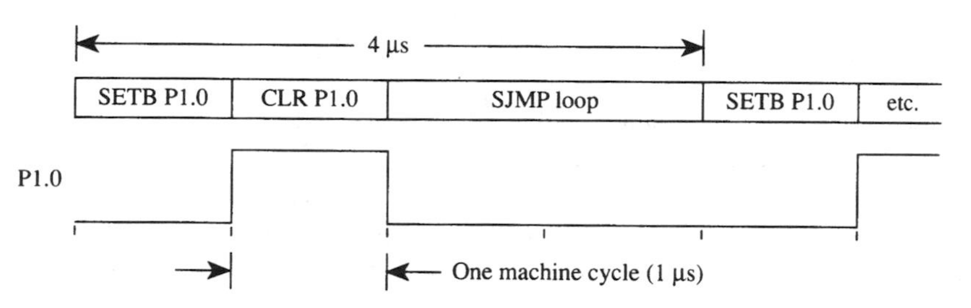

FIGURE 4–6
Waveform for example

nal. For example, adding two NOP instructions after SETB P1.0 would make the output a square wave with a period of 6 μs and a frequency of 166.7 kHz. Beyond a point, however, "software tuning" is cumbersome and a timer is the best choice for delays.

Moderate length intervals are easily obtained using 8-bit auto-reload mode, mode 2. Since the timed interval is set by an 8-bit count, the longest possible interval before overflow is $2^8 = 256$ μs.

Example 4–2: 10 kHz Square Wave

Write a program using Timer 0 to create a 10 kHz square wave on P1.0.

A 10 kHz square wave requires a high-time of 50 μs and a low-time of 50 μs. Since this interval is less than 256 μs, timer mode 2 can be used. An overflow every 50 μs requires a TH0 reload value of 50 counts less than 00H, or −50. Here's the program:

```
8100             6              ORG     8100H
8100 758902      7              MOV     TMOD,#02H    ;8-bit auto-reload mode
8103 758CCE      8              MOV     TH0,#-50     ;-50 reload value in TH0
8106 D28C        9              SETB    TR0          ;start timer
8108 308DFD     10     LOOP:    JNB     TF0,LOOP     ;wait for overflow
810B C28D       11              CLR     TF0          ;clear timer overflow flag
810D B290       12              CPL     P1.0         ;toggle port bit
810F 80F7       13              SJMP    LOOP         ;repeat
                14              END
```

This program uses a complement bit instruction (CPL) rather than the SETB and CLR bit instructions in the previous example. Between each complement operation, a delay of 1/2 the desired period (50 μs) is programmed using Timer 0 in 8-bit auto-reload mode. The reload value may be specified using decimal notation as −50, rather than using hexadecimal notation. The assembler performs the necessary conversion. Note that the timer overflow flag (TF0) must be explicitly cleared by software after each overflow.

Timed intervals longer than 256 μs must use 16-bit timer mode, mode 1. The longest delay is $2^{16} = 65{,}536$ μs or about 0.066 seconds. The inconvenience of mode 1 is that the timer registers must be reinitialized after each overflow, whereas reloading is automatic in mode 2.

Example 4–3: 1 kHz Square Wave

Write a program using Timer 0 to create a 1 kHz square wave on P1.0.

A 1 kHz square wave requires a high-time of 500 μs and a low-time of 500 μs. Since this interval is longer than 256 μs, mode 2 cannot be used. Full 16-bit timer mode, mode 1, is required. The main difference in the software is that the timer registers, TL0 and TH0, are reinitialized after each overflow.

```
8100           6          ORG    8100H
8100 758901    7          MOV    TMOD,#01H    ;16-bit timer mode
8103 758CFE    8  LOOP:   MOV    TH0,#0FEH    ;-500 (high byte)
8106 758A0C    9          MOV    TL0,#0CH     ;-500 (low byte)
8109 D28C     10          SETB   TR0          ;start timer
810B 308DFD   11  WAIT:   JNB    TF0,WAIT     ;wait for overflow
810E C28C     12          CLR    TR0          ;stop timer
8110 C28D     13          CLR    TF0          ;clear timer overflow flag
8112 B290     14          CPL    P1.0         ;toggle port bit
8114 80ED     15          SJMP   LOOP         ;repeat
              16          END
```

There is a slight error in the output frequency in the program above. This results from the extra instructions inserted after the timer overflow to reinitialize the timer. If exactly 1 kHz is required, the reload value for registers TL0/TH0 must be adjusted somewhat. Such errors do not occur in auto-reload mode, since the timer is never stopped—it overflows at a consistent rate set by the reload value in TH0.

Intervals longer than 0.066 seconds can be achieved by cascading Timer 0 and Timer 1 through software, but this ties up both timers. A more practical approach uses one of the timers in 16-bit mode with a software loop counting overflows. The desired operation is performed every *n* overflows.

Example 4–4: Buzzer Interface

A buzzer is connected to P1.7 and a debounced switch is connected to P1.6. (See Figure 4–7.) Write a program that reads the logic level provided by the switch and sounds the buzzer for 1 second for each 1-to-0 transition detected.

The buzzer in Figure 4–7 is a piezo ceramic transducer that vibrates when stimulated with a DC voltage. A typical example is the Projects Unlimited AI-430 that generates a tone of about 3 kHz at 5 volts DC. An inverter is used as a driver since the AI-430 draws 7 mA of current. As indicated in the 8051's DC Characteristics in Appendix E, Port 1 pins can sink a maximum of 1.6 mA. The AI-430 costs a few dollars.

Creating software delays is one of the most common programming tasks given to students of microprocessors. The usual method of decrementing a count within a loop is

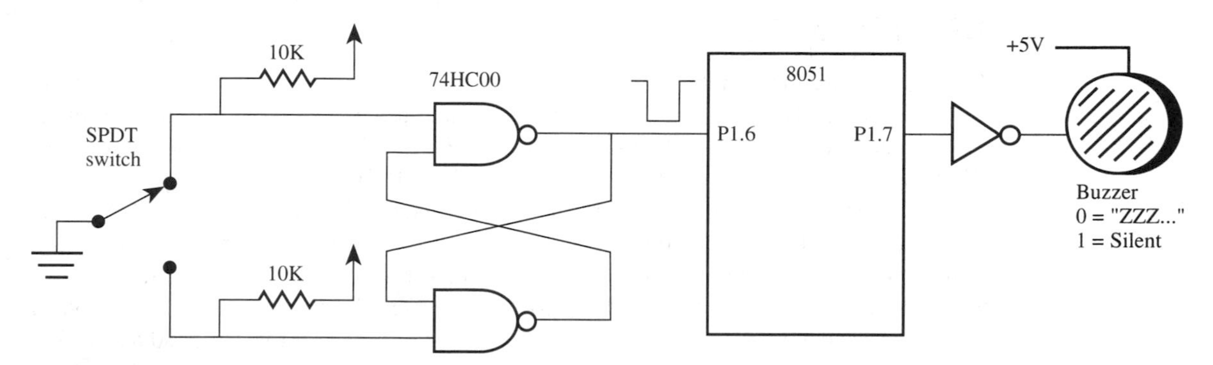

FIGURE 4–7
Buzzer example

not necessary on the 8051, since it has built-in timers. A 1-second delay subroutine using Timer 0 is shown in this example.

A 1-to-0 transition on P1.6 is detected by waiting for a 1 (JNB P1.6,LOOP) and then waiting for a 0 (JB P1.6,WAIT). Here's the program:

```
  0064           6      HUNDRED  EQU     100            ;100 x 10000 us = 1 sec.
  2710           7      COUNT    EQU     10000
8100             8               ORG     8100H
8100 758901      9               MOV     TMOD,#01H      ;use timer 0 in mode 1
8103 3096FD     10      LOOP:    JNB     P1.6,LOOP      ;wait for 1 input
8106 2096FD     11      WAIT:    JB      P1.6,WAIT      ;wait for 0 input
8109 D297       12               SETB    P1.7           ;turn buzzer on
810B 128112     13               CALL    DELAY          ;wait 1 second
810E C297       14               CLR     P1.7           ;turn buzzer off
8110 80F1       15               SJMP    LOOP
                16      ;
8112 7F64       17      DELAY:   MOV     R7,#HUNDRED
8114 758C27     18      AGAIN:   MOV     TH0,#HIGH COUNT
8117 758A10     19               MOV     TL0,#LOW COUNT
811A D28C       20               SETB    TR0
811C 308DFD     21      WAIT2:   JNB     TF0,WAIT2
811F C28D       22               CLR     TF0
8121 C28C       23               CLR     TR0
8123 DFEF       24               DJNZ    R7,AGAIN
8125 22         25               RET
                26               END
```

There are two situations not handled in the example above. First, if the input toggles during the one second that the buzzer is sounding, the transition is not detected, since the software is busy in the delay routine. Second, if the input toggles very quickly—in less than a microsecond—the transition may be missed altogether by the JNB and JB instructions. Problem 5 at the end of this chapter deals with the first situation. The second can only be handled using an interrupt input to "latch" a status flag when a 1-to-0 transition occurs. This is discussed in Chapter 6.

4.9 8052 TIMER 2

The third timer added on the 8052 IC is a powerful addition to the two just discussed. As shown earlier in Table 4–1, five extra special function registers are added to accommodate Timer 2. These include the timer registers, TL2 and TH2, the timer control register, T2CON, and the capture registers, RCAP2L and RCAP2H.

The mode for Timer 2 is set by its control register, T2CON. (See Table 4–6.) Like Timers 0 and 1, Timer 2 can operate as an interval timer or event counter. The clocking source is provided internally by the on-chip oscillator, or externally by T2, the alternate function of Port 1 bit 0 (P1.0) on the 8052 IC. The $C/\overline{T2}$ bit in T2CON selects between the internal and external clock, just as the $C/\overline{T}$ bits do in TCON for Timers 0 and 1. Regardless of the clocking source, there are three modes of operation: auto-reload, capture, and baud rate generator.

TABLE 4–6
T2CON (Timer 2 control) register summary

BIT	SYMBOL	BIT ADDRESS	DESCRIPTION
T2CON.7	TF2	CFH	Timer 2 overflow flag. (Not set when TCLK or RCLK = 1.)
T2CON.6	EXF2	CEH	Timer 2 external flag. Set when either a capture or reload is caused by 1-to-0 transition on T2EX and EXEN2 = 1; when timer interrupts are enabled, EXF2 = 1 causes CPU to vector to service routine; cleared by software
T2CON.5	RCLK	CDH	Timer 2 receiver clock. When set, Timer 2 provides serial port receive baud rate; Timer 1 provides transmit baud rate
T2CON.4	TCLK	CCH	Timer 2 transmit clock. When set, Timer 2 provides transmitter baud rate; Timer 1 provides receiver baud rate
T2CON.3	EXEN2	CBH	Timer 2 external enable. When set, capture or reload occurs on 1-to-0 transition of T2EX
T2CON.2	TR2	CAH	Timer 2 run control bit. Set/cleared by software to turn Timer 2 on/off.
T2CON.1	$C/\overline{T2}$	C9H	Timer 2 counter/interval timer select bit. 1 = event counter 0 = interval timer
T2CON.0	$CP/\overline{RL2C}$	C8H	Timer 2 capture/reload flag. When set, capture occurs on 1-to-0 transition of T2EX if EXEN2 = 1; when clear, auto reload occurs on timer overflow or T2EX transition if EXEN2 = 1; if RCLK or TCLK = 1, this bit is ignored

4.9.1 Auto-Reload Mode

The capture/reload bit in T2CON selects between the first two modes. When CP/$\overline{\text{RL2}}$ = 0, Timer 2 is in auto-reload mode with TL2/TH2 as the timer registers, and RCAP2L and RCAP2H holding the reload value. Unlike the reload mode for Timers 0 and 1, Timer 2 is always a full 16-bit timer, even in auto-reload mode.

Reload occurs on an FFFFH-to-0000H transition in TL2/TH2 and sets the Timer 2 flag, TF2. This condition is determined by software or is programmed to generate an interrupt. Either way, TF2 must be cleared by software before it is set again.

Optionally, by setting EXEN2 in T2CON, a reload also occurs on the 1-to-0 transition of the signal applied to pin T2EX, which is the alternate pin function for P1.1 on the 8052 IC. A 1-to-0 transition on T2EX also sets a new flag bit in Timer 2, EXF2. As with TF2, EXF2 is tested by software or generates an interrupt. EXF2 must be cleared by software. Timer 2 in auto-reload mode is shown in Figure 4–8.

4.9.2 Capture Mode

When CP/$\overline{\text{RL2}}$ = 1, capture mode is selected. Timer 2 operates as a 16-bit timer and sets the TF2 bit upon an FFFFH-to-0000H transition of the value in TL2/TH2. The state of TF2 is tested by software or generates an interrupt.

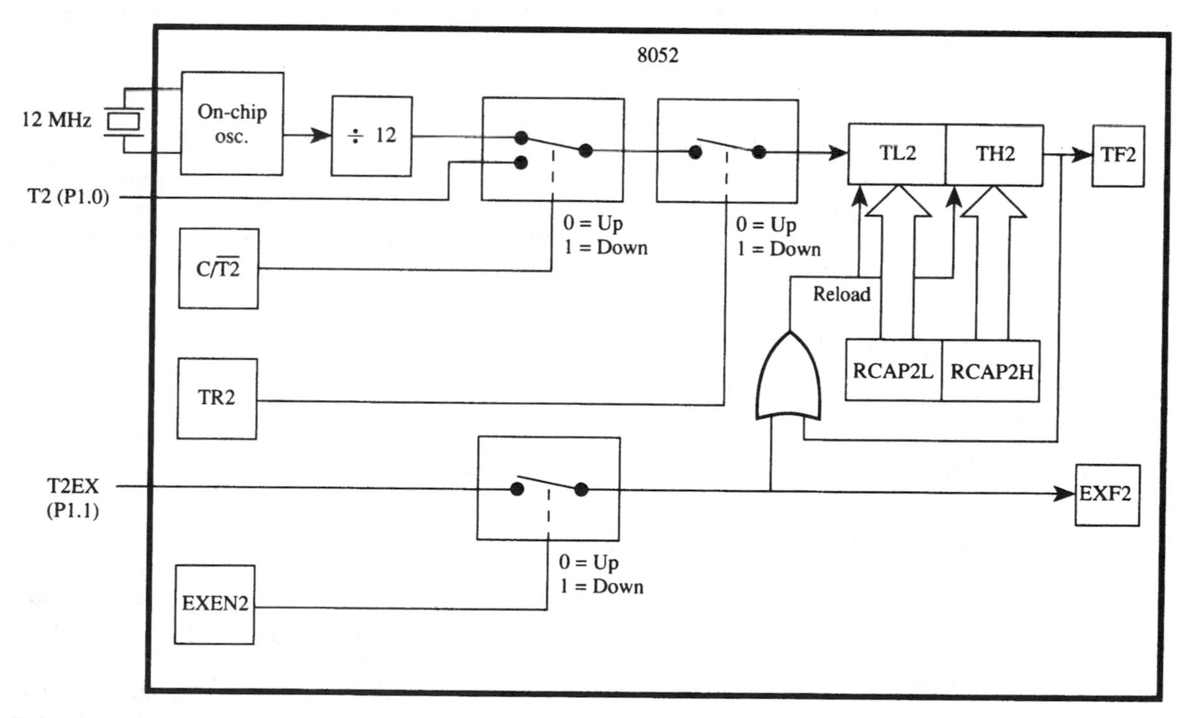

FIGURE 4–8
Timer 2 in 16-bit auto-reload mode

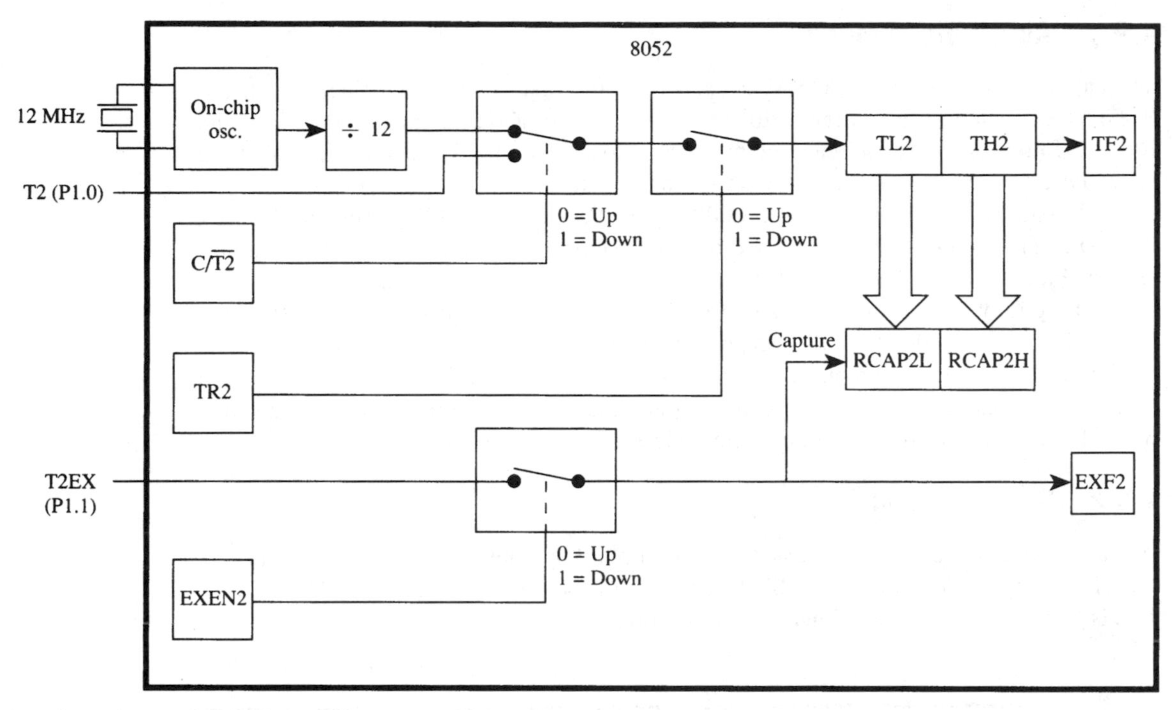

FIGURE 4–9
Timer 2 in 16-bit capture mode

To enable the capture feature, the EXEN2 bit in T2CON must be set. If EXEN2 = 1, a 1-to-0 transition on T2EX (P1.1) "captures" the value in timer registers TL2/TH2 by clocking it into registers RCAP2L and RCAP2H. The EXF2 flag in T2CON is also set and, as stated above, is tested by software or generates an interrupt. Timer 2 in capture mode is shown in Figure 4–9.

4.10 BAUD RATE GENERATION

Another use of the timers is to provide the baud rate clock for the on-chip serial port. This comes by way of Timer 1 on the 8051 IC or Timer 1 and/or Timer 2 on the 8052 IC. Baud rate generation is discussed in Chapter 5.

4.11 SUMMARY

This chapter has introduced the 8051 and 8052 timers. The software solutions for the examples presented here feature one common but rather limiting trait. They consume all of the CPU's execution time. The programs execute in wait loops, waiting for a timer overflow. This is fine for learning purposes, but for practical control-oriented applications using microcontrollers, the CPU must perform other duties and respond to external events, such as an operator entering a parameter from a keyboard. In the chapter on interrupts, we shall demonstrate how to use the timers in an "interrupt-driven" environment.

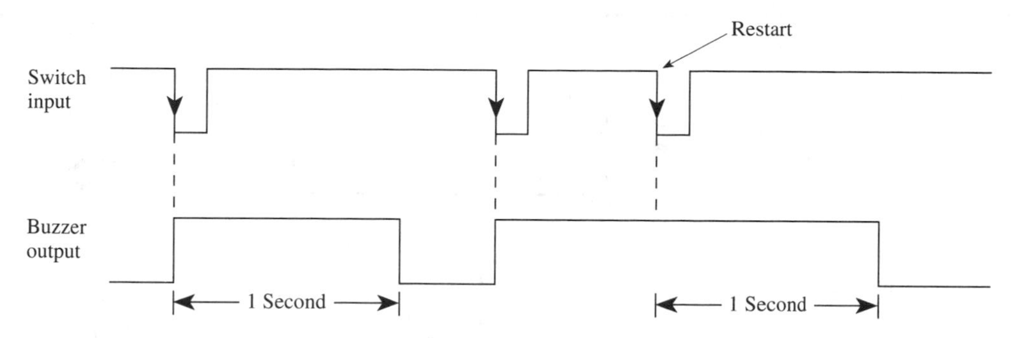

FIGURE 4–10
Timing for modified buzzer example

The timer overflow flags are not tested in a software loop, but generate an interrupt. Another program temporarily interrupts the main program while an action is performed that affects the timer interrupt (perhaps toggling a port bit). Through interrupts, the illusion of doing several things simultaneously is created.

PROBLEMS

1. Write an 8051 program that creates a square wave on P1.5 with a frequency of 100 kHz. (Hint: Don't use the timers.)
2. What is the effect of the following instruction?

```
SETB 8EH
```

3. What is the effect of the following instruction?

```
MOV  TMOD,#11010101B
```

4. Consider the three-instruction program shown in Example 4–1. What is the frequency and duty cycle of the waveform created on P1.0 for a 16 MHz 8051?
5. Rewrite the solution to Example 4–4 to include a "restart" mode. If a 1-to-0 transition occurs while the buzzer is sounding, restart the timing loop to continue the buzz for another second. This is illustrated in Figure 4–10.
6. Write an 8051 program to generate a 12 kHz square wave on P1.2 using Timer 0.
7. Design a "turnstile" application using Timer 1 to determine when the 10,000th person has entered a fairground. Assume (a) a turnstile sensor connects to T1 and generates a pulse each time the turnstile is rotated, and (b) a light is connected to P1.7 that is on when P1.7 = 1, and off otherwise. Count "events" at T1 and turn on the light at P1.7 when the 10,000th person enters the fairground. (See Figure 4–11.)
8. The international tuning standard for musical instruments is "A above middle C" at a frequency of 440 Hz. Write an 8051 program to generate this tuning frequency and sound a 440 Hz tone on a loudspeaker connected to P1.1. (See Figure 4–12.) Due to rounding of the values placed in TL1/TH1, there is a slight error in the output frequency. What is the exact output frequency and what is the percentage error? What value of crystal would yield exactly 440 Hz with the program you have written?

FIGURE 4–11
Turnstile problem

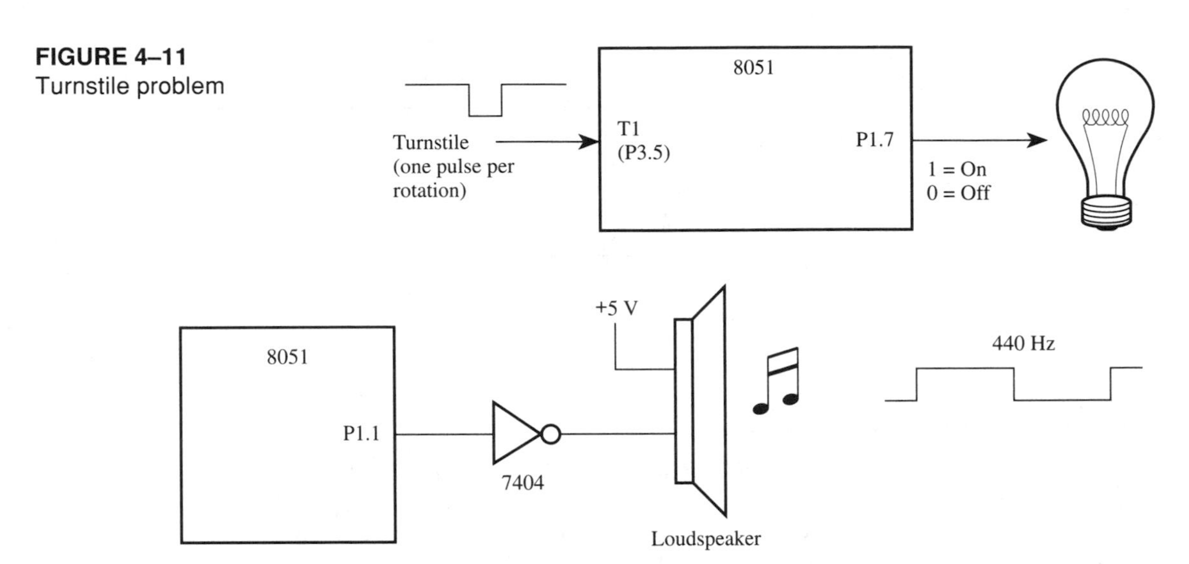

FIGURE 4–12
Loudspeaker interface

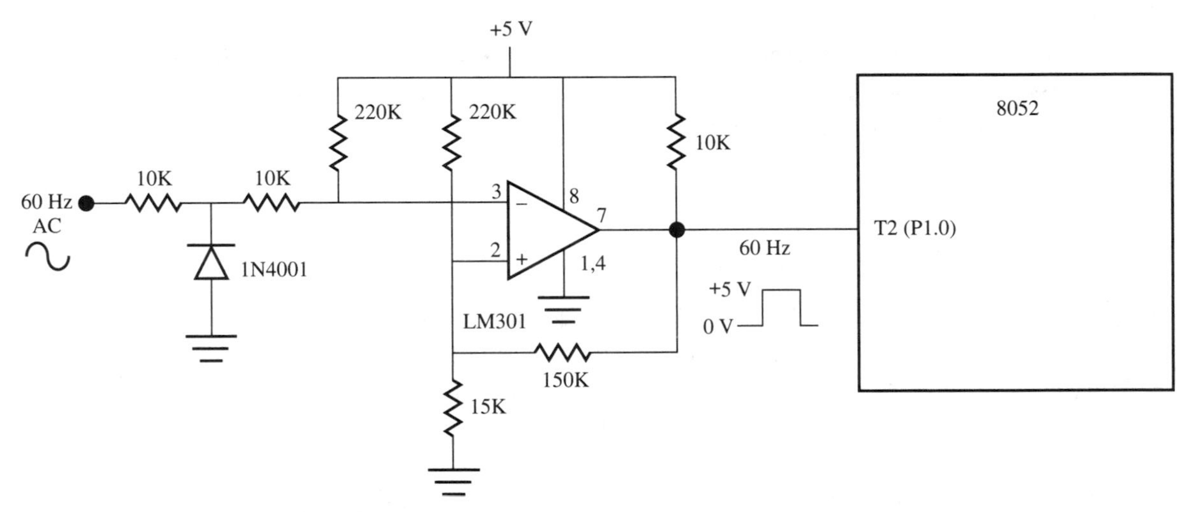

FIGURE 4–13
60 Hz time base

9. Write an 8051 program to generate a 500 Hz signal on P1.0 using Timer 0. The waveform should have a 30% duty cycle (duty cycle = high-time ÷ period).

10. The circuit shown in Figure 4–13 will provide an extremely accurate 60 Hz signal to T2 by tapping the secondary of a power supply transformer. Initialize Timer 2 such that it is clocked by T2 and overflows once per second. Upon each overflow, update a time-of-day value stored in the 8052's internal memory at locations 50H (hours), 51H (minutes), and 52H (seconds).

More timer examples and problems are found in Chapter 6.

5

SERIAL PORT OPERATION

5.1 INTRODUCTION

The 8051 includes an on-chip serial port that can operate in several modes over a wide range of frequencies. The essential function of the serial port is to perform parallel-to-serial conversion for output data, and serial-to-parallel conversion for input data.

Hardware access to the serial port is through the TXD and RXD pins introduced in Chapter 2. These pins are the alternate functions for two Port 3 bits, P3.1 on pin 11 (TXD) and P3.0 on pin 10 (RXD).

The serial port features **full duplex** operation (simultaneous transmission and reception), and **receive buffering** allowing one character to be received and held in a buffer while a second character is received. If the CPU reads the first character before the second is fully received, data are not lost.

Two special function registers provide software access to the serial port, SBUF and SCON. The serial port buffer (SBUF) at address 99H is really two buffers. Writing to SBUF loads data to be transmitted, and reading SBUF accesses received data. These are two separate and distinct registers, the transmit write-only register, and the receive read-only register. (See Figure 5–1.)

The serial port control register (SCON) at address 98H is a bit-addressable register containing status bits and control bits. Control bits set the operating mode for the serial port, and status bits indicate the end of a character transmission or reception. The status bits are tested in software or programmed to cause an interrupt.

The serial port frequency of operation, or **baud rate,** can be fixed (derived from the 8051 on-chip oscillator) or variable. If a variable baud rate is used, Timer 1 supplies the baud rate clock and must be programmed accordingly. (On the 8032/8052, Timer 2 can be programmed to supply the baud rate clock.)

5.2 SERIAL PORT CONTROL REGISTER

The mode of operation of the 8051 serial port is set by writing to the serial port mode register (SCON) at address 99H. (See Table 5–1 and Table 5–2.)

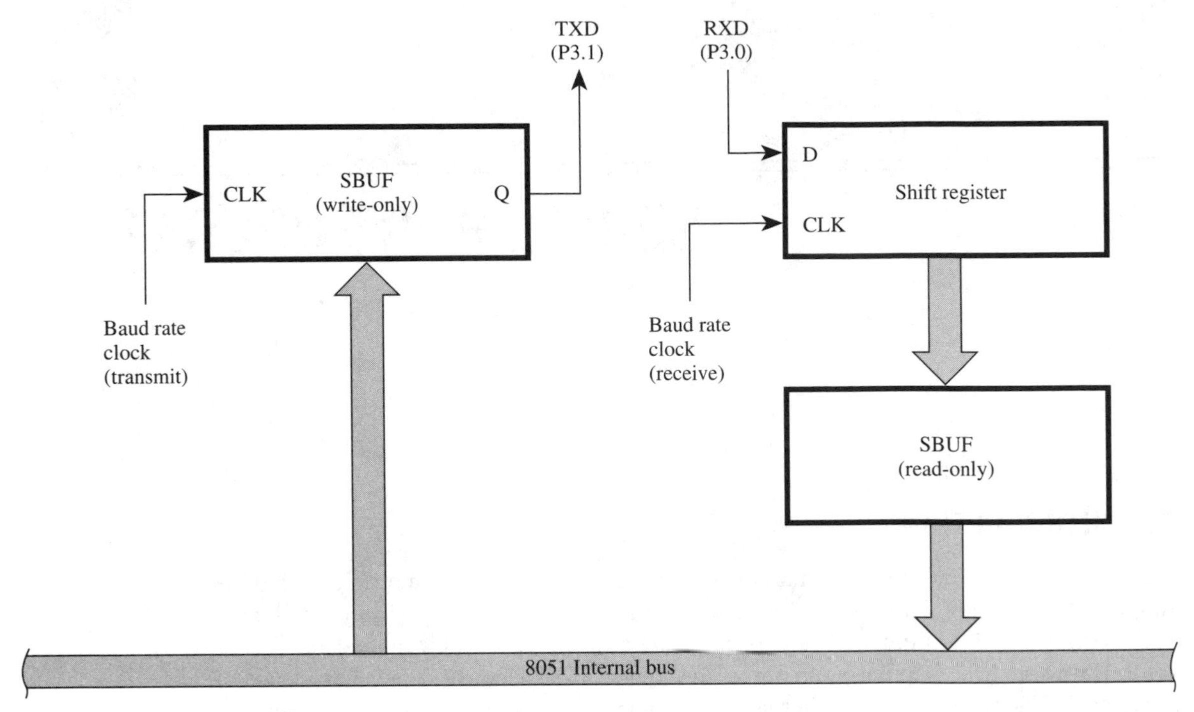

FIGURE 5–1
Serial port block diagram

Before using the serial port, SCON is initialized for the correct mode, and so on. For example, the following instruction

```
MOV  SCON,#01010010B
```

initializes the serial port for mode 1 (SM0/SM1 = 0/1), enables the receiver (REN = 1), and sets the transmit interrupt flag (T1 = 1) to indicate the transmitter is ready for operation.

5.3 MODES OF OPERATION

The 8051 serial port has four modes of operation, selectable by writing 1s or 0s into the SM0 and SM1 bits in SCON. Three of the modes enable asynchronous communications, with each character received or transmitted framed by a start bit and a stop bit. Readers familiar with the operation of a typical RS232C serial port on a microcomputer will find these modes familiar territory. In the fourth mode, the serial port operates as a simple shift register. Each mode is summarized below.

5.3.1 8-Bit Shift Register (Mode 0)

Mode 0, selected by writing 0s into bits SM1 and SM0 of SCON, puts the serial port into 8-bit shift register mode. Serial data enter and exit through RXD, and TXD outputs the shift clock. Eight bits are transmitted or received with the least-significant (LSB) first.

TABLE 5–1
SCON (serial port control) register summary

BIT	SYMBOL	ADDRESS	DESCRIPTION
SCON.7	SM0	9FH	Serial port mode bit 0 (see Table 5–2)
SCON.6	SM1	9EH	Serial port mode bit 1 (see Table 5–2)
SCON.5	SM2	9DH	Serial port mode bit 2. Enables multiprocessor communications in modes 2 & 3; RI will not be activated if received 9th bit is 0
SCON.4	REN	9CH	Receiver enable. Must be set to receive characters
SCON.3	TB8	9BH	Transmit bit 8. 9th bit transmitted in modes and 3; set/cleared by software
SCON.2	RB8	9AH	Receive bit 8. 9th bit received
SCON.1	TI	99H	Transmit interrupt flag. Set at end of character transmission; cleared by software
SCON.0	RI	98H	Receive interrupt flag. Set at end of character reception; cleared by software

The baud rate is fixed at 1/12th the on-chip oscillator frequency. The terms "RXD" and "TXD" are misleading in this mode. The RXD line is used for both data input and output, and the TXD line serves as the clock.

Transmission is initiated by any instruction that writes data to SBUF. Data are shifted out on the RXD line (P3.0) with clock pulses sent out the TXD line (P3.1). Each transmitted bit is valid on the RXD pin for one machine cycle. During each machine cycle, the clock signal goes low on S3P1 and returns high on S6P1. The timing for output data is shown in Figure 5–2.

Reception is initiated when the receiver enable bit (REN) is 1 and the receive interrupt bit (RI) is 0. The general rule is to set REN at the beginning of a program to initialize the serial port, and then clear RI to begin a data input operation. When RI is cleared, clock pulses are written out the TXD line, beginning the following machine cycle, and data are clocked in the RXD line. Obviously, it is up to the attached circuitry to provide data on the RXD line as synchronized by the clock signal on TXD. The clocking of data into the serial port occurs on the positive edge of TXD. (See Figure 5–3.)

One possible application of shift register mode is to expand the output capability of the 8051. A serial-to-parallel shift register IC can be connected to the 8051 TXD and RXD lines to provide an extra eight output lines. (See Figure 5–4.) Additional shift registers may be cascaded to the first for further expansion.

TABLE 5–2
Serial port modes

SM0	SM1	MODE	DESCRIPTION	BAUD RATE
0	0	0	Shift register	Fixed (oscillator frequency ÷ 12)
0	1	1	8-bit UART	Variable (set by timer)
1	0	2	9-bit UART	Fixed (oscillator frequency ÷ 12 or ÷ 64)
1	1	3	9-bit UART	Variable (set by timer)

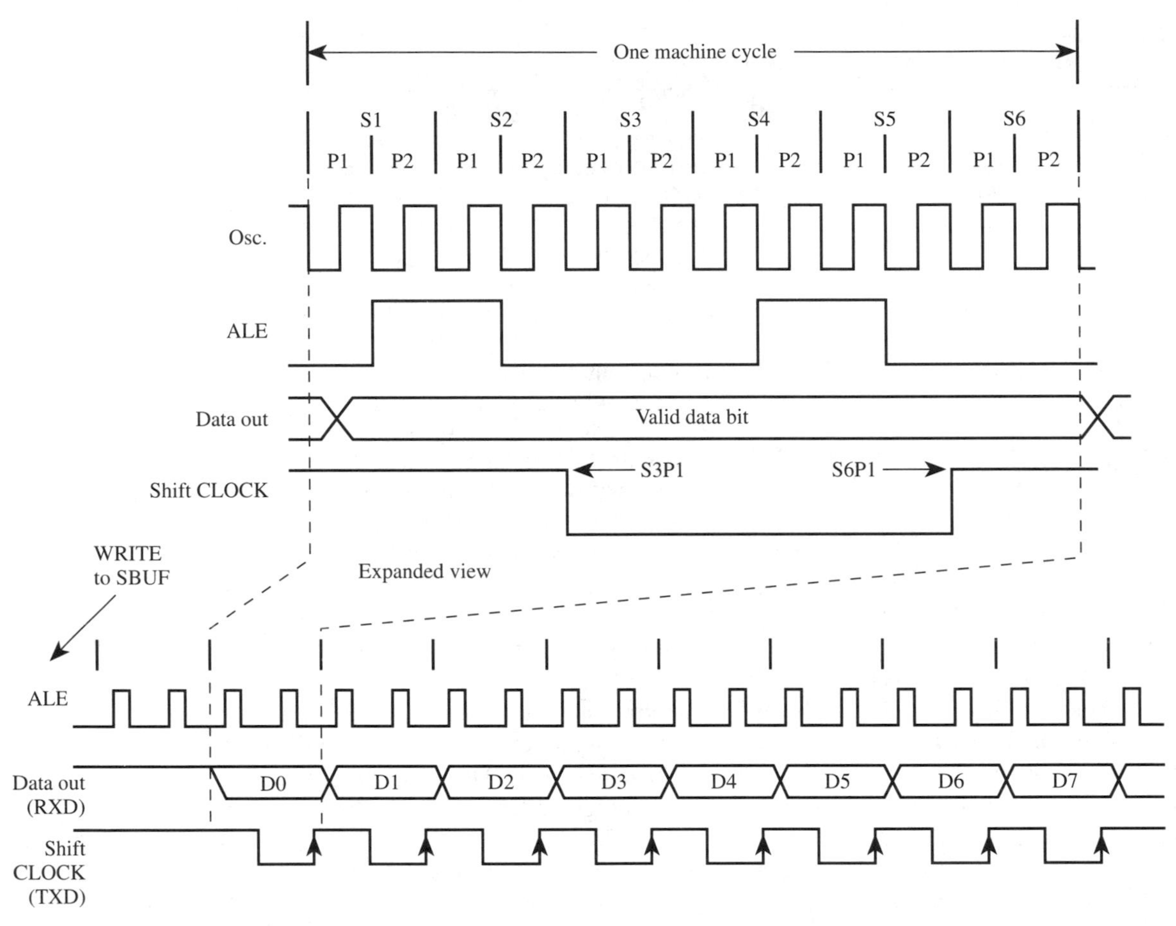

FIGURE 5–2
Serial port transmit timing for mode 0

5.3.2 8-Bit UART with Variable Baud Rate (Mode 1)

In mode 1 the 8051 serial port operates as an 8-bit UART with variable baud rate. A UART, or "universal asynchronous receiver/transmitter," is a device that receives and transmits serial data with each data character preceded by a start bit (low) and followed by a stop bit (high). A parity bit is sometimes inserted between the last data bit and the stop bit. The essential operation of a UART is parallel-to-serial conversion of output data and serial-to-parallel conversion of input data.

In mode 1, 10 bits are transmitted on TXD or received on RXD. These consist of a start bit (always 0), eight data bits (LSB first), and a stop bit (always 1). For a receive operation, the stop bit goes into RB8 in SCON. In the 8051, the baud rate is set by the Timer 1 overflow rate; the 8052 baud rate is set by the overflow rate of Timer 1 or Timer 2, or a combination of the two (one for transmit, the other for receive).

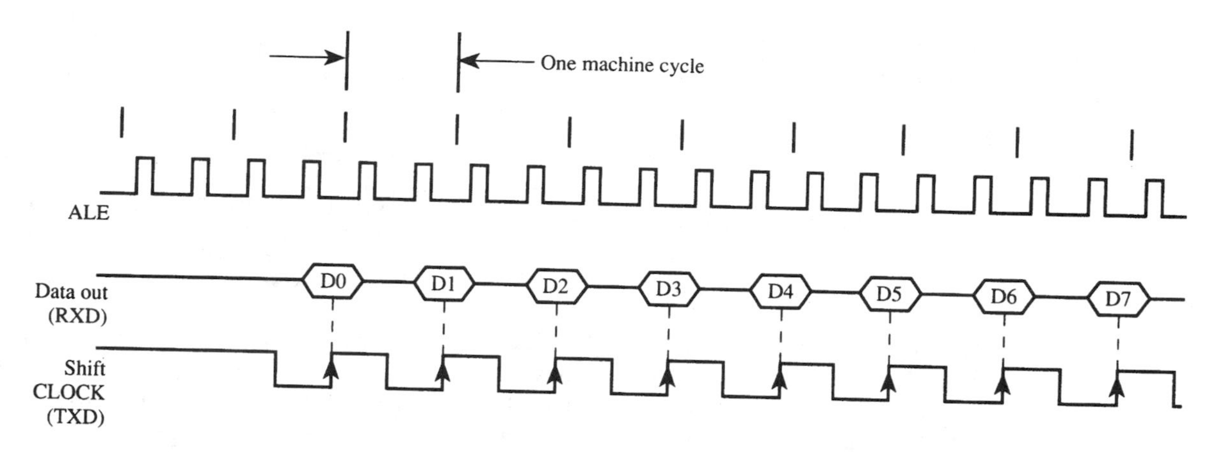

FIGURE 5–3
Serial port receive timing for mode 0

Clocking and synchronizing the serial port shift registers in modes 1, 2, and 3 is established by a 4-bit divide-by-16 counter, the output of which is the baud rate clock. (See Figure 5–5.) The input to this counter is selected through software, as discussed later.

Transmission is initiated by writing to SBUF, but does not actually start until the next rollover of the divide-by-16 counter supplying the serial port baud rate. Shifted data are outputted on the TXD line beginning with the start bit, followed by the eight data bits, then the stop bit. The period for each bit is the reciprocal of the baud rate as programmed in the timer. The transmit interrupt flag (TI) is set as soon as the stop bit appears on TXD. (See Figure 5–6.)

Reception is initiated by a 1-to-0 transition on RXD. The divide-by-16 counter is immediately reset to align the counts with the incoming bit stream (the next bit arrives on the next divide-by-16 rollover, and so on). The incoming bit stream is sampled in the middle of the 16 counts.

The receiver includes "false start bit detection" by requiring a 0 state eight counts after the first 1-to-0 transition. If this does not occur, it is assumed that the receiver was

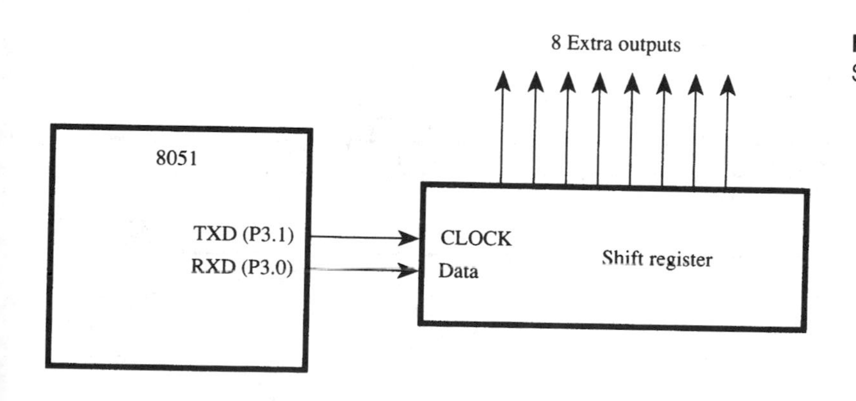

FIGURE 5–4
Serial port shift register mode

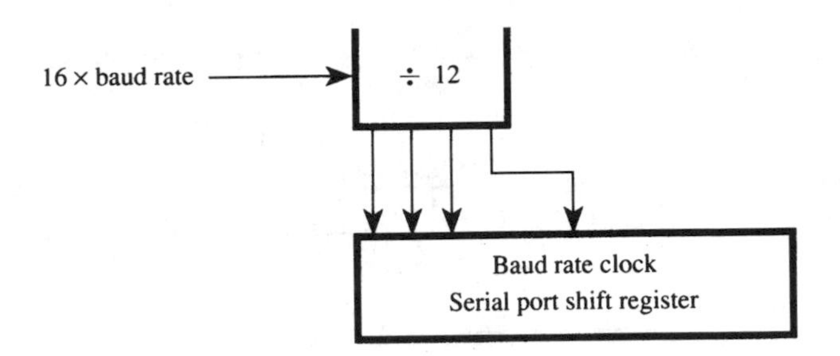

FIGURE 5–5
Serial port clocking

triggered by noise rather than by a valid character. The receiver is reset and returns to the idle state, looking for the next 1-to-0 transition.

Assuming a valid start bit was detected, character reception continues. The start bit is skipped and eight data bits are clocked into the serial port shift register. When all eight bits have been clocked in, the following occur:

1. The ninth bit (the stop bit) is clocked into RB8 in SCON,
2. SBUF is loaded with the eight data bits, and
3. The receiver interrupt flag (RI) is set.

These only occur, however, if the following conditions exist:

1. RI = 0, and
2. SM2 = 1 and the received stop bit = 1, or SM2 = 0.

The requirement that RI = 0 ensures that software has read the previous character (and cleared RI). The second condition sounds complicated, but applies only in multiprocessor communications mode (see below). It implies, "do not set RI in multiprocessor communications mode when the ninth data bit is 0."

5.3.3 9-Bit UART with Fixed Baud Rate (Mode 2)

When SM1 = 1 and SM0 = 0, the serial port operates in mode 2 as a 9-bit UART with a fixed baud rate. Eleven bits are transmitted or received: a start bit, eight data bits, a pro-

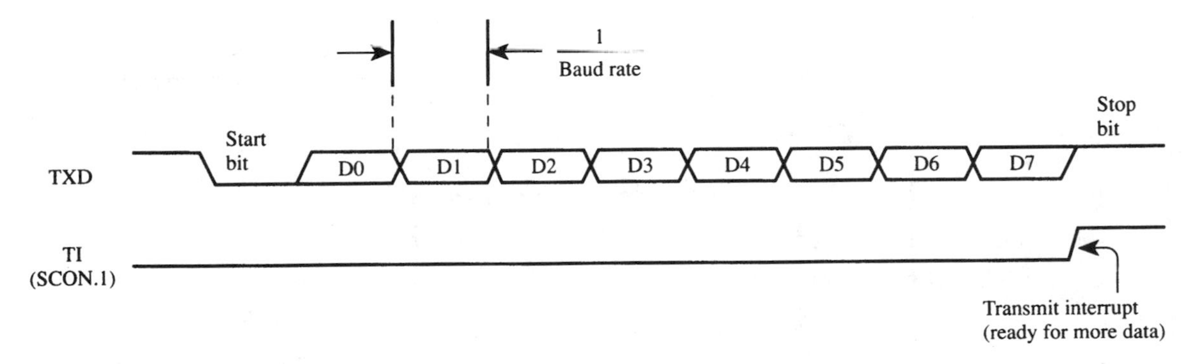

FIGURE 5–6
Setting the serial port TI flag

grammable ninth data bit, and a stop bit. On transmission, the ninth bit is whatever has been put in TB8 in SCON (perhaps a parity bit). On reception, the ninth bit received is placed in RB8. The baud rate in mode 2 is either 1/32nd or 1/64th the on-chip oscillator frequency. (See 5.6 Serial Port Baud Rates.)

5.3.4 9-Bit UART with Variable Baud Rate (Mode 3)

Mode 3, 9-bit UART with variable baud rate, is the same as mode 2 except the baud rate is programmable and provided by the timer. In fact, modes 1, 2, and 3 are very similar. The differences lie in the baud rates (fixed in mode 2, variable in modes 1 and 3) and in the number of data bits (eight in mode 1, nine in modes 2 and 3).

5.4 INITIALIZATION AND ACCESSING SERIAL PORT REGISTERS

5.4.1 Receiver Enable

The receiver enable bit (REN) in SCON must be set by software to enable the reception of characters. This is usually done at the beginning of a program when the serial port, timers, etc., are initialized. This can be done in two ways. The instruction

```
SETB REN
```

explicitly sets REN, or the instruction

```
MOV  SCON,#xxx1xxxxB
```

sets REN and sets or clears the other bits in SCON, as required. (The x's must be 1s or 0s to set the mode of operation.)

5.4.2 The 9th Data Bit

The ninth data bit transmitted in modes 2 and 3 must be loaded into TB8 by software. The ninth data bit received is placed in RB8. Software may or may not require a ninth data bit, depending on the specifications of the serial device with which communications is established. (The ninth data bit also plays an important role in multiprocessor communications. See below.)

5.4.3 Adding a Parity Bit

A common use for the ninth data bit is to add parity to a character. As discussed in Chapter 2, the P bit in the program status word (PSW) is set or cleared every machine cycle to establish even parity with the eight bits in the accumulator. If, for example, communications requires eight data bits plus even parity, the following instructions could be used to transmit the eight bits in the accumulator with even parity added in the ninth bit:

```
MOV  C,P          ; PUT EVEN PARITY BIT IN TB8
MOV  TB8,C        ; THIS BECOMES THE 9TH DATA BIT
MOV  SBUF,A       ; MOVE 8 BITS FROM ACC TO SBUF
```

If odd parity is required, then the instructions must be modified as follows:

```
MOV  C,P          ; PUT EVEN PARITY BIT IN C FLAG
CPL  C            ; CONVERT TO ODD PARITY
MOV  TB8,C
MOV  SBUF,A
```

Of course, the use of parity is not limited to modes 2 and 3. In mode 1, the eight data bits transmitted can consist of seven data bits plus a parity bit. In order to transmit a 7-bit ASCII code with even parity in bit 8, the following instructions could be used:

```
CLR  ACC.7        ; ENSURE MSB IS CLEAR
                  ; EVEN PARITY IS IN P
MOV  C,P          ; COPY TO C
MOV  ACC.7,C      ; PUT EVEN PARITY INTO MSB
MOV  SBUF,A       ; SEND CHARACTER
                  ; 7 DATA BITS PLUS EVEN PARITY
```

5.4.4 Interrupt Flags

The receive and transmit interrupt flags (RI and TI) in SCON play an important role in 8051 serial communications. Both bits are set by hardware, but must be cleared by software.

Typically, RI is set at the end of character reception and indicates "receive buffer full." This condition is tested in software or programmed to cause an interrupt. (Interrupts are discussed in Chapter 6.) If software wishes to input a character from the device connected to the serial port (perhaps a video display terminal), it must wait until RI is set, then clear RI and read the character from SBUF. This is shown below.

```
WAIT:   JNB  RI,WAIT      ; CHECK RI UNTIL SET
        CLR  RI           ; CLEAR RI
        MOV  A,SBUF       ; READ CHARACTER
```

TI is set at the end of character transmission and indicates "transmit buffer empty." If software wishes to send a character to the device connected to the serial port, it must first check that the serial port is ready. In other words, if a previous character was sent, wait until transmission is finished before sending the next character. The following instructions transmit the character in the accumulator:

```
WAIT:   JNB  TI,WAIT      ;CHECK TI UNTIL SET
        CLR  TI           ;CLEAR TI
        MOV  SBUF,A       ;SEND CHARACTER
```

The receive and transmit instruction sequences above are usually part of standard input character and output character subroutines. These are described in more detail in Example 5–2 and Example 5–3.

5.5 MULTIPROCESSOR COMMUNICATIONS

Modes 2 and 3 have a special provision for multiprocessor communications. In these modes, nine data bits are received and the ninth bit goes into RB8. The port can be programmed so that when the stop bit is received, the serial port interrupt is activated only if RB8 = 1. This feature is enabled by setting the SM2 bit in SCON. An application of this

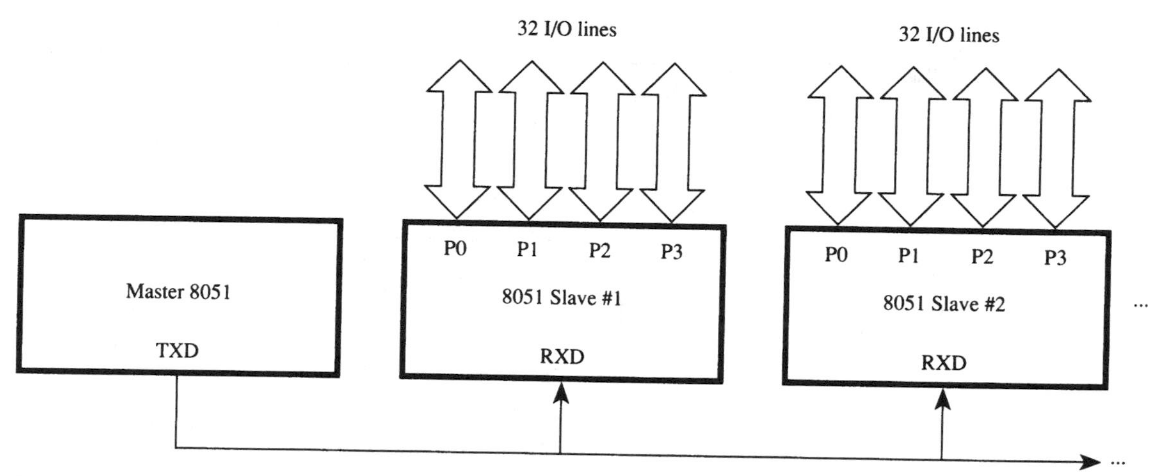

FIGURE 5–7
Multiprocessor communication

is in a networking environment using multiple 8051s in a master/slave arrangement, as shown in Figure 5–7.

When the master processor wants to transmit a block of data to one of several slaves, it first sends out an address byte that identifies the target slave. An address byte differs from a data byte in that the ninth bit is 1 in an address byte and 0 in a data byte. An address byte, however, interrupts all slaves, so that each can examine the received byte to test if it is being addressed. The addressed slave will clear its SM2 bit and prepare to receive the data bytes that follow. The slaves that weren't addressed leave their SM2 bits set and go about their business, ignoring the incoming data bytes. They will be interrupted again when the next address byte is transmitted by the master processor. Special schemes can be devised so that once a master/slave link is established, the slave can also transmit to the master. The trick is not to use the ninth data bit after a link has been established (otherwise other slaves may be inadvertently selected).

SM2 has no effect in mode 0, and in mode 1 it can be used to check the validity of the stop bit. In mode 1 reception, if SM2 = 1, the receive interrupt will not be activated unless a valid stop bit is received.

5.6 SERIAL PORT BAUD RATES

As evident in Table 5–2, the baud rate is fixed in modes 0 and 2. In mode 0 it is always the on-chip oscillator frequency divided by 12. Usually a crystal drives the 8051's on-chip oscillator, but another clock source can be used as well. (See Chapter 2.) Assuming a nominal oscillator frequency of 12 MHz, the mode 0 baud rate is 1 MHz. (See Figure 5–8a.)

By default following a system reset, the mode 2 baud rate is the oscillator frequency divided by 64. The baud rate is also affected by a bit in the power control register, PCON. Bit 7 of PCON is the SMOD bit. Setting SMOD has the effect of doubling the baud rate in modes 1, 2, and 3. In mode 2, the baud rate can be doubled from a default value of 1/64th the oscillator frequency (SMOD = 0), to 1/32nd the oscillator frequency (SMOD = 1). (See Figure 5–8b.)

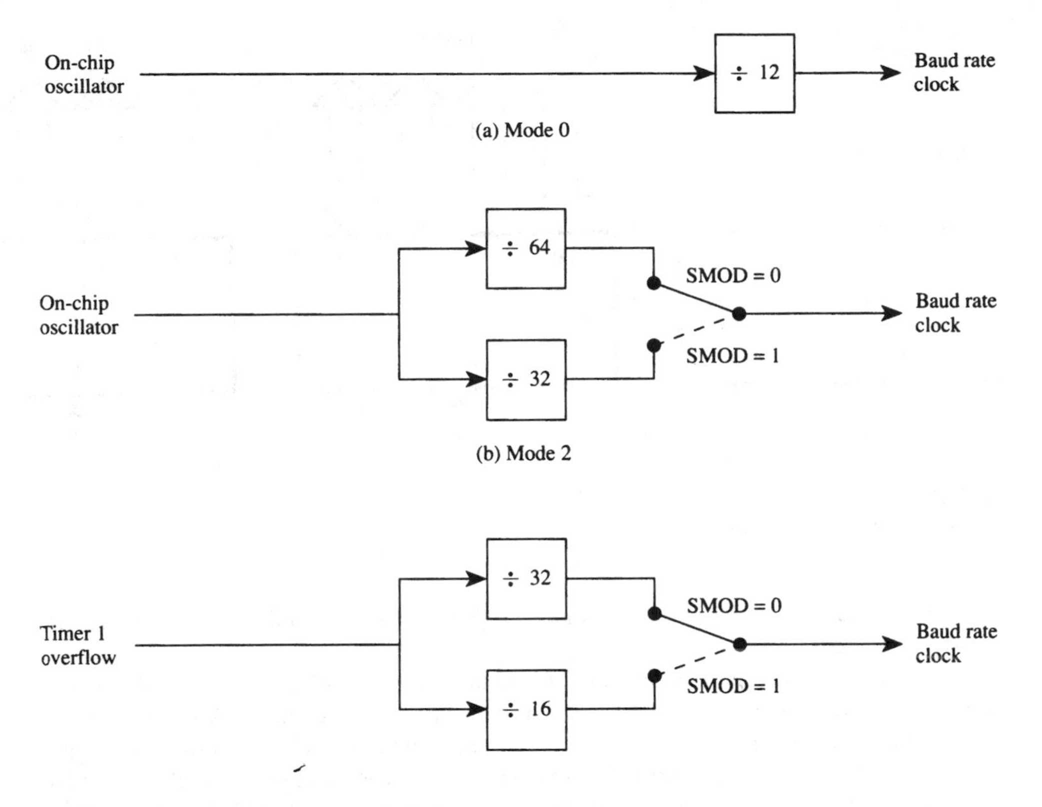

FIGURE 5–8
Serial port clocking sources. (a) Mode 0 (b) Mode 2 (c) Modes 1 and 3.

Since PCON is not bit-addressable, setting SMOD without altering the other PCON bits requires a "read-modify-write" operation. The following instructions set SMOD:

```
MOV  A,PCON          ;GET CURRENT VALUE OF PCON
SETB ACC.7           ;SET BIT 7 (SMOD)
MOV  PCON,A          ;WRITE VALUE BACK TO PCON
```

The 8051 baud rates in modes 1 and 3 are determined by the Timer 1 overflow rate. Since the timer operates at a relatively high frequency, the overflow is further divided by 32 (16 if SMOD = 1) before providing the baud rate clock to the serial port. The 8052 baud rate in modes 1 and 3 is determined by the Timer 1 or Timer 2 overflow rates, or both.

5.6.1 Using Timer 1 as the Baud Rate Clock

Considering only an 8051 for the moment, the usual technique for baud rate generation is to initialize TMOD for 8-bit auto-reload mode (timer mode 2) and put the correct reload value in TH1 to yield the proper overflow rate for the baud rate. TMOD is initialized as follows:

```
MOV  TMOD,#0010xxxxB
```

The x's are 1s or 0s as needed for Timer 0.

This is not the only possibility. Very low baud rates can be achieved by using 16-bit mode, timer mode 2 with TMOD = 0001xxxxB. There is a slight software overhead, however, since the TH1/TL1 registers must be reinitialized after each overflow. This would be performed in an interrupt service routine. Another option is to clock Timer 1 externally using T1 (P3.5). Regardless, the baud rate is the Timer 1 overflow rate divided by 32 (or divided by 16, if SMOD = 1).

The formula for determining the baud rate in modes 1 and 3, therefore, is

```
BAUD RATE = TIMER 1 OVERFLOW RATE ÷ 32
```

For example, 1200 baud operation requires an overflow rate calculated as follows:

```
1200 = TIMER 1 OVERFLOW RATE ÷ 32
TIMER 1 OVERFLOW RATE = 38.4 kHz
```

If a 12 MHz crystal drives the on-chip oscillator, Timer 1 is clocked at a rate of 1 MHz or 1000 kHz. Since the timer must overflow at a rate of 38.4 kHz and the timer is clocked at a rate of 1000 kHz, an overflow is required every 1000 ÷ 38.4 = 26.04 clocks. (Round to 26.) Since the timer counts up and overflows on the FFH-to-00H transition of the count, 26 counts less than 0 is the required reload value for TH1. The correct value is −26. The easiest way to put the reload value into TH1 is

```
MOV  TH1,#-26
```

The assembler will perform the necessary conversion. In this case −26 is converted to 0E6H; thus, the instruction above is identical to

```
MOV  TH1,#0E6H
```

Due to rounding, there is a slight error in the resulting baud rate. Generally, a 5% error is tolerable using asynchronous (start/stop) communications. Exact baud rates are possible using an 11.059 MHz crystal. Table 5–3 summarizes the TH1 reload values for the most common baud rates, using a 12.000 MHz or 11.059 MHz crystal.

TABLE 5–3
Baud rate summary

BAUD RATE	CRYSTAL FREQUENCY	SMOD	TH1 RELOAD VALUE	ACTUAL BAUD RATE	ERROR
9600	12.000 MHz	1	−7 (F9H)	8923	7%
2400	12.000 MHz	0	−13 (F3H)	2404	0.16%
1200	12.000 MHz	0	−26 (E6H)	1202	0.16%
19200	11.059 MHz	1	−3 (FDH)	19200	0
9600	11.059 MHz	0	−3 (FDH)	9600	0
2400	11.059 MHz	0	−12 (F4H)	2400	0
1200	11.059 MHz	0	−24 (E8H)	1200	0

Example 5–1: Initializing the Serial Port

Write an instruction sequence to initialize the serial port to operate as an 8-bit UART at 2400 baud. Use Timer 1 to provide the baud rate clock.

For this example, four registers must be initialized: SMOD, TMOD, TCON, and TH1. The required values are summarized below.

	SM0	SM1	SM2	REN	TB8	RB8	TI	RI
SCON:	0	1	0	1	0	0	1	0
	GTE	C/T	M1	M0	GTE	C/T	M1	M0
TMOD:	0	0	1	0	0	0	0	0
	TF1	TR1	TF0	TR0	IE1	IT1	IE0	IT0
TCON:	0	1	0	0	0	0	0	0
TH1:	1	1	1	1	0	0	1	1

Setting SM0/SM1 = 0/1 puts the serial port into 8-bit UART mode. REN = 1 enables the serial port to receive characters. Setting TI = 1 allows transmission of the first character by indicating that the transmit buffer is empty. For TMOD, setting M1/M0 = 1/0 puts Timer 1 into 8-bit auto-reload mode. Setting TR1 = 1 in TCON turns on Timer 1. The other bits are shown as 0s, since they control features or modes not used in this example.

The required TH1 value is that which provides overflows at the rate of 2400 $\times$ 32 = 76.8 kHz. Assuming the 8051 is clocked from a 12 MHz crystal, Timer 1 is clocked at a rate of 1 MHz or 1000 kHz, and the number of clocks for each overflow is 1000 $\div$ 76.8 = 13.02. (Round to 13.) The reload value is -13 or 0F3H.

The initialization instruction sequence is shown below.

Example 5–1: 8051 Serial Port Example (initialize the serial port)

```
8100              5              ORG     8100H
8100 759852       6     INIT:    MOV     SCON,#52H      ;serial port, mode 1
8103 758920       7              MOV     TMOD,#20H      ;timer 1, mode 2
8106 758DF3       8              MOV     TH1,#-13       ;reload count for 2400 baud
8109 D28E         9              SETB    TR1            ;start timer 1
                 10              END
```

Example 5–2: Output Character Subroutine

Write a subroutine called OUTCHR to transmit the 7-bit ASCII code in the accumulator out the 8051 serial port, with odd parity added as the 8th bit. Return from the subroutine with the accumulator intact, i.e., containing the same value as before the subroutine was called.

This example and the next illustrate two of the most common subroutines on microcomputer systems with an attached RS232 terminal: output character (OUTCHR) and input character (INCHAR).

```
8100             5            ORG    8100H
8100 A2D0        6   OUTCHR:  MOV    C,P          ;put parity bit in C flag
8102 B3          7            CPL    C            ;change to odd parity
8103 92E7        8            MOV    ACC.7,C      ;add to character code
8105 3099FD      9   AGAIN:   JNB    TI,AGAIN     ;Tx empty? no: check again
8108 C299       10            CLR    TI           ;yes: clear flag and
810A F599       11            MOV    SBUF,A       ;     send character
810C C2E7       12            CLR    ACC.7        ;strip off parity bit and
810E 22         13            RET                 ;  return
                14            END
```

The first three instructions place odd parity in the accumulator bit 7. Since the P bit in the PSW establishes even parity with the accumulator, it is complemented before being placed in ACC.7. The JNB instruction creates a "wait loop," repeatedly testing the transmit interrupt flag (TI) until it is set. When TI is set (because the previous character transmission is finished), it is cleared and then the character in the accumulator is written into the serial port buffer (SBUF). Transmission begins on the next rollover of the divide-by-16 counter that clocks the serial port. (See Figure 5–5.) Finally, ACC.7 is cleared so that the return value is the same as the 7-bit code passed to the subroutine.

The OUTCHR subroutine is a building block and is of little use by itself. At a "higher level," this subroutine is called to transmit a single character or a string of characters. For example, the following instructions transmit the ASCII code for the letter "Z" to the serial device attached to the 8051's serial port:

```
MOV  A,#'Z'
CALL OUTCHR
(continue)
```

As a natural extension to this idea, Problem 1 at the end of this chapter uses OUTCHR as a building block in an OUTSTR (output string) subroutine that transmits a sequence of ASCII codes (terminated by a NULL byte, 00H) to the serial device attached to the 8051's serial port.

Example 5–3: Input Character Subroutine

Write a subroutine called INCHAR to input a character from the 8051's serial port and return with the 7-bit ASCII code in the accumulator. Expect odd parity in the eighth bit received and set the carry flag if there is a parity error.

```
8100             5            ORG    8100H
8100 3098FD      6   INCHAR:  JNB    RI,$         ;wait for character
8103 C298        7            CLR    RI           ;clear flag
8105 E599        8            MOV    A,SBUF       ;read char into A
```

```
8107 A2D0      9        MOV     C,P        ;for odd parity in A,
              10                           ;   P should be set
8109 B3       11        CPL     C          ;complementing correctly
              12                           ; indicates if "error"
810A C2E7     13        CLR     ACC.7      ;strip off parity
810C 22       14        RET
              15        END
```

This subroutine begins by waiting for the receive interrupt flag (RI) to be set, indicating that a character is waiting in SBUF to be read. When RI is set, the JNB instruction falls through to the next instruction. RI is cleared and the code in SBUF is read into the accumulator. The P bit in the PSW establishes even parity with the accumulator, so it should be set if the accumulator, on its own, correctly contains odd parity in bit 7. Moving the P bit into the carry flag leaves CY = 0 if there is no error. On the other hand, if the accumulator contains a parity error, then CY = 1, correctly indicating "parity error." Finally, ACC.7 is cleared to ensure that only a 7-bit code is returned to the calling program.

5.7 SUMMARY

This chapter has presented the major details required to program the 8051 serial port. A passing mention has been made in this chapter and in the last chapter of the use of interrupts. Indeed, advanced applications using the 8051 timers or serial ports generally require input/output operations to be synchronized by interrupts. This is the topic of the next chapter.

PROBLEMS

The following problems are typical of the software routines for interfacing terminals (or other serial devices) to a microcomputer. Assume the 8051 serial port is initialized in 8-bit UART mode and the baud rate is provided by Timer 1.

1. Write a subroutine called OUTSTR that sends a null-terminated string of ASCII codes to the device (perhaps a VDT) connected to the 8051 serial port. Assume the string of ASCII codes is in external code memory and the calling program puts the address of the string in the data pointer before calling OUTSTR. A null-terminated string is a series of ASCII bytes terminated with a 00H byte.
2. Write a subroutine called INLINE that inputs a line of ASCII codes from the device connected to the 8051 serial port and places it in internal data memory beginning at address 50H. Assume the line is terminated with a carriage return code. Place the carriage return code in the line buffer along with the other codes, and then terminate the line buffer with a null byte (00H).
3. Write a program that continually sends the alphabet (lowercase) to the device attached to the 8051 serial port. Use the OUTCHR subroutine written earlier.
4. Assuming the availability of the OUTCHR subroutine, write a program that continually sends the displayable ASCII set (codes 20H to 7EH) to the device attached to the 8051 serial port.

5. Modify the solution to the above problem to suspend and resume output to the screen using XOFF and XON codes entered on the keyboard. All other codes received should be ignored. (Note: XOFF = CONTROL-S = 13H, XON = CONTROL-Q = 11H)
6. Assume the availability of the INCHAR and OUTCHR subroutines and write a program that inputs characters from the keyboard and echoes them back to the screen, converting lowercase characters to uppercase.
7. Assume the availability of the INCHAR and OUTCHR subroutines and write a program that inputs characters from the device attached to the 8051 serial port and echoes them back substituting period (.) for any control characters (ASCII codes 00H to 1FH, and 7FH).
8. Assume the availability of the OUTCHR subroutine and write a program that clears the screen on the VDT attached to the 8051 serial port and then sends your name to the VDT 10 times on 10 separate lines. The clear screen function on VDTs is accomplished by transmitting a CONTROL-Z on many terminals or <ESC> [2 J on terminals that support ANSI (American National Standards Institute) escape sequences. Use either method in your solution.
9. Figure 5–4 illustrates a technique for expanding the output capability of the 8051. Assuming such a configuration, write a program that initializes the 8051 serial port for shift register mode and then maps the contents of internal memory location 20H to the eight extra outputs, 10 times per second.

6

INTERRUPTS

6.1 INTRODUCTION

An **interrupt** is the occurrence of a condition—an event—that causes a temporary suspension of a program while the condition is serviced by another program. Interrupts play an important role in the design and implementation of microcontroller applications. They allow a system to respond asynchronously to an event and deal with the event while another program is executing. An **interrupt-driven system** gives the illusion of doing many things simultaneously. Of course, the CPU cannot execute more than one instruction at a time; but it can temporarily suspend execution of one program, execute another, then return to the first program. In a way, this is like a subroutine. The CPU executes another program—the subroutine—and then returns to the original program. The difference is that in an interrupt-driven system, the interruption is a response to an "event" that occurs asynchronously with the main program. It is not known when the main program will be interrupted.

The program that deals with an interrupt is called an **interrupt service routine (ISR)** or **interrupt handler.** The ISR executes in response to the interrupt and generally performs an input or output operation to a device. When an interrupt occurs, the main program temporarily suspends execution and branches to the ISR; the ISR executes, performs the operation, and terminates with a "return from interrupt" instruction; the main program continues where it left off. It is common to refer to the main program as executing at **base-level** and the ISRs as executing at **interrupt-level.** The terms **foreground** (base-level) and **background** (interrupt-level) are also used. This brief view of interrupts is depicted in Figure 6–1, showing (a) the execution of a program without interrupts and (b) execution at base-level with occasional interrupts and ISRs executing at interrupt-level.

A typical example of interrupts is manual input using a keyboard. Consider an application for a microwave oven. The main program (foreground) might control a microwave power element for cooking; yet, while cooking, the system must respond to manual input on the oven's door, such as a request to shorten or lengthen the cooking time. When the user depresses a key, an interrupt is generated (a signal goes from high to

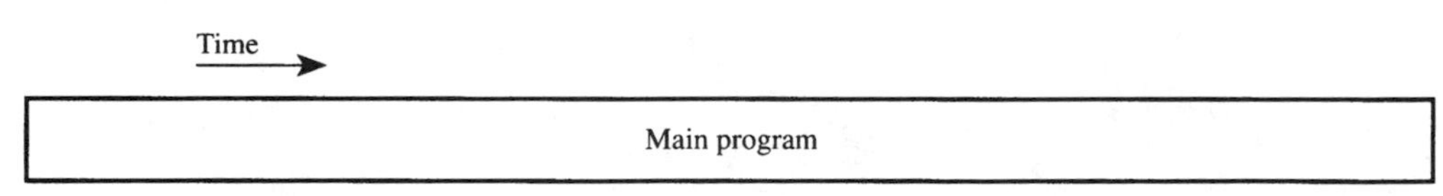

(a) Program execution without interrupts

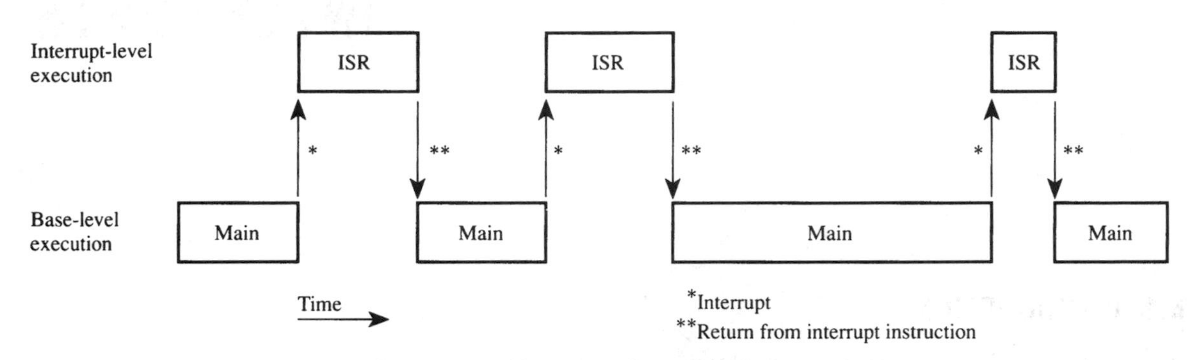

(b) Program execution with interrupts

FIGURE 6–1
Program execution with and without interrupts. (a) Without interrupts (b) With interrupts.

low, perhaps) and the main program is interrupted. The ISR takes over in the background, reads the keyboard code(s) and changes the cooking conditions accordingly, and finishes by passing control back to the main program. The main program carries on where it left off. The important point in this example is that manual input occurs "asynchronously;" that is, it occurs at intervals not predictable or controlled by the software running in the system. This is an interrupt.

6.2 8051 INTERRUPT ORGANIZATION

There are five interrupt sources on the 8051: two external interrupts, two timer interrupts, and a serial port interrupt. The 8052 adds a sixth interrupt source from the extra timer. All interrupts are disabled after a system reset and are enabled individually by software.

In the event of two or more simultaneous interrupts or an interrupt occurring while another interrupt is being serviced, there is both a polling sequence and a two-level priority scheme to schedule the interrupts. The polling sequence is fixed but the interrupt priority is programmable.

Let's begin by examining ways to enable and disable interrupts.

6.2.1 Enabling and Disabling Interrupts

Each of the interrupt sources is individually enabled or disabled through the bit-addressable special function register IE (interrupt enable) at address 0A8H. As well as individual enable bits for each interrupt source, there is a global enable/disable bit that is cleared to disable all interrupts or set to turn on interrupts. (See Table 6–1.)

Two bits must be set to enable any interrupt: the individual enable bit and the global enable bit. For example, timer 1 interrupts are enabled as follows:

```
SETB ET1                    ;ENABLE Timer 1 INTERRUPT
SETB EA                     ;SET GLOBAL ENABLE BIT
```

This could also be coded as

```
MOV  IE,#10001000B
```

Although these two approaches have exactly the same effect following a system reset, the effect is different if IE is written "on-the-fly," in the middle of a program. The first approach has no effect on the other five bits in the IE register, whereas the second approach explicitly clears the other bits. It is fine to initialize IE with a "move byte" instruction at the beginning of a program (i.e., following a power-up or system reset), but enabling and disabling interrupts on-the-fly within a program should use "set bit" and "clear bit" instructions to avoid side effects with other bits in the IE register.

6.2.2 Interrupt Priority

Each interrupt source is individually programmed to one of two priority levels through the bit-addressable special function register IP (interrupt priority) at address 0B8H. (See Table 6–2.)

IP is cleared after a system reset to place all interrupts at the lower priority level by default. The idea of "priorities" allows an ISR to be interrupted by an interrupt if the new interrupt is of higher priority than the interrupt currently being serviced. This is straightforward on the 8051, since there are only two priority levels. If a low-priority ISR is executing when a high-priority interrupt occurs, the ISR is interrupted. A high-priority ISR cannot be interrupted.

The main program, executing at base level and not associated with any interrupt, can always be interrupted regardless of the priority of the interrupt. If two interrupts of different priorities occur simultaneously, the higher priority interrupt will be serviced first.

TABLE 6–1
IE (interrupt enable) register summary

BIT	SYMBOL	BIT ADDRESS	DESCRIPTION (1 = ENABLE, 0 = DISABLE)
IE.7	EA	AFH	Global enable/disable
IE.6	-	AEH	Undefined
IE.5	ET2	ADH	Enable Timer 2 Interrupt (8052)
IE.4	ES	ACH	Enable serial port interrupt
IE.3	ET1	ABH	Enable Timer 1 interrupt
IE.2	EX1	AAH	Enable external 1 interrupt
IE.1	ET0	A9H	Enable Timer 0 interrupt
IE.0	EX0	A8H	Enable external 0 interrupt

TABLE 6–2
IP (interrupt priority) register summary

BIT	SYMBOL	BIT ADDRESS	DESCRIPTION (1 = HIGHER LEVEL, 0 = LOWER LEVEL)
IP.7	-	-	Undefined
IP.6	-	-	Undefined
IP.5	PT2	0BDH	Priority for Timer 2 interrupt (8052)
IP.4	PS	0BCH	Priority for serial port interrupt
IP.3	PT1	0BBH	Priority for Timer 1 interrupt
IP.2	PX1	0BAH	Priority for external 1 interrupt
IP.1	PT0	0B9H	Priority for Timer 0 interrupt
IP.0	PX0	0B8H	Priority for external 0 interrupt

6.2.3. Polling Sequence

If two interrupts of the same priority occur simultaneously, a fixed polling sequence determines which is serviced first. The polling sequence is external 0, Timer 0, external 1, Timer 1, serial port, Timer 2.

Figure 6–2 illustrates the five interrupt sources, the individual and global enable mechanism, the polling sequence, and the priority levels. The state of all interrupt sources is available through the respective flag bits in the SFRs. Of course, if any interrupt is disabled, an interrupt does not occur, but software can still test the interrupt flag. The timer and serial port examples in the previous two chapters used the interrupt flags extensively without actually using interrupts.

A serial port interrupt results from the logical OR of a receive interrupt (RI) or a transmit interrupt (TI). Likewise, Timer 2 interrupts are generated by a timer overflow (TF2) or by the external input flag (EXF2). The flag bits that generate interrupts are summarized in Table 6–3.

6.3 PROCESSING INTERRUPTS

When an interrupt occurs and is accepted by the CPU, the main program is interrupted. The following actions occur:

- The current instruction completes execution
- The PC is saved on the stack
- The current interrupt status is saved internally
- Interrupts are blocked at the level of the interrupt
- The PC is loaded with the vector address of the ISR
- The ISR executes

The ISR executes and takes action in response to the interrupt. The ISR finishes with a RETI (return from interrupt) instruction. This retrieves the old value of the PC from the stack and restores the old interrupt status. Execution of the main program continues where it left off.

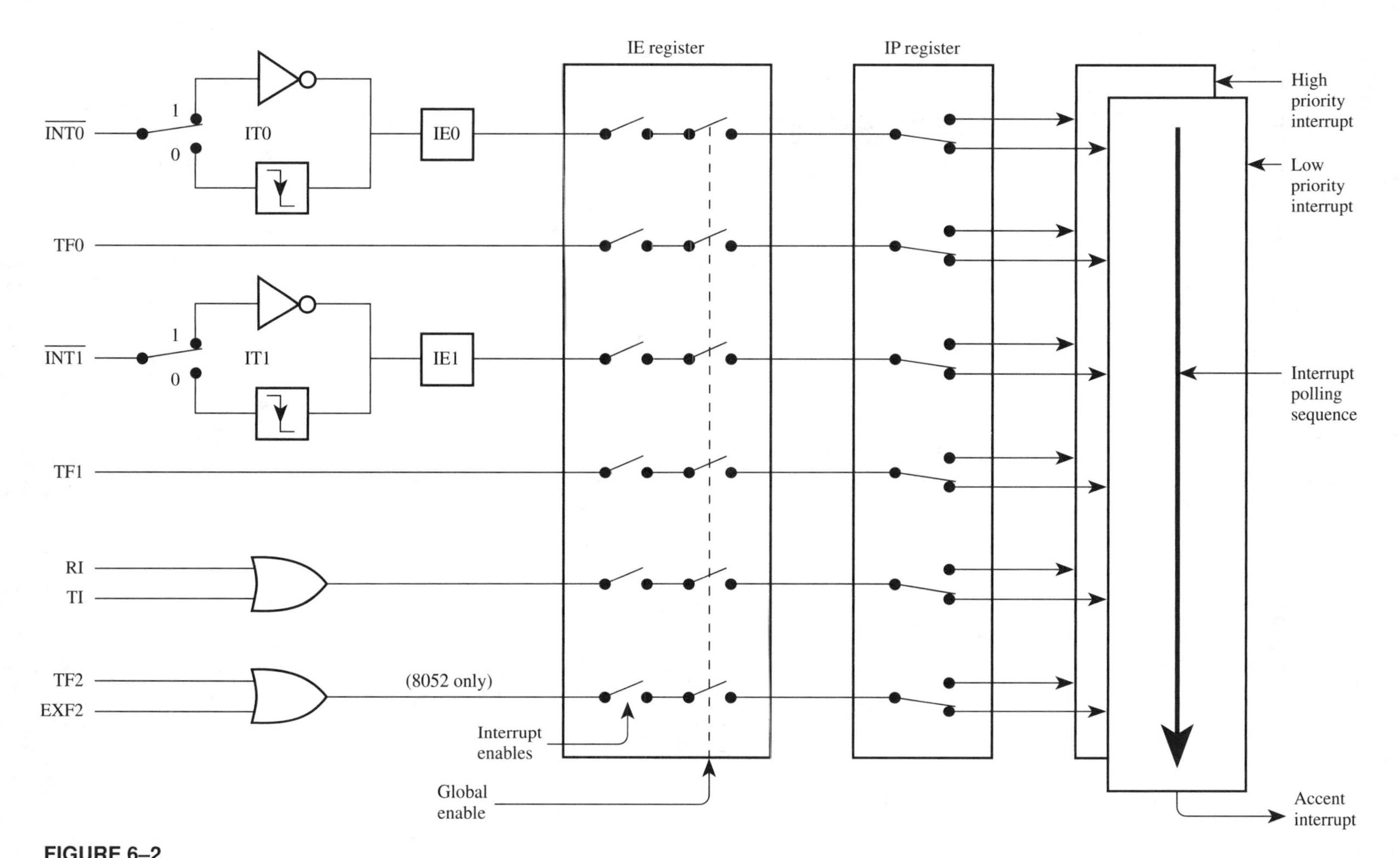

FIGURE 6–2
Overview of 8051 interrupt structure

TABLE 6–3
Interrupt flag bits

INTERRUPT	FLAG	SFR REGISTER AND BIT POSITION
External 0	IE0	TCON.1
External 1	IE1	TCON.3
Timer 1	TF1	TCON.7
Timer 0	TF0	TCON.5
Serial port	T1	SCON.1
Serial port	RI	SCON.0
Timer 2	TF2	T2CON.7 (8052)
Timer 2	EXF2	T2CON.6 (8052)

6.3.1 Interrupt Vectors

When an interrupt is accepted, the value loaded into the PC is called the **interrupt vector.** It is the address of the start of the ISR for the interrupting source. The interrupt vectors are given in Table 6–4.

The system reset vector (RST at address 0000H) is included in this table, since, in this sense, it is like an interrupt: it interrupts the main program and loads the PC with a new value.

When "vectoring to an interrupt," the flag that caused the interrupt is automatically cleared by hardware. The exceptions are RI and TI for serial port interrupts, and TF2 and EXF2 for Timer 2 interrupts. Since there are two possible sources for each of these interrupts, it is not practical for the CPU to clear the interrupt flag. These bits must be tested in the ISR to determine the source of the interrupt, and then the interrupting flag is cleared by software. Usually a branch occurs to the appropriate action, depending on the source of the interrupt.

Since the interrupt vectors are at the bottom of code memory, the first instruction of the main program is often a jump above this area of memory, such as LJMP 0030H.

6.4 PROGRAM DESIGN USING INTERRUPTS

The examples in Chapter 3 and Chapter 4 did not use interrupts but made extensive use of "wait loops" to test the timer overflow flags (TF0, TF1, or TF2) or the serial port transmit and receive flags (TI or RI). The problem in this approach is that the CPU's valuable execution time is fully consumed waiting for flags to be set. This is inappropri-

TABLE 6–4
Interrupt vectors

INTERRUPT	FLAG	VECTOR ADDRESS
System reset	RST	0000H
External 0	IE0	0003H
Timer 0	TF0	000BH
External 1	IE1	0013H
Timer 1	TF1	001BH
Serial port	RI or TI	0023H
Timer 2	TF2 or EXF2	002BH

ate for control-oriented applications where a microcontroller must interact with many input and output devices simultaneously.

In this section, examples are developed to demonstrate practical methods for implementing software for control-oriented applications. The key ingredient is the interrupt. Although the examples are not necessarily bigger, they are more complex, and in recognition of this, we proceed one step at a time. The reader is advised to follow the examples slowly and to examine the software meticulously. Some of the most difficult bugs in system designs often involve interrupts. The details must be understood thoroughly.

Since we are using interrupts, the examples will be complete and self-contained. Each program starts at address 0000H with the assumption that it begins execution following a system reset. The idea is that eventually these programs develop into full-fledged applications that reside in ROM or EPROM.

The suggested framework for a self-contained program using interrupts is shown below.

```
            ORG  0000H                 ;RESET ENTRY POINT
            LJMP MAIN
            .                          ;ISR ENTRY POINTS
            .
            .
            ORG  0030H                 ;MAIN PROGRAM ENTRY POINT
MAIN:       .                          ;MAIN PROGRAM BEGINS
            .
            .
```

The first instruction jumps to address 0030H, just above the vector locations where the ISRs begin, as given in Table 6–4. As shown in Figure 6–3, the main program begins at address 0030H.

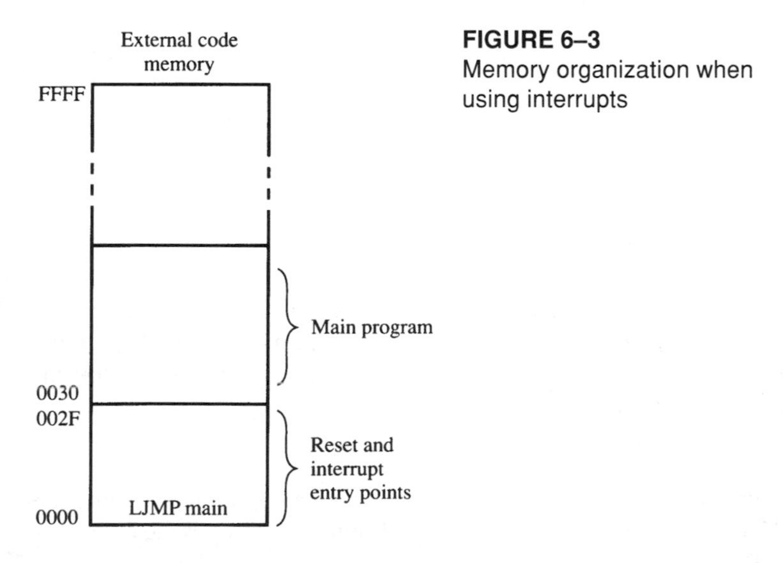

FIGURE 6–3
Memory organization when using interrupts

6.4.1 Small Interrupt Service Routines

Interrupt service routines must begin near the bottom of code memory at the addresses shown in Table 6–4. Although there are only eight bytes between each interrupt entry point, this is often enough memory to perform the desired operation and return from the ISR to the main program.

If only one interrupt source was used, say Timer 0, then the following framework could be used:

```
            ORG  0000H              ;RESET
            LJMP MAIN
            ORG  000BH              ;Timer 0 ENTRY POINT
T0ISR:      .                       ;Timer 0 ISR BEGINS
            .
            RETI                    ;RETURN TO MAIN PROGRAM
MAIN:       .                       ;MAIN PROGRAM
            .
            .
```

If more interrupts are used, care must be taken to ensure they start at the correct location (see Table 6–4) and do not overrun the next ISR. Since only one interrupt is used in the example above, the main program can begin immediately after the RETI instruction.

6.4.2 Large Interrupt Service Routines

If an ISR is longer than eight bytes, it may be necessary to move it elsewhere in code memory or it may trespass on the entry point for the next interrupt. Typically, the ISR begins with a jump to another area of code memory where the ISR can stretch out. Considering only Timer 0 for the moment, the following framework could be used:

```
            ORG  0000H              ;RESET ENTRY POINT
            LJMP MAIN
            ORG  000BH              ;Timer 0 ENTRY POINT
            LJMP T0ISR
            ORG  0030H              ;ABOVE INTERRUPT VECTORS
MAIN:       .
            .
            .
T0ISR:      .
            .                       ;Timer 0 ISR
            .
            RETI                    ;RETURN TO MAIN PROGRAM
```

To keep it simple, our programs will only do one thing at a time initially. The main or foreground program initializes the timer, serial port, and interrupt registers as appropriate, and then does nothing. The work is done totally in the ISR. After the initialize instructions, the main program consists of the following instruction:

```
HERE:       SJMP HERE
```

When an interrupt occurs, the main program is interrupted temporarily while the ISR executes. The RETI instruction at the end of the ISR returns control to the main program, and it continues doing nothing. This is not as farfetched as one might think. In many control-oriented applications, the bulk of the work is in fact done in the interrupt service routines.

Example 6–1 A Square Wave Using Timer Interrupts

Write a program using Timer 0 and interrupts to create a 10 kHz square wave on P1.0.

Timer interrupts occur when the timer registers (TLx/THx) overflow and set the overflow flag (TFx). This example appears in Chapter 4 without using interrupts (see example). The bulk of the program is the same except it is now organized into the framework for interrupts. Here's the program:

```
0000            5              ORG     0           ;reset entry point
0000 020030     6              LJMP    MAIN        ;jump above interrupt vectors
000B            7              ORG     000BH       ;Timer 0 interrupt vector
000B B290       8     T0ISR:   CPL     P1.0        ;toggle port bit
000D 32         9              RETI
0030           10              ORG     0030H       ;Main program entry point
0030 758902    11     MAIN:    MOV     TMOD,#02H   ;timer 0, mode 2
0033 758CCE    12              MOV     TH0,#-50    ;50 us delay
0036 D28C      13              SETB    TR0         ;start timer
0038 75A882    14              MOV     IE,#82H     ;enable timer 0 interrupt
003B 80FE      15              SJMP    $           ;do nothing
               16              END
```

This is a complete program, which could be burned into EPROM and installed in an 8051 single-board computer for execution. Immediately after reset, the program counter is loaded with 0000H. The first instruction executed is LJMP MAIN, which branches over the timer ISR to address 0030H in code memory. The next three instructions (lines 11–13) initialize Timer 0 for 8-bit auto-reload mode with overflows every 50 μs. The MOV IE,#82H instruction enables Timer 0 interrupts, so each overflow of the timer generates an interrupt. Of course, the first overflow will not occur for 50 μs, so the main program falls through to the "do-nothing" loop. Each 50 μs an interrupt occurs; the main program is interrupted and the Timer 0 ISR executes. The ISR simply complements the port bit and returns to the main program where the do-nothing loop executes for another 50 μs.

Note that the timer flag, TF0, is not explicitly cleared by software. When interrupts are enabled, TF0 is automatically cleared by hardware when the CPU vectors to the interrupt.

Incidentally, the return address in the main program is the location of the SJMP instruction. This address gets pushed on the 8051's internal stack prior to vectoring to each interrupt, and gets popped from the stack when the RETI instruction executes at the end of the ISR. Since the SP was not initialized, it defaults to its reset value of 07H. The push operation leaves the return address in internal RAM locations 08H (PC_L) and 09H (PC_H).

Example 6–2: Two Square Waves Using Interrupts

Write a program using interrupts to simultaneously create 7 kHz and 500 Hz square waves on P1.7 and P1.6.

The hardware configuration with the timings for the desired waveforms is shown in Figure 6–4.

This combination of outputs would be extremely difficult to generate on a non-interrupt-driven system. Timer 0, providing synchronization for the 7 kHz signal, operates in mode 2, as in the previous example; and timer 1, providing synchronization for the 500 Hz signal, operates in mode 1, 16-bit timer mode. Since 500 Hz requires a high-time of 1 ms and low-time of 1 ms, mode 2 cannot be used. (Recall that 256 μs is the maximum timed interval in mode 2 when the 8051 is operating at 12 MHz.) Here's the program:

```
0000            5             ORG     0
0000 020030     6             LJMP    MAIN
000B            7             ORG     000BH          ;Timer 0 vector address
000B 02003F     8             LJMP    T0ISR
001B            9             ORG     001BH          ;Timer 1 vector address
001B 020042    10             LJMP    T1ISR
0030           11             ORG     0030H
0030 758912    12   MAIN:     MOV     TMOD,#12H      ;Timer 1 = mode 1
               13                                    ;Timer 0 = mode 2
0033 758CB9    14             MOV     TH0,#-71       ;7 kHz using timer 0
0036 D28C      15             SETB    TR0
0038 D28F      16             SETB    TF1            ;force timer 1 interrupt
003A 75A88A    17             MOV     IE,#8AH        ;enable both timer intrrpts
003D 80FE      18             SJMP    $
               19   ;
003F B297      20   T0ISR:    CPL     P1.7
0041 32        21             RETI
0042 C28E      22   T1ISR:    CLR     TR1
0044 758DFC    23             MOV     TH1,#HIGH(-1000)  ;1 ms high time &
0047 758B18    24             MOV     TL1,#LOW(-1000)   ; low time
004A D28E      25             SETB    TR1
004C B296      26             CPL     P1.6
004E 32        27             RETI
               28             END
```

Again, the framework is for a complete program that could be installed in EPROM or ROM on an 8051-based product. The main program and the ISRs are located above the vector locations for the system reset and interrupts. Both waveforms are created by "CPL bit" instructions; however, the timed intervals necessitate a slightly different approach for each.

Since the TL1/TH1 registers must be reloaded after each overflow (i.e., after each interrupt), Timer 1 ISR (a) stops the timer, (b) reloads TL1/TH1, (c) starts the timer, then

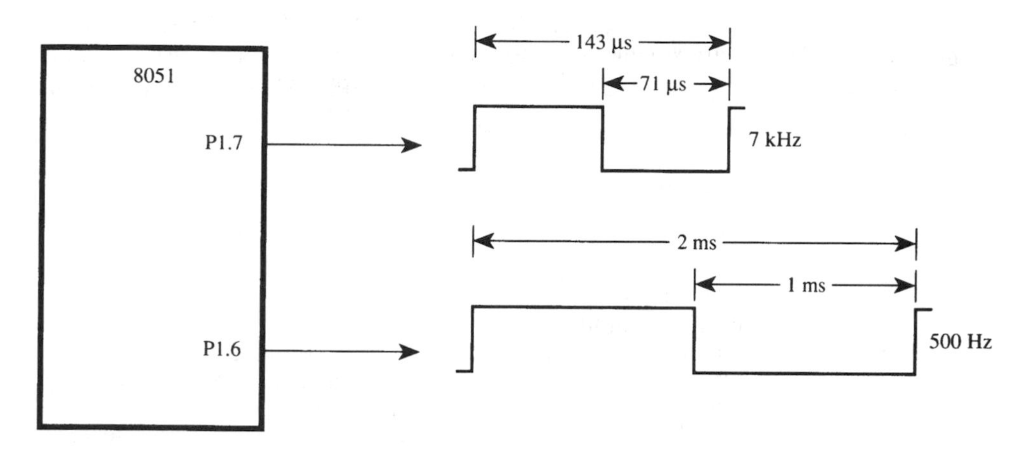

FIGURE 6–4
Waveform example

(d) complements the port bit. Note also that TL1/TH1 are *not* initialized at the beginning of the main program, unlike TH0. Since TL1/TH1 must be reinitialized after each overflow, TF1 is set in the main program by software to "force" an initial interrupt as soon as interrupts are turned on. This effectively gets the 500 Hz waveform started.

The Timer 0 ISR, as in the previous example, simply complements the port bit and returns to the main program. SJMP $ is used in the main program as the abbreviated form of HERE: SJMP HERE. The two forms are functionally equivalent. (See "Special Assembler Symbols" in Chapter 7.)

6.5 SERIAL PORT INTERRUPTS

Serial port interrupts occur when either the transmit interrupt flag (TI) or the receive interrupt flag (RI) is set. A transmit interrupt occurs when transmission of the previous character written to SBUF has finished. A receive interrupt occurs when a character has been completely received and is waiting in SBUF to be read.

Serial port interrupts are slightly different from timer interrupts. The flag that causes a serial port interrupt is not cleared by hardware when the CPU vectors to the interrupt. The reason is that there are two sources for a serial port interrupt, TI or RI. The source of the interrupt must be determined in the ISR and the interrupting flag cleared by software. Recall that with timer interrupts the interrupting flag is cleared by hardware when the processor vectors to the ISR.

Example 6–3: Character Output Using Interrupts

Write a program using interrupts to continually transmit the ASCII code set (excluding control codes) to a terminal attached to the 8051's serial port.

There are 128 7-bit codes in the ASCII chart. (See Appendix F.) These consist of 95 graphic codes (20H to 7EH) and 33 control codes (00H to 1FH, and 7FH). The program shown below is self-contained and executable from EPROM or ROM immediately after a system reset.

```
0000             5                ORG     0
0000 020030      6                LJMP    MAIN
0023             7                ORG     0023H          ;serial port interrupt entry
0023 020042      8                LJMP    SPISR
0030             9                ORG     0030H
0030 758920     10      MAIN:     MOV     TMOD,#20H      ;Timer 1, mode 2
0033 758DE6     11                MOV     TH1,#-26       ;12000 baud reload value
0036 D28E       12                SETB    TR1            ;start timer
0038 759842     13                MOV     SCON,#42H      ;mode 1, set TI to force 1st
                14                                       ; interrupt; send 1st char.
003B 7420       15                MOV     A,#20H         ;send ASCII space first
003D 75A890     16                MOV     IE,#90H        ;enable serial port interrupt
0040 80FE       17                SJMP    $              ;do nothing
                18      ;
0042 B47F02     19      SPISR:    CJNE    A,#7FH,SKIP    ;if finished ASCII set,
0045 7420       20                MOV     A,#20H         ; reset to SPACE
0047 F599       21      SKIP:     MOV     SBUF,A         ;send char. to serial port
0049 04         22                INC     A              ;increment ASCII code
004A C299       23                CLR     TI             ;clear interrupt flag
004C 32         24                RETI
                25                END
```

After jumping to MAIN at code address 0030H, the first three instructions initialize Timer 1 to provide a 1200 baud clock to the serial port (lines 10–12). MOV SCON,#42H initializes the serial port for mode 1 (8-bit UART) and sets the TI flag to force an interrupt as soon as interrupts are enabled. Then, the first ASCII graphic code (20H) is loaded into A and serial port interrupts are enabled. Finally, the main body of the program enters a do-nothing loop (SJMP $).

The serial port interrupt service routine does all the work once the main program sets up initial conditions. The first two instructions check the accumulator, and if the ASCII code has reached 7FH (i.e., the last code transmitted was 7EH), reset the accumulator to 20H (lines 19–20). Then, the ASCII code is sent to the serial port buffer (MOV SBUF,A), the code is incremented (INC A), the transmit interrupt flag is cleared (CLR TI), and the ISR is terminated (RETI). Control returns to the main program and SJMP $ executes until TI is set at the end of the next character transmission.

If we compare the CPU's speed to the rate of character transmission, we see that SJMP $ executes for a very large percentage of the time for this program. What is this percentage? At 1200 baud, each bit transmitted takes 1/1200 = 0.833 ms. Eight data bits plus a start and stop bit, therefore, take 8.33 ms or 8333 μs. The worst-case execution time for the SPISR is found by totaling the number of cycles for each instruction and multiplying by 1 μs (assuming 12 MHz operation). This turns out to be 8 μs. So, of the

8333 μs for each character transmission, only 8 μs are for the interrupt service routine. The SJMP $ instruction executes about 8325 ÷ 8333 × 100 = 99.90% of the time. Since interrupts are used, the SJMP $ instruction could be replaced with other instructions performing other tasks required in the application. Interrupts would still occur every 8.33 ms, and characters would still be transmitted out the serial port as they are in the above program.

6.6 EXTERNAL INTERRUPTS

External interrupts occur as a result of a low-level or negative edge on the $\overline{\text{INT0}}$ or $\overline{\text{INT1}}$ pin on the 8051 IC. These are the alternate functions for Port 3 bits P3.2 (pin 12) and P3.3 (pin 13) respectively.

The flags that actually generate these interrupts are bits IE0 and IE1 in TCON. When an external interrupt is generated, the flag that generated it is cleared by hardware when vectoring to the ISR only if the interrupt was transition-activated. If the interrupt was level-activated, then the external requesting source controls the level of the request flag, rather than the on-chip hardware.

The choice of low-level-activated interrupts versus negative-edge-activated interrupts is programmable through the IT0 and IT1 bits in TCON. For example, if IT1 = 0, external interrupt 1 is triggered by a detected low at the $\overline{\text{INT1}}$ pin. If IT1 =1, external interrupt 1 is edge-triggered. In this mode, if successive samples of the $\overline{\text{INT1}}$ pin show a high in one cycle and a low in the next, the interrupt request flag IE1 in TCON is set. Flag bit IE1 then requests the interrupt.

Since the external interrupt pins are sampled once each machine cycle, an input should be held for at least 12 oscillator periods to ensure proper sampling. If the external interrupt is transition-activated, the external source must hold the request pin high for at least 1 cycle, and then hold it low for at least 1 more cycle to ensure the transition is detected. IE0 and IE1 are automatically cleared when the CPU vectors to the interrupt.

If the external interrupt is level-activated, the external source must hold the request active until the requested interrupt is actually generated. Then it must deactivate the request before the interrupt service routine is completed, or another interrupt will be generated. Usually, an action taken in the ISR causes the requesting source to return the interrupting signal to the inactive state.

Example 6–4: Furnace Controller

Using interrupts, design an 8051 furnace controller that keeps a building at 20°C ±1°C.

The following interface is assumed for this example. The furnace ON/OFF solenoid is connected to P1.7 such that

```
P1.7 = 1 for solenoid engaged (furnace ON)
P1.7 = 0 for solenoid disengaged (furnace OFF)
```

Temperature sensors are connected to $\overline{\text{INT0}}$ and $\overline{\text{INT1}}$ and provide $\overline{\text{HOT}}$ and $\overline{\text{COLD}}$ signals, respectively, such that

```
HOT  = 0 if T > 21°C
COLD = 0 if T < 19°C
```

The program should turn on the furnace for T < 19°C and turn it off for T > 21°C. The hardware configuration and a timing diagram are shown in Figure 6–5.

```
0000            5            ORG     0
0000 020030     6            LJMP    MAIN
                7                              ;EXT 0 vector at 0003H
0003 C297       8   EX0ISR:  CLR     P1.7      ;turn furnace off
0005 32         9            RETI
0013           10            ORG     0013H
0013 D297      11   EX1ISR:  SETB    P1.7      ;turn furnace on
0015 32        12            RETI
0030           13            ORG     30H
0030 75A885    14   MAIN:    MOV     IE,#85H   ;enable external interrupts
0033 D288      15            SETB    IT0       ;negative edge triggered
0035 D28A      16            SETB    IT1
0037 D297      17            SETB    P1.7      ;turn furnace off
0039 20B202    18            JB      P3.2,SKIP ;if T > 21 degrees,
003C C297      19            CLR     P1.7      ; turn furnace off
003E 80FE      20   SKIP:    SJMP    $         ;do nothing
               21            END
```

The first three instructions in the main program (lines 14–16) turn on external interrupts and make both $\overline{\text{INT0}}$ and $\overline{\text{INT1}}$ negative-edge triggered. Since the current state of the $\overline{\text{HOT}}$ (P3.3) and $\overline{\text{COLD}}$ (P3.3) inputs is not known, the next three instructions (lines 17–19) are required to turn the furnace ON or OFF, as appropriate. First, the furnace is turned ON (SETB P1.7), and then the $\overline{\text{HOT}}$ input is sampled (JB P3.2,SKIP). If $\overline{\text{HOT}}$ is high, then T < 21°C, so the next instruction is skipped and the furnace is left ON. If, however, $\overline{\text{HOT}}$ is low, then T > 21°C. In this case the jump does not take place. The next instruction turns the furnace OFF (CLR P1.7) before entering the do-nothing loop.

Once everything is set up properly in the main program, little remains to be done. Each time the temperature rises above 21°C or falls below 19°C, an interrupt occurs. The ISRs simply turn the furnace ON (SETB P1.7) or OFF (CLR P1.7), as appropriate, and return to the main program.

Note that an ORG 0003H statement is not necessary immediately before the EX0ISR label. Since the LJMP MAIN instruction is three bytes long, EX0ISR is certain to start at 0003H, the correct entry point for external 0 interrupts.

Example 6–5: Intrusion Warning System

Design an intrusion warning system using interrupts that sounds a 400 Hz tone for 1 second (using a loudspeaker connected to P1.7) whenever a door sensor connected $\overline{\text{INT0}}$ makes a high-to-low transition.

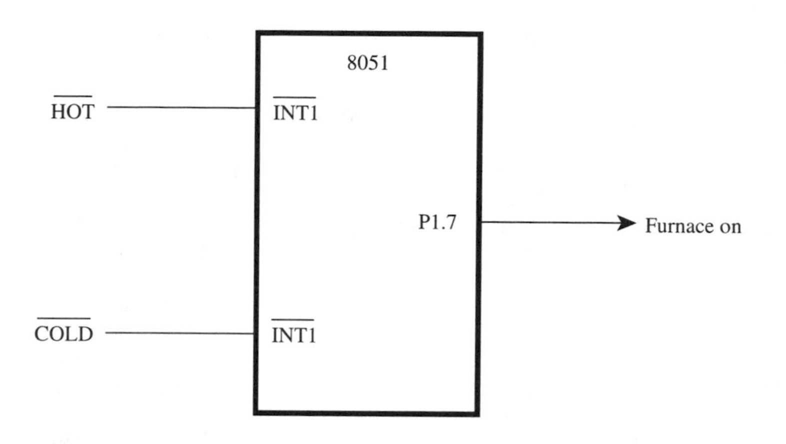

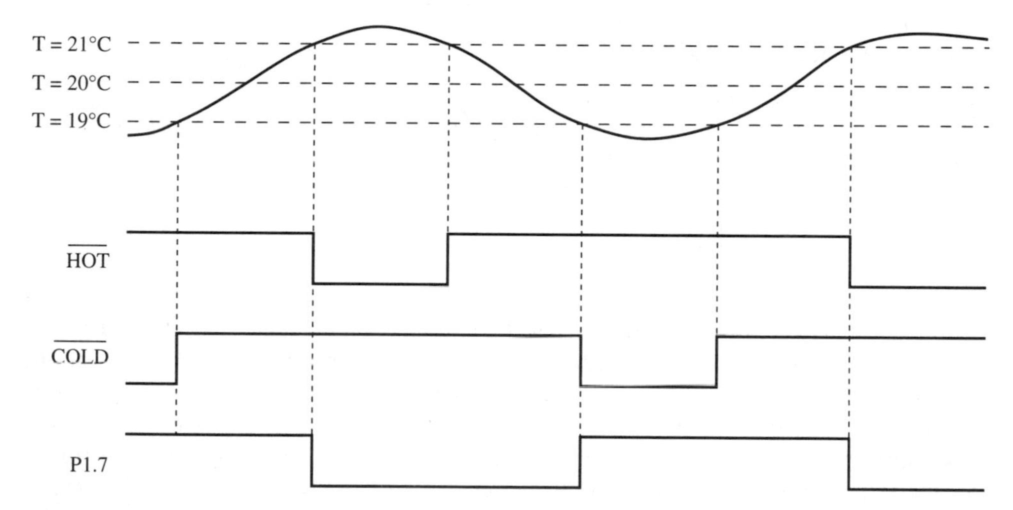

FIGURE 6–5
Furnace example. (a) Hardware connections (b) Timing.

The solution to this example uses three interrupts: external 0 (door sensor), Timer 0 (400 Hz tone), and Timer 1 (1 second timeout). The hardware configuration and timings are shown in Figure 6–6.

```
0000             5              ORG     0
0000 020030      6              LJMP    MAIN        ;3-byte instruction
0003 02003A      7              LJMP    EX0ISR      ;EXT 0 vector address
000B             8              ORG     000BH       ;Timer 0 vector
000B 020045      9              LJMP    T0ISR
001B            10              ORG     001BH       ;Timer 1 vector
001B 020059     11              LJMP    T1ISR
0030            12              ORG     0030H
0030 D288       13     MAIN:    SETB    IT0         ;negative edge activated
0032 758911     14              MOV     TMOD,#11H   ;16-bit timer mode
```

```
0035 75A881  15              MOV     IE,#81H          ;enable EXT 0 only
0038 80FE    16              SJMP    $                ;now relax
             17    ;
003A 7F14    18    EX0ISR:   MOV     R7,#20           ;20 x 5000 us = 1 second
003C D28D    19              SETB    TF0              ;force timer 0 interrupt
003E D28F    20              SETB    TF1              ;force timer 1 interrupt
0040 D2A9    21              SETB    ET0              ;begin tone for 1 second
0042 D2AB    22              SETB    ET1              ;enable timer interrupts
0044 32      23              RETI                     ;timer ints will do the work
             24    ;
0045 C28C    25    T0ISR:    CLR     TR0              ;stop timer
0047 DF07    26              DJNZ    R7,SKIP          ;if not 20th time, exit
0049 C2A9    27              CLR     ET0              ;if 20th, disable tone
004B C2AB    28              CLR     ET1              ;disable itself
004D 020058  29              LJMP    EXIT
0050 758C3C  30    SKIP:     MOV     TH0,#HIGH(-50000)  ;0.05 sec. delay
0053 758AB0  31              MOV     TL0,#LOW(-50000)
0056 D28C    32              SETB    TR0
0058 32      33    EXIT:     RETI
             34    ;
0059 C28E    35    T1ISR:    CLR     TR1
005B 758DFB  36              MOV     TH1,#HIGH(-1250)   ;count for 400 Hz
005E 758B1E  37              MOV     TL1,#LOW(-1250)
0061 B297    38              CPL     P1.7             ;music maestro!
0063 D28E    39              SETB    TR1
0065 32      40              RETI
             41              END
```

This is our largest program thus far. Five distinct sections are the interrupt vector locations, the main program, and the three interrupt service routines. All vector locations contain LJMP instructions to the respective routines. The main program, starting at code address 0030H, contains only four instructions. SETB IT0 configures the door-sensing interrupt input as negative-edge triggered. MOV TMOD,#11H configures both timers for mode 1, 16-bit timer mode. Only the external 0 interrupt is enabled initially (MOV IE,#81H), so a "door open" condition is needed before any interrupt is accepted. Finally, SJMP $ puts the main program in a do-nothing loop.

When a door-open condition is sensed (by a high-to-low transition of $\overline{\text{INT0}}$), an external 0 interrupt is generated. EX0ISR begins by putting the constant 20 in R7 (see below), then sets the overflow flags for both timers to force timer interrupts to occur. Timer interrupts will only occur, however, if the respective bits are enabled in the IE register. The next two instructions (SETB ET0 and SETB ET1) enable timer interrupts. Finally, EX0ISR terminates with a RET1 to the main program.

Timer 0 creates the 1 second timeout, and Timer 1 creates the 400 Hz tone. After EX0ISR returns to the main program, timer interrupts are immediately generated (and accepted after one execution of SJMP $). Because of the fixed polling sequence (see Figure 6–2), the Timer 0 interrupt is serviced first. A 1 second timeout is created by programming 20 repetitions of a 50,000 μs timeout. R7 serves as the counter. Nineteen times out of 20, T0ISR operates as follows. First, Timer 0 is turned off and R7 is decremented. Then, TH0/TL0 is reloaded with −50000, the timer is turned back on and the interrupt is terminated. On the 20th Timer 0 interrupt, R7 is decremented to 0 (1 second

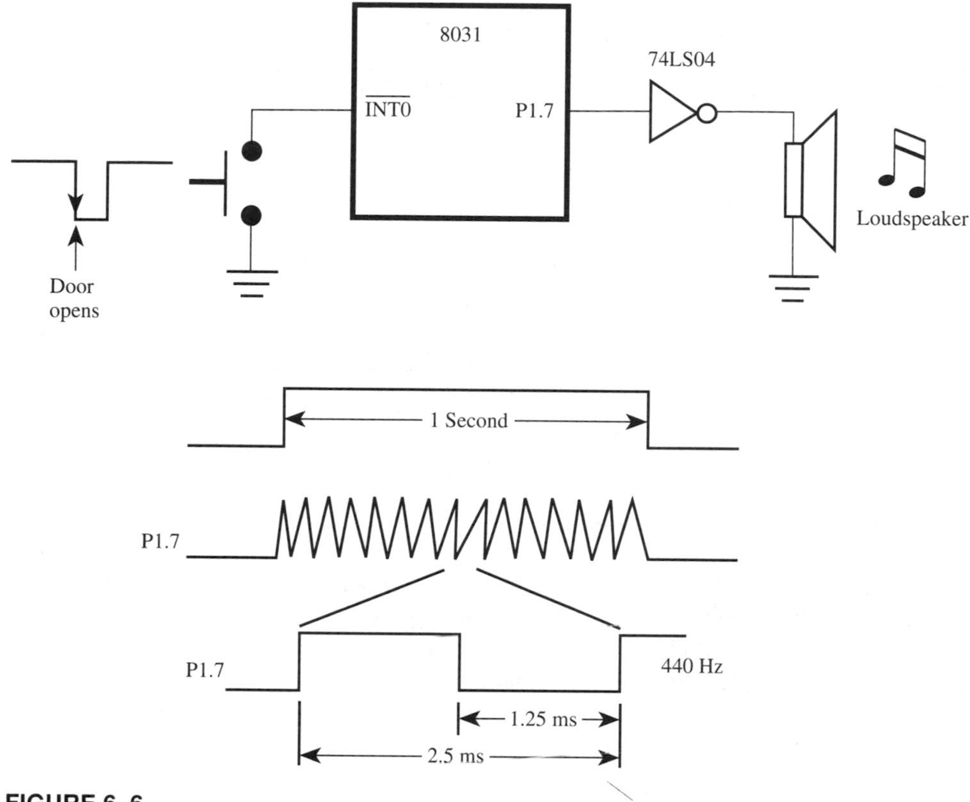

FIGURE 6–6
Loudspeaker interface using interrupts. (a) Hardware connections (b) Timing.

has elapsed). Both timer interrupts are disabled (CLR, ET0, CLR ET1) and the interrupt is terminated. No further timer interrupts will be generated until the next "door-open" condition is sensed.

The 400 Hz tone is programmed using Timer 1 interrupts. 400 Hz requires a period of 1/400 = 2,500 μs, or 1,250 μs high-time and 1,250 μs low-time. Each timer 1 ISR simply puts −1250 in TH1/TL1, complements the port bit driving the loudspeaker, then terminates.

6.7 INTERRUPT TIMINGS

Interrupts are sampled and latched on S5P2 of each machine cycle. (See Figure 6–7.) They are polled on the next machine cycle and if an interrupt condition exists, it is accepted if (a) no other interrupt of equal or higher priority is in progress, (b) the polling cycle is the last cycle in an instruction, and (c) the current instruction is not a RETI or any access to IE or IP. During the next two cycles, the processor pushes the PC on the stack and loads the PC with the interrupt vector address. The ISR begins.

The stipulation that the current instruction is not RETI ensures that at least one instruction executes after each interrupt service routine. The timing is shown in Figure 6–8.

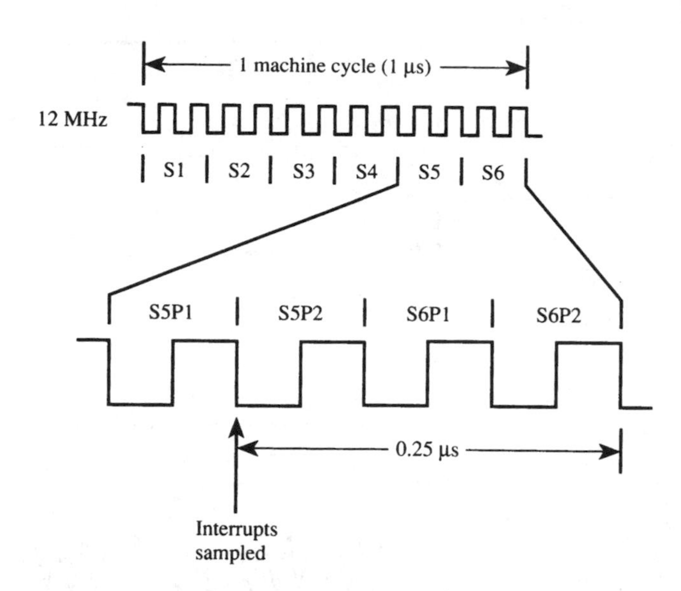

FIGURE 6–7
Sampling of interrupts on S5P2

The time between an interrupt condition occurring and the ISR beginning is called **interrupt latency.** Interrupt latency is critical in many control applications. With a 12 MHz crystal, the interrupt latency can be as short as 3.25 μs on the 8051. An 8051 system that uses one high-priority interrupt will have a worst-case interrupt latency of 9.25 μs (assuming the high-priority interrupt is always enabled). This occurs if the interrupt condition happens just before the RETI of a level 0 ISR that is followed by a multiply instruction (see Figure 6–9).

6.8 SUMMARY

This chapter has presented the major details required to embark on the design of interrupt-driven systems with the 8051 microcontroller. Readers are advised to begin programming with interrupts in increments. The examples in this chapter serve as a good first contact with 8051 interrupts.

8051 single-board computers usually contain a monitor program in EPROM residing at the bottom of code memory. If interrupts are not used in the monitor program, the vector locations probably contain LJMP instructions to an area of CODE RAM where

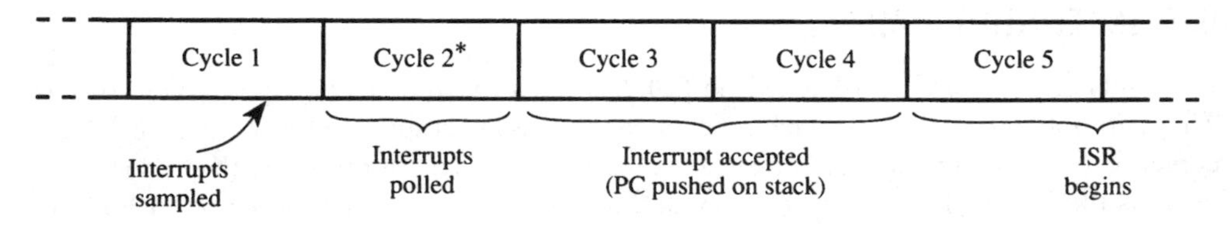

FIGURE 6–8
Polling of interrupts

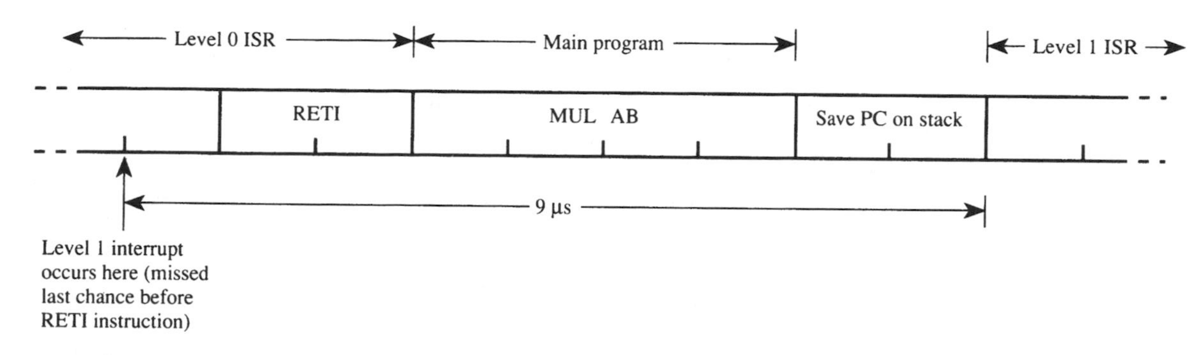

FIGURE 6–9
Interrupt latency

user applications are loaded for execution and debugging. The manufacturer's literature will provide the addresses for programmers to use as entry points for interrupt service routines. Alternatively, users can simply "look" in the interrupt vector locations using the monitor program's commands for examining code memory locations. The content of code memory address 0003H, for example, will contain the opcode of the first instruction to execute for an external 0 interrupt. If this is an LJMP opcode (22H; see Appendix B), then the next two addresses (0003H and 0004H) contain the address of the ISR, and so on.

Alternately, users can develop self-contained interrupt applications, as shown in the examples. The object bytes can be burned into EPROM and installed in the target system at code address 0000H. When the system is powered up or reset, the application begins execution without the need of a monitor program for loading and starting the application.

PROBLEMS

1. Modify Example 1 to shut off interrupts and terminate if any key is hit on the terminal.
2. Create a 1 kHz square wave on P1.7 using interrupts.
3. Create a 7 kHz pulse wave with a 30% duty cycle on P1.6 using interrupts.
4. Combine Example 6–1 and Example 6–3 (earlier in the chapter) into one program.
5. Modify Example 6–3 to send one character per second. (Hint: use a timer and output the character in the timer ISR.)

7

ASSEMBLY LANGUAGE PROGRAMMING

7.1 INTRODUCTION

This chapter introduces assembly language programming for the 8051 microcontroller. Assembly language is a computer language lying between the extremes of machine language and high-level language. Typical high-level languages like Pascal or C use words and statements that are easily understood by humans, although still a long way from "natural" language. Machine language is the binary language of computers. A machine language program is a series of binary bytes representing instructions the computer can execute.

Assembly language replaces the binary codes of machine language with easy to remember "mnemonics" that facilitate programming. For example, an addition instruction in machine language might be represented by the code "10110011." It might be represented in assembly language by the mnemonic "ADD." Programming with mnemonics is obviously preferable to programming with binary codes.

Of course, this is not the whole story. Instructions operate on data, and the location of the data is specified by various "addressing modes" embedded in the binary code of the machine language instruction. So, there may be several variations of the ADD instruction depending on what is added. The rules for specifying these variations are central to the theme of assembly language programming.

An assembly language program is not executable by a computer. Once written, the program must undergo translation to machine language. In the example above, the mnemonic "ADD" must be translated to the binary code "10110011." Depending on the complexity of the programming environment, this translation may involve one or more steps before an executable machine language program results. As a minimum, a program called an "assembler" is required to translate the instruction mnemonics to machine language binary codes. A further step may require a "linker" to combine portions of programs from separate files and to set the address in memory at which the program may execute. We begin with a few definitions.

An **assembly language program** is a program written using labels, mnemonics, and so on, in which each statement corresponds to a machine instruction. Assembly language programs, often called source code or symbolic code, cannot be executed by a computer.

A **machine language program** is a program containing binary codes that represent instructions to a computer. Machine language programs, often called object code, are executable by a computer.

An **assembler** is a program that translates an assembly language program into a machine language program. The machine language program (object code) may be in "absolute" form or in "relocatable" form. In the latter case, "linking" is required to set the absolute address for execution.

A **linker** is a program that combines relocatable object programs (modules) and produces an absolute object program that is executable by a computer. A linker is sometimes called a "linker/locator" to reflect its separate functions of combining relocatable modules (linking) and setting the address for execution (locating).

A **segment** is a unit of code or data memory. A segment may be relocatable or absolute. A relocatable segment has a name, type, and other attributes that allow the linker to combine it with other partial segments, if required, and to correctly locate the segment. An absolute segment has no name and cannot be combined with other segments.

A **module** contains one or more segments or partial segments. A module has a name assigned by the user. The module definitions determine the scope of local symbols. An object file contains one or more modules. A module may be thought of as a "file" in many instances.

A **program** consists of a single absolute module, merging all absolute and relocatable segments from all input modules. A program contains only the binary codes for instructions (with addresses and data constants) that are understood by a computer.

7.2 ASSEMBLER OPERATION

There are many assembler programs and other support programs available to facilitate the development of applications for the 8051 microcontroller. Intel's original MCS-51™family assembler, ASM51, is the standard to which the others are compared. In this chapter, we focus on assembly language programming as undertaken using the most common features of ASM51. Although many features are standardized, some may not be implemented in assemblers from other companies.

ASM51 is a powerful assembler with all the bells and whistles. It is available on Intel development systems and on the IBM *PC* family of microcomputers. Since these "host" computers contain a CPU chip other than the 8051, ASM51 is called a **cross assembler.** An 8051 source program may be written on the host computer (using any text editor) and may be assembled to an object file and listing file (using ASM51), but the program may not be executed. Since the host system's CPU chip is not an 8051, it does not understand the binary instructions in the object file. Execution on the host computer requires either hardware emulation or software simulation of the target CPU. A third possibility is to download the object program to an 8051-based target system for execution. Hardware emulation, software simulation, downloading, and other development techniques are discussed in Chapter 9.

ASM51 is invoked from the system prompt by

```
ASM51 source_file [assembler_controls]
```

The source file is assembled and any assembler controls specified take effect. (Assembler controls, which are optional, are discussed later in this chapter.) The assembler re-

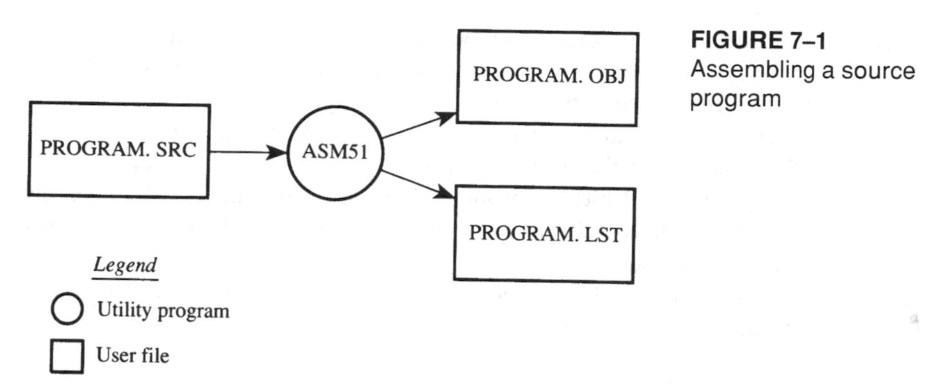

FIGURE 7–1
Assembling a source program

ceives a source file as input (e.g., PROGRAM.SRC) and generates an object file (PROGRAM.OBJ) and listing file (PROGRAM.LST) as output. This is illustrated in Figure 7–1.

Since most assemblers scan the source program twice in performing the translation to machine language, they are described as **two-pass assemblers.** The assembler uses a **location counter** as the address of instructions and the values for labels. The action of each pass is described below.

7.2.1 Pass One

During the first pass, the source file is scanned line-by-line and a **symbol table** is built. The location counter defaults to 0 or is set by the ORG (set origin) directive. As the file is scanned, the location counter is incremented by the length of each instruction. Define data directives (DB or DW) increment the location counter by the number of bytes defined. Reserve memory directives (DS) increment the location counter by the number of bytes reserved.

Each time a label is found at the beginning of a line, it is placed in the symbol table along with the current value of the location counter. Symbols that are defined using equate directives (EQU) are placed in the symbol table along with the "equated" value. The symbol table is saved and then used during pass two.

7.2.2 Pass Two

During pass two, the object and listing files are created. Mnemonics are converted to opcodes and placed in the output files. Operands are evaluated and placed after the instruction opcodes. Where symbols appear in the operand field, their values are retrieved from the symbol table (created during pass one) and used in calculating the correct data or addresses for the instructions.

Since two passes are performed, the source program may use "forward references," that is, use a symbol before it is defined. This would occur, for example, when branching ahead in a program.

The object file, if it is absolute, contains only the binary bytes (00H–FFH of the machine language program. A relocatable object file will also contain a symbol table and other information required for linking and locating. The listing file contains ASCII text codes (20H–7EH) for both the source program and the hexadecimal bytes in the machine language program.

A good demonstration of the distinction between an object file and a listing file is to display each on the host computer's CRT display (using, for example, the TYPE command on MS-DOS systems). The listing file clearly displays, with each line of output containing an address, opcode, and perhaps data, followed by the program statement from the source file. The listing file displays properly because it contains only ASCII text codes. Displaying the object file is a problem, however. The output will appear as "garbage," since the object file contains binary codes of an 8051 machine language program, rather than ASCII text codes.

A sketch of a two-pass assembler is shown in Figure 7–2 written in a pseudo computer language (similar to Pascal or C) to enhance readability.

7.3 ASSEMBLY LANGUAGE PROGRAM FORMAT

Assembly language programs contain the following:

- Machine instructions
- Assembler directives
- Assembler controls
- Comments

Machine instructions are the familiar mnemonics of executable instructions (e.g., ANL). Assembler directives are instructions to the assembler program that define program structure, symbols, data, constants, and so on (e.g., ORG). Assembler controls set assembler modes and direct assembly flow (e.g., $TITLE). Comments enhance the readability of programs by explaining the purpose and operation of instruction sequences.

Those lines containing machine instructions or assembler directives must be written following specific rules understood by the assembler. Each line is divided into "fields" separated by space or tab characters. The general format for each line is as follows:

```
[label:]    mnemonic [operand][,operand][. . .][;comment]
```

Only the mnemonic field is mandatory. Many assemblers require the label field, if present, to begin on the left in column 1, and subsequent fields to be separated by space or tab characters. With ASM51, the label field needn't begin in column 1 and the mnemonic field needn't be on the same line as the label field. The operand field must, however, begin on the same line as the mnemonic field. The fields are described below.

7.3.1 Label Field

A label represents the address of the instruction (or data) that follows. When branching to this instruction, this label is used in the operand field of the branch or jump instruction (e.g., SJMP SKIP).

Whereas the term "label" always represents an address, the term "symbol" is more general. Labels are one type of symbol and are identified by the requirement that they must terminate with a colon (:). Symbols are assigned values or attributes using directives such as EQU, SEGMENT, BIT, DATA, etc. Symbols may be addresses, data con-

```
ASM(input_file) /* assemble source program in input_file */

BEGIN

   /* pass 1: build the symbol table */

   [lc = 0]              /* lc = location counter; default to 0 */
   [mnemonic = null]
   [open input_file]
   WHILE [mnemonic != end] DO BEGIN
      [get line from input_file]
      [scan line and get label/symbol and mnemonic]
      IF [label] THEN [enter "label = lc" into symbol table]
      CASE [mnemonic] OF
         null, comment, END:
               [do nothing]
         ORG:  [lc = operand]
         EQU:  [enter "symbol = operand" into symbol table]
         DB:   [increment lc by number of bytes defined]
         DW:   [increment lc by twice the number of words defined]
         DS:   [lc = lc + operand]
         1_byte_instruction:   [lc = lc + 1]
         2_byte_instruction:   [lc = lc + 2]
         3_byte_instruction:   [lc = lc + 3]
   END

   /* pass 2: create the object program */

   [rewind input_file pointer]
   [lc = 0]
   [mnemonic = null]
   [open output_file]
   WHILE [mnemonic != end] DO BEGIN
      [get line from input_file]
      [scan line and determine mnemonic op code and value(s) of operand(s)]

      /*  Note: If symbols are used in operand field,
       *  their values are looked-up in the symbol table
       *  created during pass one.
       */

      CASE [mnemonic] OF
         null, comment, EQU, END:
               [do nothing]
         ORG:  [lc = operand]
         DB:   [put bytes into object_file and increment lc by # of bytes]
         DW:   [put words into object_file and inc lc by twice # of words]
         DS:   [lc = lc + operand]
         1_byte_instruction:   [put opcode into output_file]
         2_byte_instruction:   [put opcode into output_file]
                               [put low-byte of operand into output_file]
         3_byte_instruction:   [put opcode into output_file]
                               [put high-byte of operand into output_file]
                               [put low-byte of operand into output_file]
   END
   [close input_file]
   [close output_file]

END
```

FIGURE 7–2
Pseudo code sketch of a two-pass assembler

stants, names of segments, or other constructs conceived by the programmer. Symbols do not terminate with a colon. In the example below, PAR is a symbol and START is a label (which is a type of symbol).

```
PAR          EQU  500              ;"PAR" IS A SYMBOL WHICH
                                   ; REPRESENTS THE VALUE 500
START:       MOV  A,#0FFH          ;"START" IS A LABEL WHICH
                                   ; REPRESENTS THE ADDRESS OF
                                   ; THE MOV INSTRUCTION
```

A symbol (or label) must begin with a letter, question mark, or underscore (_); must be followed by letters, digits, "?", or "_"; and can contain up to 31 characters.[1] Symbols may use upper- or lower-case characters, but they are treated the same. Reserved words (mnemonics, operators, predefined symbols, and directives) may not be used.

7.3.2 Mnemonic Field

Instruction mnemonics or assembler directives go in the mnemonic field, which follows the label field. Examples of instruction mnemonics are ADD, MOV, DIV, or INC. Examples of assembler directives are ORG, EQU, or DB. Assembler directives are described later in this chapter.

7.3.3 Operand Field

The operand field follows the mnemonic field. This field contains the address or data used by the instruction. A label may be used to represent the address of the data, or a symbol may be used to represent a data constant. The possibilities for the operand field are largely dependent on the operation. Some operations have no operand (e.g., the RET instruction), while others allow for multiple operands separated by commas. Indeed, the possibilities for the operand field are numerous, and we shall elaborate on these at length. But first, the comment field.

7.3.4 Comment Field

Remarks to clarify the program go in the comment field at the end of each line. Comments must begin with a semicolon (;). Entire lines may be comment lines by beginning them with a semicolon. Subroutines and large sections of a program generally begin with a comment block—several lines of comments that explain the general properties of the section of software that follows.

7.3.5 Special Assembler Symbols

Special assembler symbols are used for the register-specific addressing modes. These include A, R0 through R7, DPTR, PC, C, and AB. As well, a dollar sign ($) can be used to refer to the current value of the location counter. Some examples follow.

---

[1]The reader is reminded that the rules specified in this chapter apply to Intel's ASM51. Other assemblers may have different requirements.

```
            SETB C
            INC  DPTR
            JNB  TI,$
```

The last instruction above makes effective use of ASM51's location counter to avoid using a label. It could also be written as

```
HERE:       JNB  TI,HERE
```

7.3.6 Indirect Address

For certain instructions, the operand field may specify a register that contains the address of the data. The commercial "at" sign (@) indicates address indirection and may only be used with R0, R1, the DPTR, or the PC, depending on the instruction. For example,

```
            ADD  A,@R0
            MOVC A,@A+PC
```

The first instruction above retrieves a byte of data from internal RAM at the address specified in R0. The second instruction retrieves a byte of data from external code memory at the address formed by adding the contents of the accumulator to the program counter. Note that the value of the program counter, when the add takes place, is the address of the instruction following MOVC. For both instructions above, the value retrieved is placed into the accumulator.

7.3.7 Immediate Data

Instructions using immediate addressing provide data in the operand field that become part of the instruction. Immediate data are preceded with a pound sign (#). For example,

```
CONSTANT    EQU  100
            MOV  A,#0FEH
            ORL  40H,#CONSTANT
```

All immediate data operations (except MOV DPTR,#data) require eight bits of data. The immediate data are evaluated as a 16-bit constant and then the low-byte is used. All bits in the high-byte must be the same (00H or FFH) or the error message "value will not fit in a byte" is generated. For example, the following instructions are syntactically correct:

```
            MOV  A,#0FF00H
            MOV  A,#00FFH
```

But the following two instructions generate error messages:

```
            MOV  A,#0FE00H
            MOV  A,#01FFH
```

If using signed decimal notation, constants from −256 to +256 may be used. For example, the following two instructions are equivalent (and syntactically correct):

```
            MOV  A,#-256
            MOV  A,#0FF00H
```

Both instructions above put 00H into accumulator A.

7.3.8 Data Address

Many instructions access memory locations using direct addressing and require an on-chip data memory address (00H to 7FH) or an SFR address (80H to 0FFH) in the operand field. Predefined symbols may be used for the SFR addresses. For example,

```
MOV  A,45H
MOV  A,SBUF          ;SAME AS MOV A,99H
```

7.3.9 Bit Address

One of the most powerful features of the 8051 is the ability to access individual bits without the need for masking operations on bytes. Instructions accessing bit-addressable locations must provide a bit address in internal data memory (00H to 7FH) or a bit address in the SFRs (80H to 0FFH).

There are three ways to specify a bit address in an instruction: (a) explicitly by giving the address, (b) using the **dot operator** between the byte address and the bit position, and (c) using a predefined assembler symbol. Some examples follow.

```
SETB 0E7H            ;EXPLICIT BIT ADDRESS
SETB ACC.7           ;DOT OPERATOR (SAME AS ABOVE)
JNB  TI,$            ;"TI" IS A PRE-DEFINED SYMBOL
JNB  99H,$           ;(SAME AS ABOVE)
```

7.3.10 Code Address

A code address is used in the operand field for jump instructions, including relative jumps (SJMP and conditional jumps), absolute jumps and calls (ACALL, AJMP), and long jumps and calls (LJMP, LCALL).

The code address is usually given in the form of a label. For example,

```
HERE:           .
                .
                .
                SJMP HERE
```

ASM51 will determine the correct code address and insert into the instruction the correct 8-bit signed offset, 11-bit page address, or 16-bit long address, as appropriate.

7.3.11 Generic Jumps and Calls

ASM51 allows programmers to use a generic JMP or CALL mnemonic. "JMP" can be used instead of SJMP, AJMP, or LJMP; and "CALL" can be used instead of ACALL or LCALL. The assembler converts the generic mnemonic to a "real" instruction following a few simple rules. The generic mnemonic converts to the short form (for JMP only) if no forward references are used and the jump destination is within −128 locations, or to the absolute form if no forward references are used and the instruction following the JMP or CALL instruction is in the same 2K block as the destination instruction. If short or absolute forms cannot be used, the conversion is to the long form.

```
MCS-51 MACRO ASSEMBLER     GENERIC


DOS 3.31 (038-N) MCS-51 MACRO ASSEMBLER, V2.2
OBJECT MODULE PLACED IN GENERIC.OBJ
ASSEMBLER INVOKED BY:  C:\ASM51\ASM51.EXE GENERIC.SRC EP


LOC  OBJ            LINE     SOURCE

1234                   1               ORG     1234H
1234 04                2     START:    INC     A
1235 80FD              3               JMP     START          ;ASSEMBLES AS SJMP
12FC                   4               ORG     START + 200
12FC 4134              5               JMP     START          ;ASSEMBLES AS AJMP
12FE 021301            6               JMP     FINISH         ;ASSEMBLES AS LJMP
1301 04                7     FINISH:   INC     A
                       8               END
```

FIGURE 7–3
Use of the generic JMP mnemonic

The conversion is not necessarily the best programming choice. For example, if branching ahead a few instructions, the generic JMP will always convert to LJMP even though an SJMP is probably better. Consider the assembled instruction sequence in Figure 7–3 using three generic jumps. The first jump (line 3) assembles as SJMP because the destination is before the jump (i.e., no forward reference) and the offset is less than -128. The ORG directive in line 4 creates a gap of 200 locations between the label START and the second jump, so the conversion on line 5 is to AJMP because the offset is too great for SJMP. Note also that the address following the second jump (12FCH) and the address of START (1234H) are within the same 2K page, which, for this instruction sequence, is bounded by 1000H and 17FFH. This criterion must be met for absolute addressing. The third jump assembles as LJMP because the destination (FINISH) is not yet defined when the jump is assembled (i.e., a forward reference is used). The reader can verify that the conversion is as stated by examining the object field for each jump instruction. Verify the hexadecimal codes with those found in Appendix C for SJMP, AJMP, and LJMP.

7.4 ASSEMBLE-TIME EXPRESSION EVALUATION

Values and constants in the operand field may be expressed three ways: (a) explicitly (e.g., 0EFH), (b) with a predefined symbol (e.g., ACC), or (c) with an expression (e.g., 2 + 3). The use of expressions provides a powerful technique for making assembly language programs more readable and more flexible. When an expression is used, the assembler calculates a value and inserts it into the instruction.

All expression calculations are performed using 16-bit arithmetic; however, either 8 or 16 bits are inserted into the instruction as needed. For example, the following two instructions are the same:

```
MOV  DPTR,#04FFH + 3
MOV  DPTR,#0502H            ;ENTIRE 16-BIT RESULT USED
```

If the same expression is used in a "MOV A,#data" instruction, however, the error message "value will not fit in a byte" is generated by ASM51. An overview of the rules for evaluating expressions follows.

7.4.1 Number Bases

The base for numeric constants is indicated in the usual way for Intel microprocessors. Constants must be followed with "B" for binary, "O" or "Q" for octal, "D" or nothing for decimal, or "H" for hexadecimal. For example, the following instructions are the same:

```
MOV  A,#15
MOV  A,#1111B
MOV  A,#0FH
MOV  A,#17Q
MOV  A,#15D
```

Note that a digit must be the first character for hexadecimal constants in order to differentiate them from labels (i.e., "0A5H" not "A5H").

7.4.2 Character Strings

Strings using one or two characters may be used as operands in expressions. The ASCII codes are converted to the binary equivalent by the assembler. Character constants are enclosed in single quotes ('). Some examples follow.

```
CJNE A,#'Q',AGAIN
SUBB A,#'0'                ;CONVERT ASCII DIGIT TO
                           ; BINARY DIGIT
MOV  DPTR,#'AB'
MOV  DPTR,#4142H           ;SAME AS ABOVE
```

7.4.3 Arithmetic Operators

The arithmetic operators are

```
+      addition
-      subtraction
*      multiplication
/      division
MOD    modulo (remainder after division)
```

For example, the following two instructions are the same:

```
MOV  A,#10 + 10H
MOV  A,#1AH
```

The following two instructions are also the same:

```
MOV  A,#25 MOD 7
MOV  A,#4
```

Since the MOD operator could be confused with a symbol, it must be separated from its operands by at least one space or tab character, or the operands must be enclosed in parentheses. The same applies for the other operators composed of letters.

7.4.4 Logical Operators

The logical operators are

```
OR     logical OR
AND    logical AND
XOR    logical Exclusive OR
NOT    logical NOT (complement)
```

The operation is applied on the corresponding bits in each operand. The operator must be separated from the operands by space or tab characters. For example, the following two instructions are the same:

```
MOV  A,#'9' AND 0FH
MOV  A,#9
```

The NOT operator only takes one operand. The following three MOV instructions are the same:

```
THREE          EQU  3
MINUS_THREE    EQU  -3
               MOV  A,#(NOT THREE) + 1
               MOV  A,#MINUS_THREE
               MOV  A,#11111101B
```

7.4.5 Special Operators

The special operators are

```
SHR     shift right
SHL     shift left
HIGH    high-byte
LOW     low-byte
()      evaluate first
```

For example, the following two instructions are the same:

```
MOV  A,#8 SHL 1
MOV  A,#10H
```

The following two instructions are also the same:

```
MOV  A,#HIGH 1234H
MOV  A,#12H
```

7.4.6 Relational Operators

When a relational operator is used between two operands, the result is always false (0000H) or true (FFFFH). The operators are

```
EQ    =     equals
NE    <>    not equals
LT    <     less than
LE    <=    less than or equal to
```

```
GT     >      greater than
GE     >=     greater than or equal to
```

Note that for each operator, two forms are acceptable (e.g., "EQ" or "="). In the following examples, all relational tests are "true":

```
MOV  A,#5 = 5
MOV  A,#5 NE 4
MOV  A,#'X' LT 'Z'
MOV  A,#'X' >= 'X'
MOV  A,#$ > 0
MOV  A,#100 GE 50
```

So, the assembled instructions are all equal to

```
MOV  A,#0FFH
```

Even though expressions evaluate to 16-bit results (i.e., 0FFFFH), in the examples above only the low-order eight bits are used, since the instruction is a move byte operation. The result is not considered too big in this case, because as signed numbers the 16-bit value FFFFH and the 8-bit value FFH are the same (−1).

7.4.7 Expression Examples

The following are examples of expressions and the values that result:

Expression	Result
'B' − 'A'	0001H
8/3	0002H
155 MOD 2	0001H
4 * 4	0010H
8 AND 7	0000H
NOT 1	FFFEH
'A' SHL 8	4100H
LOW 65535	00FFH
(8 + 1) * 2	0012H
5 EQ 4	0000H
'A' LT 'B'	FFFFH
3 <= 3	FFFFH

A practical example that illustrates a common operation for timer initialization follows: Put −500 into Timer 1 registers TH1 and TL1. Using the HIGH and LOW operators, a good approach is

```
VALUE       EQU  -500
            MOV  TH1,#HIGH VALUE
            MOV  TL1,#LOW VALUE
```

The assembler converts −500 to the corresponding 16-bit value (FE0CH); then the HIGH and LOW operators extract the high (FEH) and low (0CH) bytes, as appropriate for each MOV instruction.

7.4.8 Operator Precedence

The precedence of expression operators from highest to lowest is

```
()
HIGH LOW
* / MOD SHL SHR
+ -
EQ NE LT LE GT GE = <> < <= > >=
NOT
AND
OR XOR
```

When operators of the same precedence are used, they are evaluated left-to-right. Examples:

Expression	Value
HIGH ('A' SHL 8)	0041H
HIGH 'A' SHL 8	0000H
NOT 'A' -1	FFBFH
'A' OR 'A' SHL 8	4141H

7.5 ASSEMBLER DIRECTIVES

Assembler directives are instructions to the assembler program. They are *not* assembly language instructions executable by the target microprocessor. However, they are placed in the mnemonic field of the program. With the exception of DB and DW, they have no direct effect on the contents of memory.

ASM51 provides several categories of directives:

- Assembler state control (ORG, END, USING)
- Symbol definition (SEGMENT, EQU, SET, DATA, IDATA, XDATA, BIT, CODE)
- Storage initialization/reservation (DS, DBIT, DB, DW)
- Program linkage (PUBLIC, EXTRN, NAME)
- Segment selection (RSEG, CSEG, DSEG, ISEG, BSEG, XSEG)

Each assembler directive is presented below, ordered by category.

7.5.1 Assembler State Control

7.5.1.1 ORG (Set Origin)

The format for the ORG (set origin) directive is

```
ORG  expression
```

The ORG directive alters the location counter to set a new program origin for statements that follow. A label is not permitted. Two examples follow.

```
ORG    100H                      ;SET LOCATION COUNTER TO 100H
ORG    ($ + 1000H) AND 0F000H    ;SET TO NEXT 4K BOUNDARY
```

The ORG directive can be used in any segment type. If the current segment is absolute, the value will be an absolute address in the current segment. If a relocatable segment is active, the value of the ORG expression is treated as an offset from the base address of the current instance of the segment.

7.5.1.2 END

The format for the END directive is

```
END
```

END should be the last statement in the source file. No label is permitted and nothing beyond the END statement is processed by the assembler.

7.5.1.3 USING

The format for the USING directive is

```
USING                   expression
```

This directive informs ASM51 of the currently active register bank. Subsequent uses of the predefined symbolic register addresses AR0 to AR7 will convert to the appropriate direct address for the active register bank. Consider the following sequence:

```
USING      3
PUSH       AR7
USING      1
PUSH       AR7
```

The first push above assembles to PUSH 1FH (R7 in bank 3), whereas the second push assembles to PUSH 0FH (R7 in bank 1).

Note that USING does not actually switch register banks; it only informs ASM51 of the active bank. Executing 8051 instructions is the only way to switch register banks. This is illustrated by modifying the example above as follows:

```
MOV        PSW,#00011000B   ;SELECT REGISTER BANK 3
USING      3
PUSH       AR7              ;ASSEMBLE TO PUSH 1FH
MOV        PSW,#00001000B   ;SELECT REGISTER BANK 1
USING      1
PUSH       AR7              ;ASSEMBLE TO PUSH 0FH
```

7.5.2 Symbol Definition

The symbol definition directives create symbols that represent segments, registers, numbers, and addresses. None of these directives may be preceded by a label. Symbols defined by these directives may not have been previously defined and may not be redefined by any means. The SET directive is the only exception. Symbol definition directives are described below.

7.5.2.1 Segment

The format for the SEGMENT directive is shown below.

```
symbol        SEGMENT segment_type
```

The symbol is the name of a relocatable segment. In the use of segments, ASM51 is more complex than conventional assemblers, which generally support only "code" and "data" segment types. However, ASM51 defines additional segment types to accommodate the diverse memory spaces in the 8051. The following are the defined 8051 segment types (memory spaces):

- □ CODE (the code segment)
- □ XDATA (the external data space)
- □ DATA (the internal data space accessible by direct addressing, 00H–7FH)
- □ IDATA (the entire internal data space accessible by indirect addressing, 00H–7FH, 00H–FFH on the 8052)
- □ BIT (the bit space; overlapping byte locations 20H–2FH of the internal data space)

For example, the statement

```
EPROM         SEGMENT   CODE
```

declares the symbol EPROM to be a SEGMENT of type CODE. Note that this statement simply declares what EPROM is. To actually begin using this segment, the RSEG directive is used (see below).

7.5.2.2 EQU (Equate)

The format for the EQU directive is

```
symbol        EQU  expression
```

The EQU directive assigns a numeric value to a specified symbol name. The symbol must be a valid symbol name and the expression must conform to the rules described earlier.

The following are examples of the EQU directive:

```
N27           EQU  27                      ;SET N27 TO THE VALUE 27
HERE          EQU  $                       ;SET "HERE" TO THE VALUE
                                           ; OF THE LOCATION COUNTER
CR            EQU  0DH                     ;SET CR (CARRIAGE RETURN) TO 0DH
MESSAGE:      DB   'This is a message'
LENGTH        EQU  $ - MESSAGE             ;"LENGTH" EQUALS LENGTH OF "MESSAGE"
```

7.5.2.3 Other Symbol Definition Directives

The SET directive is similar to the EQU directive except the symbol may be redefined later, using another SET directive.

The DATA, IDATA, XDATA, BIT, and CODE directives assign addresses of the corresponding segment type to a symbol. These directives are not essential. A similar effect can be achieved using the EQU directive; however, if used they evoke powerful type-checking by ASM51. Consider the following two directives and four instructions:

```
FLAG1          EQU  05H
FLAG2          BIT  05H
               SETB FLAG1
               SETB FLAG2
               MOV  FLAG1,#0
               MOV  FLAG2,#0
```

The use of FLAG2 in the last instruction in this sequence will generate a "data segment address expected" error message from ASM51. Since FLAG2 is defined as a bit address (using the BIT directive), it can be used in a set bit instruction, but it cannot be used in a move byte instruction. Hence, the error. Even though FLAG1 represents the same value (05H), it was defined using EQU and does not have an associated address space. This is not an advantage of EQU, but, rather, a disadvantage. By properly defining address symbols for use in a specific memory space (using the directives BIT, DATA, XDATA, etc.), the programmer takes advantage of ASM51's powerful type-checking and avoids bugs from the misuse of symbols.

7.5.3 Storage Initialization/Reservation

The storage initialization and reservation directives initialize and reserve space in either word, byte, or bit units. The space reserved starts at the location indicated by the current value of the location counter in the currently active segment. These directives may be preceded by a label. The storage initialization/reservation directives are described below.

7.5.3.1 DS (Define Storage)

The format for the DS (define storage) directive is

```
[label:]       DS   expression
```

The DS directive reserves space in byte units. It can be used in any segment type except BIT. The expression must be a valid assemble-time expression with no forward references and no relocatable or external references. When a DS statement is encountered in a program, the location counter of the current segment is incremented by the value of the expression. The sum of the location counter and the specified expression should not exceed the limitations of the current address space.

The following statements create a 40-byte buffer in the internal data segment:

```
               DSEG AT 30H ;PUT IN DATA SEGMENT (ABSOLUTE, INTERNAL)
LENGTH         EQU  40
BUFFER:        DS   LENGTH ;40 BYTES RESERVED
```

The label BUFFER represents the address of the first location of reserved memory. For this example, the buffer begins at address 30H because "AT 30H" is specified with DSEG. (See 7.5.5.2 Selecting Absolute Segments.) This buffer could be cleared using the following instruction sequence:

```
              MOV  R7,#LENGTH
              MOV  R0,#BUFFER
LOOP:         MOV  @R0,#0
              DJNZ R7,LOOP
              (continue)
```

To create a 1000-byte buffer in external RAM starting at 4000H, the following directives could be used:

```
XSTART        EQU  4000H
XLENGTH       EQU  1000
              XSEG AT XSTART
XBUFFER:      DS   XLENGTH
```

This buffer could be cleared with the following instruction sequence:

```
              MOV  DPTR,#XBUFFER
LOOP:         CLR  A
              MOVX @DPTR,A
              INC  DPTR
              MOV  A,DPL
              CJNE A,#LOW(XBUFFER + XLENGTH + 1),LOOP
              MOV  A,DPH
              CJNE A,#HIGH(XBUFFER + XLENGTH + 1), LOOP
              (continue)
```

This is an excellent example of a powerful use of ASM51's operators and assemble-time expressions. Since an instruction does not exist to compare the data pointer with an immediate value, the operation must be fabricated from available instructions. Two compares are required, one each for the high- and low-bytes of the DPTR. Furthermore, the compare-and-jump-if-not-equal instruction works only with the accumulator or a register, so the data pointer bytes must be moved into the accumulator before the CJNE instruction. The loop terminates only when the data pointer has reached XBUFFER + LENGTH + 1. (The "+1" is needed because the data pointer is incremented *after* the last MOVX instruction.)

7.5.3.2 DBIT

The format for the DBIT (define bit) directive is,

```
[label:]      DBIT expression
```

The DBIT directive reserves space in bit units. It can be used only in a BIT segment. The expression must be a valid assemble-time expression with no forward references. When the DBIT statement is encountered in a program, the location counter of the current (BIT) segment is incremented by the value of the expression. Note that in a BIT segment, the basic unit of the location counter is bits rather than bytes. The following directives create three flags in an absolute bit segment:

```
              BSEG          ;BIT SEGMENT (ABSOLUTE)
KBFLAG:       DBIT 1        ;KEYBOARD STATUS
PRFLAG:       DBIT 1        ;PRINTER STATUS
DKFLAG:       DBIT 1        ;DISK STATUS
```

Since an address is not specified with BSEG in the example above, the address of the flags defined by DBIT could be determined (if one wishes to do so) by examining the symbol table in the .LST or .M51 files. (See Figure 7–1 and Figure 7–6.) If the definitions above were the first use of BSEG, then KBFLAG would be at bit address 00H (bit 0 of byte address 20H; see Figure 2–6.) If other bits were defined previously using BSEG, then the definitions above would follow the last bit defined. (See 7.5.5.2. Selecting Absolute Segments.)

7.5.3.3 DB (Define Byte)

The format for the DB (define byte) directive is

```
[label:]     DB   expression [,expression][. . . ]
```

The DB directive initializes code memory with byte values. Since it is used to actually place data constants in code memory, a CODE segment must be active. The expression list is a series of one or more byte values (each of which may be an expression) separated by commas.

The DB directive permits character strings (enclosed in single quotes) longer than two characters as long as they are not part of an expression. Each character in the string is converted to the corresponding ASCII code. If a label is used, it is assigned the address of the first byte. For example, the following statements

```
             CSEG AT 0100H
SQUARES:     DB   0,1,4,9,16,25  ;SQUARES OF NUMBERS 0-5
MESSAGE:     DB   'Login:',0     ;NULL-TERMINATED CHARACTER STRING
```

when assembled, result in the following hexadecimal memory assignments for external code memory:

Address	Contents
0100	00
0101	01
0102	04
0103	09
0104	10
0105	19
0106	4C
0107	6F
0108	67
0109	69
010A	6E
010B	3A
010C	00

7.5.3.4 DW (Define Word)

The format for the DW (define word) directive is

```
[label:]     DW    expression    [,expression][. . . ]
```

The DW directive is the same as the DB directive except two memory locations (16 bits) are assigned for each data item. For example, the statements

```
            CSEG AT 200H
            DW   $,'A',1234H,2,'BC'
```

result in the following hexadecimal memory assignments:

Address	Contents
0200	02
0201	00
0202	00
0203	41
0204	12
0205	34
0206	00
0207	02
0208	42
0209	43

7.5.4 Program Linkage

Program linkage directives allow the separately assembled modules (files) to communicate by permitting intermodule references and the naming of modules. In the following discussion, a "module" can be considered a "file." (In fact, a module may encompass more than one file.)

7.5.4.1 PUBLIC

The format for the PUBLIC (public symbol) directive is

```
PUBLIC symbol [,symbol][. . . ]
```

The PUBLIC directive allows the list of specified symbols to be known and used outside the currently assembled module. A symbol declared PUBLIC must be defined in the current module. Declaring it PUBLIC allows it to be referenced in another module. For example,

```
PUBLIC INCHAR, OUTCHR, INLINE, OUTSTR
```

7.5.4.2 EXTRN

The format for the EXTRN (external symbol) directive is,

```
EXTRN segment_type(symbol [,symbol][. . . ], . . . )
```

The EXTRN directive lists symbols to be referenced in the current module that are defined in other modules. The list of external symbols must have a segment type associated with each symbol in the list. (The segment types are CODE, XDATA, DATA, IDATA, BIT, and NUMBER. NUMBER is a type-less symbol defined by EQU.) The segment type indicates the way a symbol may be used. The information is important at link-time to ensure symbols are used properly in different modules.

The PUBLIC and EXTRN directives work together. Consider the two files shown below, MAIN.SRC and MESSAGES.SRC. The subroutines HELLO and GOOD_BYE are defined in the module MESSAGES but are made available to other modules using the PUBLIC directive. The subroutines are called in the module MAIN even though they are not defined there. The EXTRN directive declares that these symbols are defined in another module.

MAIN.SRC

```
                        EXTRN   CODE(HELLO,GOOD_BYE)
                        . . .
                        CALL    HELLO
                        . . .
                        CALL    GOOD_BYE
                        . . .
                        END
```

MESSAGES.SRC

```
                        PUBLIC HELLO,GOOD_BYE
                        . . .
            HELLO:      (begin subroutine)
                        . . .
                        RET
            GOOD_BYE:   (begin subroutine)
                        . . .
                        RET
                        . . .
                        END
```

Neither MAIN.SRC nor MESSAGES.SRC is a complete program; they must be assembled separately and linked together to form an executable program. During linking, the external references are resolved with correct addresses inserted as the destination for the CALL instructions.

7.5.4.3 NAME

The format for the NAME directive is

```
NAME module_name
```

All the usual rules for symbol names apply to module names. If a name is not provided, the module takes on the file name (without a drive or subdirectory specifier and without an extension). In the absence of any use of the NAME directive, a program will contain one module for each file. The concept of "modules," therefore, is somewhat cumbersome, at least for relatively small programming problems. Even programs of moderate size (encompassing, for example, several files complete with relocatable segments) needn't use the NAME directive and needn't pay any special attention to the concept of "modules." For this reason, it was mentioned in the definition that a module may be considered a "file," to simplify learning ASM51. However, for very large programs (several thousand lines of code, or more), it makes sense to partition the problem into modules, where, for example, each module may encompass several files containing routines having a common purpose.

7.5.5 Segment Selection Directives

When the assembler encounters a segment selection directive, it diverts the following code or data into the selected segment until another segment is selected by a segment selection directive. The directive may select a previously defined relocatable segment, or optionally create and select absolute segments.

7.5.5.1 RSEG (Relocatable Segment)

The format for the RSEG (relocatable segment) directive is

```
RSEG segment_name
```

where "segment_name" is the name of a relocatable segment previously defined with the SEGMENT directive. RSEG is a "segment selection" directive that diverts subsequent code or data into the named segment until another segment selection directive is encountered.

7.5.5.2 Selecting Absolute Segments

RSEG selects a relocatable segment. An "absolute" segment, on the other hand, is selected using one of the following directives:

```
CSEG [AT address]
DSEG [AT address]
ISEG [AT address]
BSEG [AT address]
XSEG [AT address]
```

These directives select an absolute segment within the code, internal data, indirect internal data, bit, or external data address spaces, respectively. If an absolute address is provided (by indicating "AT address"), the assembler terminates the last absolute address segment, if any, of the specified segment type and creates a new absolute segment starting at that address. If an absolute address is not specified, the last absolute segment of the specified type is continued. If no absolute segment of this type was previously selected and the absolute address is omitted, a new segment is created starting at location 0. Forward references are not allowed and start addresses must be absolute.

Each segment has its own location counter, which is always set to 0 initially. The default segment is an absolute code segment; therefore, the initial state of the assembler is location 0000H in the absolute code segment. When another segment is chosen for the first time, the location counter of the former segment retains the last active value. When that former segment is reselected, the location counter picks up at the last active value.

```
LOC  OBJ             LINE    SOURCE

                        1    ONCHIP      SEGMENT     DATA     ;A RELOCATABLE DATA SEGMENT
                        2    EPROM       SEGMENT     CODE     ;A RELOCATABLE CODE SEGMENT
                        3
----                    4                BSEG        AT 70H   ;BEGIN ABSOLUTE BIT SEGMENT
0070                    5    FLAG1:      DBIT        1
0071                    6    FLAG2:      DBIT        1
                        7
----                    8                RSEG        ONCHIP   ;BEGIN RELOCATABLE DATA SEGMENT
0000                    9    TOTAL:      DS          1
0001                   10    COUNT:      DS          1
0002                   11    SUM16:      DS          2
                       12
----                   13                RSEG        EPROM    ;BEGIN RELOCATABLE EPROM SEGMENT
0000 750000      F     14    BEGIN:      MOV         TOTAL,#0
                       15    ;           (continue program)
                       16                END
```

FIGURE 7–4
Defining and initiating absolute and relocatable segments

The ORG directive may be used to change the location counter within the currently selected segment. Figure 7–4 shows examples of defining and initiating relocatable and absolute segments.

The first two lines in Figure 7–4 declare the symbols ONCHIP and EPROM to be segments of type DATA (internal data RAM) and CODE respectively. Line 4 begins an absolute bit segment starting at bit address 70H (bit 0 of byte address 2EH; see Figure 2–6). Next, FLAG1 and FLAG2 are created as labels corresponding to bit-addressable locations 70H and 71H. RSEG in line 8 begins the relocatable ONCHIP segment for internal data RAM. TOTAL and COUNT are labels corresponding to byte locations. SUM16 is a label corresponding to a word (2-byte) location. The next occurrence of RSEG in line 13 begins the relocatable EPROM segment for code memory. The label BEGIN is the address of the first instruction in this instance of the EPROM. Note that it is not possible to determine the address of the labels TOTAL, COUNT, SUM16, and BEGIN from Figure 7–4. Since these labels occur in relocatable segments, the object file must be processed by the linker/locator (see 7.7 Linker Operation) with starting addresses specified for the ONCHIP and EPROM segments. The .M51 listing file created by the linker/locator gives the absolute addresses for these labels. FLAG1 and FLAG2, however, always correspond to bit addresses 70H and 71H because they are defined in an absolute BIT segment.

7.6 ASSEMBLER CONTROLS

Assembler controls establish the format of the listing and object files by regulating the actions of ASM51. For the most part, assembler controls affect the look of the listing file, without having any effect on the program itself. They can be entered on the invocation line when a program is assembled, or they can be placed in the source file. Assembler controls appearing in the source file must be preceded with a dollar sign and must begin in column one.

There are two categories of assembler controls: primary and general. Primary controls can be placed in the invocation line or at the beginning of the source program. Only other primary controls may precede a primary control. General controls may be placed anywhere in the source program. Figure 7–5 shows the assembler controls supported by ASM51.

NAME	PRIMARY/ GENERAL	DEFAULT	ABBREV.	MEANING
DATE(date)	P	DATE()	DA	Places string in header (9 char. max.)
DEBUG	P	NODEBUG	DB	Outputs debug symbol information to object file
NODEBUG	P	NODEBUG	NODB	Symbol information not placed in object file
EJECT	G	not applicable	EJ	Continue listing on next page
ERRORPRINT(file)	P	NOERRORPRINT	EP	Designates a file to receive error messages in addition to the listing file (defaults to console)
NOERRORPRINT	P	NOERRORPRINT	NOEP	Designates that error messages will be printed in listing file only
GEN	G	GENONLY	GE	Generates a full listing of the macro expansion process including macro calls in the listing file
GENONLY	G	GENONLY	GO	List only the fully expanded source as if all lines generated by a macro call were already in the source file
NOGEN	G	GENONLY	NOGE	List only the original source text in the listing file
INCLUDE(file)	G	not applicable	IC	Designates a file to be included as part of the program
LIST	G	LIST	LI	Print subsequent lines of source code in listing file
NOLIST	G	LIST	NOLI	Do not print subsequent lines of source code in listing file
MACRO[(mem_percent)	P	MACRO(50)	MR	Evaluate and expand all macro calls. Allocate percentage of free memory for macro processing
NOMACRO	P	MACRO(50)	NOMR	Do evaluate macro calls
MOD51	P	MOD51	MO	Recognize the 8051-specific predefined special registers
NOMOD51	P	MOD51	NOMO	Do not recognize the 8051-specific predefined special registers
OBJECT[(file)]	P	OBJECT(source.OBJ)	OJ	Designate file to receive object code
NOOBJECT	P	OBJECT(source.OBJ)	NOOJ	Designates that no object file will be created
PAGING	P	PAGING	PI	Designates that listing will be broken into pages and each will have a header
NOPAGING	P	PAGING	NOPI	Designates that listing file will contain no page breaks
PAGELENGTH(n)	P	PAGELENGTH(60)	PL	Sets maximum number of lines in each page of listing file (range = 10 to 65,536)
PAGEWIDTH(n)	P	PAGEWIDTH(120)	PW	Sets maximum number of characters in each line of listing file (range = 72 to 132)
PRINT[(file)]	P	PRINT(source.LST)	PR	Designates file to receive source listing
NOPRINT	P	PRINT(source.LST)	NOPR	Designates that no listing file will be created
SAVE	G	not applicable	SA	Stores current control setting for LIST and GEN
RESTORE	G	not applicable	RS	Restores control setting from SAVE stack
REGISTERBANK(rb,...)	P	REGISTERBANK(0)	RB	Indicates one or more banks used in program module
NOREGISTERBANK	P	REGISTERBANK(0)	NORB	Indicates that no banks are used
SYMBOLS	P	SYMBOLS	SB	Creates a formatted table of all symbols used in program
NOSYMBOLS	P	SYMBOLS	NOSB	No symbol table created
TITLE(string)	G	TITLE()	TT	Places a string in all subsequent page headers (maximum 60 characters)
WORKFILES(path)	P	same as source	WF	Designates alternate path for temporary workfiles
XREF	P	NOXREF	XR	Creates a cross reference listing of all symbols used in program
NOXREF	P	NOXREF	NOXR	No cross reference list created

FIGURE 7–5
Assembler controls supported by ASM51

7.7 LINKER OPERATION

When developing large application programs, it is common to divide tasks into subprograms or modules containing sections of code (usually subroutines) that can be written separately from the overall program. The term "modular programming" refers to this programming strategy. Generally, modules are relocatable, meaning they are not intended for a specific address in the code or data space. A linking and locating program is needed to combine the modules into one absolute object module that can be executed.

Intel's RL51 is a typical linker/locator. It processes a series of relocatable object modules as input and creates an executable machine language program (PROGRAM, perhaps) and a listing file containing a memory map and symbol table (PROGRAM.M51). This is illustrated in Figure 7–6.

As relocatable modules are combined, all values for external symbols are resolved with values inserted into the output file. The linker is invoked from the system prompt by

```
RL51 input_list [TO output_file][location_controls]
```

The input_list is a list of relocatable object modules (files) separated by commas. The output_file is the name of the output absolute object module. If none is supplied, it defaults to the name of the first input file without any suffix. The location__controls set start addresses for the named segments.

For example, suppose three modules or files (MAIN.OBJ, MESSAGES.OBJ, and SUBROUTINES.OBJ) are to be combined into an executable program (EXAMPLE), and that these modules each contain 2 relocatable segments, one called EPROM of type CODE, and the other called ONCHIP of type DATA. Suppose further that the code segment is to be executable at address 4000H and the data segment is to reside starting at address 30H (in internal RAM). The following linker invocation could be used:

```
RL51 MAIN.OBJ,MESSAGES.OBJ,SUBROUTINES.OBJ TO EXAMPLE &
CODE (EPROM(4000H)) DATA(ONCHIP(30H))
```

Note that the ampersand character "&" is used as the line continuation character.

If the program begins at the label START, and this is the first instruction in the MAIN module, then execution begins at address 4000H. If the MAIN module was not linked first, or if the label START is not at the beginning of MAIN, then the program's

FIGURE 7–6
Linker operation

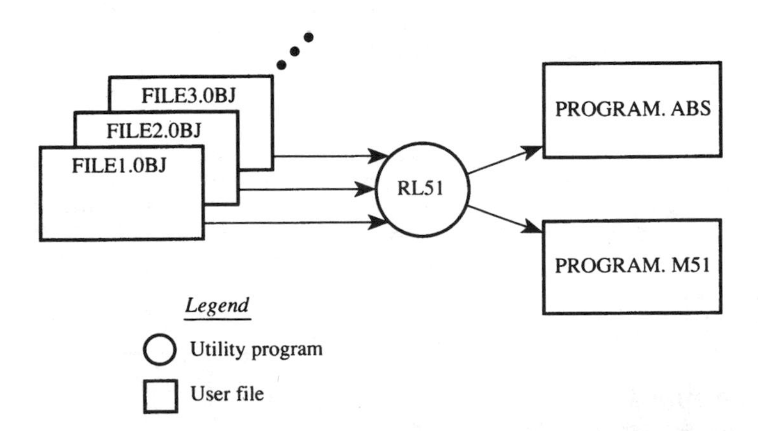

entry point can be determined by examining the symbol table in the listing file EXAMPLE.M51 created by RL51. By default, EXAMPLE.M51 will contain only the link map. If a symbol table is desired, then each source program must have used the $DEBUG control. (See Figure 7–5.)

7.8 ANNOTATED EXAMPLE: LINKING RELOCATABLE SEGMENTS AND MODULES

Many of the concepts just introduced are now brought together in an annotated example of a simple 8051 program. The source code is split over two files and uses symbols declared as EXTRN or PUBLIC to allow inter-file communication. Each file is a module—one named MAIN, the other named SUBROUTINES. The program uses a relocatable code segment named EPROM and a relocatable internal data segment named ONCHIP. Working with multiple files, modules, and segments is essential for large programming projects. A careful examination of the example that follows will strengthen these core concepts and prepare the reader to embark on practical 8051-based designs.

Our example is a simple input/output program using the 8051's serial port and a VDT's keyboard and CRT display. The program does the following:

- Initialize the serial port (once)
- Output the prompt "Enter a command:"
- Input a line from the keyboard, echoing each character as it is received
- Echo back the entire line
- Repeat

Figure 7–7 shows (a) the listing file (ECHO.LST) for the first source file, (b) the listing file (IO.LST) for the second source file, and (c) the listing file (EXAMPLE.M51) created by the linker/locator.

7.8.1 ECHO.LST

Figure 7–7a shows the contents of the file ECHO.LST created by ASM51 when the source file (ECHO.SRC) was assembled. The first several lines in the listing file provide general information on the programming environment. Among other things, the invocation line is restated in an expanded form showing the path to the files. Note the use of the assembler control EP (for ERRORPRINT) on the invocation line. This causes error messages to be sent to the console as well as the listing file. (See Figure 7–5.)

The original source file is shown under the column heading SOURCE, just to the right of the column LINE. As evident, ECHO.SRC contains 22 lines. Lines 1 to 4 contain assembler controls. (See Figure 7–5.) $DEBUG in line 1 instructs ASM51 to place a symbol table in the object file, ECHO.OBJ. This is necessary for hardware emulation or for the linker/locator to create a symbol table in its listing file. $TITLE defines a string to be placed at the top of each page of the listing file. $PAGEWIDTH specifies the maximum width of each line in the listing file. $NOPAGING prevents page breaks (form feeds) from being inserted in the listing file. Most assembler controls affect the look of the output listing file. Some trial-and-error will usually produce the desired output for printing.

(a)

```
MCS-51 MACRO ASSEMBLER    *** ANNOTATED EXAMPLE (MAIN MODULE) ***           03/17/91


DOS 3.31 (038-N) MCS-51 MACRO ASSEMBLER, V2.2
OBJECT MODULE PLACED IN ECHO.OBJ
ASSEMBLER INVOKED BY:  C:\ASM51\ASM51.EXE ECHO.SRC EP


LOC  OBJ            LINE     SOURCE

                       1     $DEBUG
                       2     $TITLE(*** ANNOTATED EXAMPLE (MAIN MODULE) ***)
                       3     $PAGEWIDTH(98)
                       4     $NOPAGING
                       5
                       6             NAME      MAIN            ;MODULE NAME IS "MAIN"
                       7             EXTRN CODE(INIT,OUTSTR)   ;DECLARE EXTERNAL SYMBOLS
                       8             EXTRN CODE(INLINE,OUTLINE)
                       9
 000D                 10     CR      EQU       0DH             ;CARRIAGE RETURN CODE
                      11     EPROM   SEGMENT   CODE            ;DEFINE SYMBOL "EPROM"
                      12
----                  13             RSEG      EPROM           ;BEGIN CODE SEGMENT
0000 120000     F     14     MAIN:   CALL      INIT            ;INITIALIZE SERIAL PORT
0003 900000     F     15     LOOP:   MOV       DPTR,#PROMPT    ;SEND PROMPT
0006 120000     F     16             CALL      OUTSTR
0009 120000     F     17             CALL      INLINE          ;GET A COMMAND LINE AND
000C 120000     F     18             CALL      OUTLINE         ; ECHO IT BACK
000F 80F2             19             JMP       LOOP            ;REPEAT
                      20
0011 0D               21     PROMPT: DB        CR,'Enter a command: ',0
0012 456E7465
0016 72206120
001A 636F6D6D
001E 616E643A
0022 20
0023 00
                      22             END



SYMBOL TABLE LISTING
------ ----- -------


N A M E     T Y P E  V A L U E      A T T R I B U T E S

CR . . . .    NUMB   000DH   A
EPROM. . .  C SEG    0024H          REL=UNIT
INIT . . .  C ADDR   ----       EXT
INLINE . .  C ADDR   ----       EXT
LOOP . . .  C ADDR   0003H   R      SEG=EPROM
MAIN . . .  C ADDR   0000H   R      SEG=EPROM
OUTLINE. .  C ADDR   ----       EXT
OUTSTR . .  C ADDR   ----       EXT
PROMPT . .  C ADDR   0011H   R      SEG=EPROM


REGISTER BANK(S) USED: 0

ASSEMBLY COMPLETE, NO ERRORS FOUND
```

FIGURE 7–7
Annotated example: linking relocatable segments and modules. (a) ECHO.LST (b) IO.LST (c) EXAMPLE.M51.

```
DOS 3.31 (038-N) MCS-51 MACRO ASSEMBLER, V2.2
OBJECT MODULE PLACED IN IO.OBJ
ASSEMBLER INVOKED BY:  C:\ASM51\ASM51.EXE IO.SRC EP

LOC  OBJ           LINE    SOURCE

                      1    $DEBUG
                      2    $TITLE(*** ANNOTATED EXAMPLE (SUBROUTINES MODULE) ***)
                      3    $PAGEWIDTH(98)
                      4    $NOPAGING
                      5
                      6                NAME     SUBROUTINES         ;MODULE NAME
                      7                PUBLIC   INIT,OUTCHR,INCHAR ;DECLARE PUBLIC SYMBOLS
                      8                PUBLIC   INLINE,OUTLINE,OUTSTR
                      9
                     10    ;*****************************************************************
                     11    ; DEFINE SYMBOLS                                                 *
                     12    ;*****************************************************************
  000D               13    CR          EQU      0DH         ;CARRIAGE RETURN
  0028               14    LENGTH      EQU      40          ;40-CHARACTER BUFFER
                     15    EPROM       SEGMENT  CODE        ;"EPROM" IS A CODE SEGMENT
                     16    ONCHIP      SEGMENT  DATA        ;"ONCHIP" IS A DATA SEGMENT
                     17
----                 18                RSEG     EPROM       ;BEGIN RELOCATABLE CODE SEGMENT
                     19
                     20    ;*****************************************************************
                     21    ; INITIALIZE THE SERIAL PORT                                     *
                     22    ;*****************************************************************
0000 759852          23    INIT:       MOV      SCON,#52H   ;8-BIT UART MODE
0003 758920          24                MOV      TMOD,#20H   ;TIMER 1 SUPPLIES BAUD RATE CLOCK
0006 758DF3          25                MOV      TH1,#-13    ;2400 BAUD
0009 D28E            26                SETB     TR1         ;START TIMER
000B 22              27                RET
                     28
                     29    ;*****************************************************************
                     30    ; OUTPUT CHARACTER IN ACC (NOTE: VDT MUST CONVERT CR INTO CR/LF) *
                     31    ;*****************************************************************
000C 3099FD          32    OUTCHR:     JNB      TI,$        ;WAIT FOR TRANSMIT BUFFER EMPTY
000F C299            33                CLR      TI          ;WHEN EMPTY, CLEAR FLAG AND
0011 F599            34                MOV      SBUF,A      ; SEND CHARACTER
0013 22              35                RET
                     36
                     37    ;*****************************************************************
                     38    ; INPUT CHARACTER TO ACC                                         *
                     39    ;*****************************************************************
0014 3098FD          40    INCHAR:     JNB      RI,$        ;WAIT FOR RECEIVE BUFFER FULL
0017 C298            41                CLR      RI          ;WHEN CHAR ARRIVES, CLEAR FLAG &
0019 E599            42                MOV      A,SBUF      ; INPUT CHAR TO ACC
001B 22              43                RET
                     44
                     45    ;*****************************************************************
                     46    ; OUTPUT NULL-TERMINATED STRING                                  *
                     47    ;*****************************************************************
001C E4              48    OUTSTR:     CLR      A           ;DPTR POINTS TO STRING OF CHAR
001D 93              49                MOVC     A,@A+DPTR   ;GET CHARACTER
001E 6006            50                JZ       EXIT        ;IF NULL BYTE, DONE
0020 120000     F    51                CALL     OUTCHR      ;OTHERWISE, SEND IT
0023 A3              52                INC      DPTR        ;POINT TO NEXT CHARACTER
0024 80F6            53                JMP      OUTSTR      ; AND SEND IT TOO
0026 22              54    EXIT:       RET
```

FIGURE 7–7
continued

```
                        55
                        56    ;*****************************************************************
                        57    ; INPUT CHARACTERS TO BUFFER                                      *
                        58    ;*****************************************************************
0027 7800      F        59    INLINE:     MOV     R0,#BUFFER  ;USE R0 AS POINTER TO BUFFER
0029 120000    F        60    AGAIN:      CALL    INCHAR      ;GET A CHARACTER
002C 120000    F        61                CALL    OUTCHR      ; ECHO IT BACK
002F F6                 62                MOV     @R0,A       ;PUT IT IN BUFFER
0030 08                 63                INC     R0          ;INCREMENT POINTER TO BUFFER
0031 B40DF5             64                CJNE    A,#CR,AGAIN ;IF NOT CR, GET ANOTHER CHAR
0034 7600               65                MOV     @R0,#0      ;PUT NULL BYTE AT END
0036 22                 66                RET
                        67
                        68    ;*****************************************************************
                        69    ; OUTPUT CONTENTS OF BUFFER                                       *
                        70    ;*****************************************************************
0037 7800      F        71    OUTLINE:    MOV     R0,#BUFFER  ;USE R0 AS POINTER TO BUFFER
0039 E6                 72    AGAIN2:     MOV     A,@R0       ;GET CHARACTER FROM BUFFER
003A 6006               73                JZ      EXIT2       ;IF NULL BYTE, DONE
003C 120000    F        74                CALL    OUTCHR      ;OTHERWISE, SEND IT
003F 08                 75                INC     R0          ;POINT TO NEXT CHAR IN BUFFER
0040 80F7               76                JMP     AGAIN2      ; AND SEND IT TOO
0042 22                 77    EXIT2:      RET
                        78
                        79    ;*****************************************************************
                        80    ; CREATE A BUFFER IN ONCHIP RAM                                   *
                        81    ;*****************************************************************
----                    82                RSEG    ONCHIP      ;BEGIN RELOCATABLE DATA SEGMENT
0000                    83    BUFFER:     DS      LENGTH      ;ALLOCATE INTERNAL RAM AS BUFFER
                        84                END

SYMBOL TABLE LISTING
------ ----- -------

N A M E       T Y P E    V A L U E       A T T R I B U T E S

AGAIN . . .   C ADDR     0029H    R      SEG=EPROM
AGAIN2. . .   C ADDR     0039H    R      SEG=EPROM
BUFFER. . .   D ADDR     0000H    R      SEG=ONCHIP
CR. . . . .     NUMB     000DH    A
EPROM . . .   C SEG      0043H           REL=UNIT
EXIT. . . .   C ADDR     0026H    R      SEG=EPROM
EXIT2 . . .   C ADDR     0042H    R      SEG=EPROM
INCHAR. . .   C ADDR     0014H    R PUB  SEG=EPROM
INIT. . . .   C ADDR     0000H    R PUB  SEG=EPROM
INLINE. . .   C ADDR     0027H    R PUB  SEG=EPROM
LENGTH. . .     NUMB     0028H    A
ONCHIP. . .   D SEG      0028H           REL=UNIT
OUTCHR. . .   C ADDR     000CH    R PUB  SEG=EPROM
OUTLINE . .   C ADDR     0037H    R PUB  SEG=EPROM
OUTSTR. . .   C ADDR     001CH    R PUB  SEG=EPROM
RI. . . . .   B ADDR     0098H.0  A
SBUF. . . .   D ADDR     0099H    A
SCON. . . .   D ADDR     0098H    A
SUBROUTINES     ----     ----
TH1 . . . .   D ADDR     008DH    A
TI. . . . .   B ADDR     0098H.1  A
TMOD. . . .   D ADDR     0089H    A
TR1 . . . .   B ADDR     0088H.6  A

REGISTER BANK(S) USED: 0

ASSEMBLY COMPLETE, NO ERRORS FOUND
```

FIGURE 7–7
continued

(c)

```
DATE : 03/17/91
DOS 3.31 (038-N) MCS-51 RELOCATOR AND LINKER V3.0, INVOKED BY:
C:\ASM51\RL51.EXE ECHO.OBJ,IO.OBJ TO EXAMPLE CODE(EPROM(8000H))DATA(ONCHIP(30H
>> ))


INPUT MODULES INCLUDED
  ECHO.OBJ(MAIN)
  IO.OBJ(SUBROUTINES)


LINK MAP FOR EXAMPLE(MAIN)


            TYPE    BASE      LENGTH    RELOCATION   SEGMENT NAME
            ----    ----      ------    ----------   ------------

            REG     0000H     0008H                  "REG BANK 0"
                    0008H     0028H                  *** GAP ***
            DATA    0030H     0028H     UNIT         ONCHIP

                    0000H     8000H                  *** GAP ***
            CODE    8000H     0067H     UNIT         EPROM


SYMBOL TABLE FOR EXAMPLE(MAIN)

VALUE          TYPE           NAME
-----          ----           ----

-------        MODULE         MAIN
N:000DH        SYMBOL         CR
C:8000H        SEGMENT        EPROM
C:8003H        SYMBOL         LOOP
C:8000H        SYMBOL         MAIN
C:8011H        SYMBOL         PROMPT
-------        ENDMOD         MAIN

-------        MODULE         SUBROUTINES
C:804DH        SYMBOL         AGAIN
C:805DH        SYMBOL         AGAIN2
D:0030H        SYMBOL         BUFFER
N:000DH        SYMBOL         CR
C:8000H        SEGMENT        EPROM
C:804AH        SYMBOL         EXIT
C:8066H        SYMBOL         EXIT2
C:8038H        PUBLIC         INCHAR
C:8024H        PUBLIC         INIT
C:804BH        PUBLIC         INLINE
N:0028H        SYMBOL         LENGTH
D:0030H        SEGMENT        ONCHIP
C:8030H        PUBLIC         OUTCHR
C:805BH        PUBLIC         OUTLINE
C:8040H        PUBLIC         OUTSTR
B:0098H        SYMBOL         RI
D:0099H        SYMBOL         SBUF
D:0098H        SYMBOL         SCON
D:008DH        SYMBOL         TH1
B:0098H.1      SYMBOL         TI
D:0089H        SYMBOL         TMOD
B:0088H.6      SYMBOL         TR1
-------        ENDMOD         SUBROUTINES
```

FIGURE 7–7
continued

The NAME assembler directive in line 6 defines the current file as part of the module MAIN. For this example, no further instance of the MAIN module is used; however, larger projects may include other files also defined as part of the MAIN module. It may help the reader for the rest of this example to read "file" for the term "module."

Lines 7 and 8 identify the symbols used in the current module but defined elsewhere. Without these EXTRN directives, ASM51 will generate the message "undefined symbol" on each line in the source program where one of these symbols is used. The "segment type" must also be defined for each symbol to ensure its proper use. All of the external symbols defined in this example are of type CODE.

Symbol definitions come next. Line 10 defines the symbol CR as the carriage return ASCII code 0DH. Line 11 defines the symbol EPROM as a segment of type CODE. Recall that the SEGMENT directive defines only what the symbol is—nothing more, nothing less.

The RSEG directive in line 13 begins the relocatable segment named EPROM. Subsequent instructions, data constant definitions, and so on, will be placed in the EPROM code segment.

The program begins on line 14 at the label MAIN. The first instruction in the program is a call to the subroutine INIT, which will initialize the 8051's serial port. The assembled code under the OBJ column contains the correct opcode (12H for LCALL); however, bytes 2 and 3 of the instruction (the address of the subroutine) appear as 0000H followed by the letter "F." The linker/locator must "fix" this when the program modules are linked together and addresses are set for the relocatable segments. Note, too, that the address under the LOC column is also entered as 0000H. Since the EPROM segment is relocatable, it is not known at assemble-time where the segment will start. All relocatable segments will display 0000H as the starting address in the listing file.

The rest of the program instructions are on lines 15 to 19. A prompt message is sent to the VDT by loading the DPTR with a starting address of the prompt and calling the subroutine OUTSTR. Since the OUTSTR, INLINE, and OUTLINE subroutines are not defined in ECHO.SRC, one can only guess at their operation from the name of the subroutine and the comment lines.

The prompt is a null-terminated ASCII string, which is placed in the EPROM code segment using the DB (define byte) directive on line 21. Since the prompt bytes are constant (i.e., unchanging) it is correct to place them in code memory (even though they are data bytes). The prompt begins with a carriage return to ensure it displays on a new line. (In this example, it is assumed the VDT converts CR to CR/LF.)

All the symbols and labels in ECHO.SRC appear in the symbol table at the bottom of ECHO.LST. Since the EPROM segment is relocatable and the subroutines are external, the VALUE column is not of much use. The value for the symbol EPROM, however, gives the length of the segment, which in this case is 24H or 36 bytes.

7.8.2 IO.LST

Figure 7–7b shows the contents of the file IO.LST—the file containing the input/output subroutines. This module is named SUBROUTINES in line 6. Lines 7 and 8 declare all subroutine names as PUBLIC symbols. This makes these symbols available to other modules. Note that all the subroutines are made public even though only four of them were used in the MAIN module. Perhaps, as the program grows, other modules will be added that may need these subroutines. So, they are all made public.

Lines 13 to 16 define several symbols. Once again, EPROM is used as the name of the code segment. Another segment is used in this module. ONCHIP is defined in line 17 as an internal data segment.

The subroutines are each written in turn beginning at line 20. The comment block beginning each is deliberately brief in this example; however, a more detailed description of a subroutine is usually given. It is useful to provide, for example, entry and exit conditions for each subroutine.

After the last subroutine, a buffer in internal RAM is created using the ONCHIP segment. The segment is started using RSEG (line 82), and the buffer is created using the DS (define storage) directive (line 83). The length of the buffer is assigned to the symbol LENGTH "equated" at the top of the program (line 14) as 40. The placement in the source file of the definition of the symbol LENGTH and of the instance of the segment ONCHIP is largely a matter of taste. Both could also be positioned just before or after the INLINE subroutine, where they are used.

As with the EPROM segment, ONCHIP is given an initial address of 0000H under the LOC column at line 83. Again, the actual location of the ONCHIP segment will not be determined until link-time (see below). The letter "F" appears in numerous locations in IO.LST. Each line so identified contains an instruction using a symbol whose value cannot be determined at assemble-time. The zeros placed in the object file at these locations will be replaced with "absolute" values by the linker/locator.

7.8.3 EXAMPLE.M51

Figure 7–7c shows the contents of the file EXAMPLE.M51 created by the linker/locator program, RL51. The invocation line is repeated near the top of EXAMPLE.M51 and should be examined carefully. Here it is again (leaving out the path):

```
RL51 ECHO.OBJ,IO.OBJ TO EXAMPLE CODE (EPROM (8000H)) &
DATA (ONCHIP(30H))
```

Following the command, the object modules are listed separated by commas in the order they are to be linked. Following the input list, the optional control TO EXAMPLE is specified providing the name for the absolute object module created by RL51. If omitted, the name of the first file in the input list is used (without any file extension). The listing file, in this example, automatically takes on the name EXAMPLE.M51. Finally, the locating controls CODE and DATA specify the names of segments of the associated type and the absolute address at which the segment is to begin. In this example the EPROM code segment begins at address 8000H and the ONCHIP data segment begins at byte address 30H in the 8051's internal RAM.

Following the restatement of the invocation line, EXAMPLE.M51 contains a list of the input modules included by RL51. In this example only two files (ECHO.OBJ and IO.OBJ) and two modules (MAIN and SUBROUTINES) are listed. If the NAME directive had not been used in the source files, the module names would be the same as the file names.

The link map appears next. Both the ONCHIP and EPROM segments are identified, and the starting address and the length (in hexadecimal) are given for each.

ONCHIP is identified as a data segment starting at address 30H and 28H (40) bytes in length. EPROM is identified as a code segment starting at address 8000H and 67H (103) bytes in length.

Finally, EXAMPLE.M51 contains a symbol table. All symbols (including labels) used in the program are listed, sorted on a module-by-module basis. All values are "absolute." Remember that the symbol table in the .M51 file can only be created if the $DEBUG assembler control is placed at the top of each source file. The INIT subroutine address (which we noted earlier was absent in ECHO.LST) is identified under the SUBROUTINES module as 8024H. This address is substituted as the code address in any object module using the instruction CALL INIT, as noted earlier in the MAIN module. Knowing the absolute value of labels is important when debugging. When a bug is found, often a temporary "patch" can be made by modifying the program bytes and re-executing the program. If the patch fixes the bug, the appropriate change is made to the source program.

7.9 MACROS

For the final topic in this chapter, we return to ASM51. The macro processing facility (MPL) of ASM51 is a "string replacement" facility. Macros allow frequently used sections of code to be defined once using a simple mnemonic and used anywhere in the program by inserting the mnemonic. Programming using macros is a powerful extension of the techniques described thus far. Macros can be defined anywhere in a source program and subsequently used like any other instruction. The syntax for a macro definition is

```
%*DEFINE (call_pattern) (macro_body)
```

Once defined, the call pattern is like a mnemonic; it may be used like any assembly language instruction by placing it in the mnemonic field of a program. Macros are made distinct from "real" instructions by preceding them with a percent sign, "%." When the source program is assembled, everything within the macro-body, on a character-by-character basis, is substituted for the call-pattern. The mystique of macros is largely unfounded. They provide a simple means for replacing cumbersome instruction patterns with primitive, easy-to-remember mnemonics. The substitution, we reiterate, is on a character-by-character basis—nothing more, nothing less.

For example, if the following macro definition appears at the beginning of a source file,

```
%*DEFINE(PUSH_DPTR)
              (PUSH DPH
              PUSH DPL
              )
```

then the statement

```
              %PUSH_DPTR
```

will appear in the .LST file as

```
              PUSH DPH
              PUSH DPL
```

The example above is a typical macro. Since the 8051 stack instructions operate only on direct addresses, pushing the data pointer requires two PUSH instructions. A similar macro can be created to POP the data pointer.

There are several distinct advantages in using macros:

- A source program using macros is more readable, since the macro mnemonic is generally more indicative of the intended operation than the equivalent assembler instructions.
- The source program is shorter and requires less typing.
- Using macros reduces bugs.
- Using macros frees the programmer from dealing with low-level details.

The last two points above are related. Once a macro is written and debugged, it is used freely without the worry of bugs. In the PUSH_DPTR example above, if PUSH and POP instructions are used rather than push and pop macros, the programmer may inadvertently reverse the order of the pushes or pops. (Was it the high-byte or low-byte that was pushed first?) This would create a bug. Using macros, however, the details are worked out once—when the macro is written—and the macro is used freely thereafter, without the worry of bugs.

Since the replacement is on a character-by-character basis, the macro definition should be carefully constructed with carriage returns, tabs, etc., to ensure proper alignment of the macro statements with the rest of the assembly language program. Some trial-and-error is required.

There are advanced features of ASM51's macro-processing facility that allow for parameter passing, local labels, repeat operations, assembly flow control, and so on. These are discussed below.

7.9.1 Parameter Passing

A macro with parameters passed from the main program has the following modified format:

```
%*DEFINE (macro_name (parameter_list)) (macro_body)
```

For example, if the following macro is defined,

```
%*DEFINE(CMPA# (VALUE))
                (CJNE   A,#%VALUE,$ + 3
                )
```

then the macro call

```
%CMPA#(20H)
```

will expand to the following instruction in the .LST file:

```
CJNE   A,#20H,$ + 3
```

Although the 8051 does not have a "compare accumulator" instruction, one is easily created using the CJNE instruction with "$+3" (the next instruction) as the destination for the conditional jump. The CMPA# mnemonic may be easier to remember for many programmers. Besides, use of the macro unburdens the programmer from remembering notational details, such as "$+3."

Let's develop another example. It would be nice if the 8051 had instructions such as

```
JUMP IF ACCUMULATOR GREATER THAN X
JUMP IF ACCUMULATOR GREATER THAN OR EQUAL TO X
JUMP IF ACCUMULATOR LESS THAN X
JUMP IF ACCUMULATOR LESS THAN OR EQUAL TO X
```

but it does not. These operations can be created using CJNE followed by JC or JNC, but the details are tricky. Suppose, for example, it is desired to jump to the label GREATER_THAN if the accumulator contains an ASCII code greater than "Z" (5AH). The following instruction sequence would work:

```
CJNE    A,#5BH,$+3
JNC     GREATER_THAN
```

The CJNE instruction subtracts 5BH (i.e., "Z" + 1) from the content of A and sets or clears the carry flag accordingly. CJNE leaves C = 1 for accumulator values 00H up to and including 5AH. (Note: 5AH − 5BH < 0, therefore C = 1; but 5BH − 5BH = 0, therefore C = 0.) Jumping to GREATER_THAN on the condition "not carry" correctly jumps for accumulator values 5BH, 5CH, 5DH, and so on, up to FFH. Once details such as these are worked out, they can be simplified by inventing an appropriate mnemonic, defining a macro, and using the macro instead of the corresponding instruction sequence. Here's the definition for a "jump if greater than" macro:

```
%*DEFINE(JGT(VALUE, LABEL))
            (CJNE  A,#%VALUE+1,$+3       ;JGT
            JNC    %LABEL
            )
```

To test if the accumulator contains an ASCII code greater than "Z," as just discussed, the macro would be called as

```
            %JGT('Z',GREATER_THAN)
```

ASM51 would expand this into

```
            CJNE  A,#5BH,$+3       ;JGT
            JNC   GREATER_THAN
```

The JGT macro is an excellent example of a relevant and powerful use of macros. By using macros, the programmer benefits by using a meaningful mnemonic and avoiding messy and potentially bug-ridden details.

7.9.2 Local Labels

Local labels may be used within a macro using the following format:

```
%*DEFINE(macro_name [(parameter_list)])
            [LOCAL list_of_local_labels](macro_body)
```

For example, the following macro definition

```
%*DEFINE (DEC_DPTR) LOCAL SKIP
             (DEC DPL                    ;DECREMENT DATA POINTER
             MOV  A,DPL
             CJNE A,#0FFH,%SKIP
             DEC  DPH
%SKIP:       )
```

would be called as

```
             %DEC_DPTR
```

and would be expanded by ASM51 into

```
             DEC  DPL                    ;DECREMENT DATA POINTER
             MOV  A,DPL
             CJNE A,#0FFH,SKIP00
             DEC  DPH
SKIP00:
```

Note that a local label generally will not conflict with the same label used elsewhere in the source program, since ASM51 appends a numeric code to the local label when the macro is expanded. Furthermore, the next use of the same local label receives the next numeric code, and so on.

The macro above has a potential "side effect." The accumulator is used as a temporary holding place for DPL. If the macro is used within a section of code that uses A for another purpose, the value in A would be lost. This side effect probably represents a bug in the program. The macro definition could guard against this by saving A on the stack. Here's an alternate definition for the DEC__DPTR macro:

```
%*DEFINE (DEC_DPTR) LOCAL SKIP
             (PUSH  ACC
             DEC    DPL                  ;DECREMENT DATA POINTER
             MOV    A,DPL
             CJNE   A,#0FFH,%SKIP
             DEC    DPH
%SKIP:       POP    ACC
             )
```

7.9.3 Repeat Operations

This is one of several built-in (predefined) macros. The format is

```
             %REPEAT (expression) (text)
```

For example, to fill a block of memory with 100 NOP instructions,

```
             %REPEAT (100)
             (NOP
             )
```

7.9.4 Control Flow Operations

The conditional assembly of sections of code is provided by ASM51's control flow macro definition. The format is

```
%IF(expression) THEN (balanced_text)
[ELSE (balanced_text)]FI
```

For example,

```
INTERNAL     EQU  1                      ;1 = 8051 SERIAL I/O DRIVERS
                                         ;0 = 8251 SERIAL I/O DRIVERS
             .
             .
             %IF (INTERNAL) THEN
(INCHAR:     .                           ;8051 DRIVERS
             .
OUTCHR:      .
             .
             ) ELSE
(INCHAR:     .                           ;8251 DRIVERS
             .
OUTCHR:      .
             .
             )
```

In this example, the symbol INTERNAL is given the value 1 to select I/O subroutines for the 8051's serial port, or the value 0 to select I/O subroutines for an external UART, in this case the 8251. The IF macro causes ASM51 to assemble one set of drivers and skip over the other. Elsewhere in the program, the INCHAR and OUTCHR subroutines are used without consideration for the particular hardware configuration. As long as the program was assembled with the correct value for INTERNAL, the correct subroutine is executed.

PROBLEMS

1. Recast the following instructions with the operand expressed in binary.

```
MOV  A,#255
MOV  A,#11Q
MOV  A,#1AH
MOV  A,#'A'
```

2. What is wrong with the coding of the following instruction?

```
ORL  80H,#F0H
```

3. Identify the error in the following symbols.

```
?byte.bit
@GOOD_bye
1ST_FLAG
MY_PROGRAM
```

4. Recast the following instructions with the expression evaluated as a 16-bit hexadecimal constant.

```
MOV  DPTR,#'0' EQ 48
MOV  DPTR,#HIGH 'AB'
MOV  DPTR,#-1
MOV  DPTR,#NOT (257 MOD 256)
```

5. What are the "segment types" defined by ASM51 for the 8051, and what memory spaces do they represent?
6. How could a relocatable segment in external data memory be defined, selected, and a 100-byte buffer created? (Give the segment the name "OFFCHIP" and give the buffer the name "XBUFFER.")
7. A certain application requires five status bits (FLAG1 to FLAG5). How could a 5-bit buffer be defined in an absolute BIT segment starting at bit address 08H? At what byte address do these bits reside?
8. What are two good reasons for making generous use of the EQU directive in assembly language programs?
9. What is the difference between the DB and DW directives?
10. What are the memory assignments for the following assembler directives:

```
ORG  OFH
DW   $ SHL 4
DB   65535
DW   '0'
```

11. What directive is used to select an absolute code segment?
12. A file called "ASCII" contains 33 equate directives, 1 for each control code:

```
NUL        EQU  00H          ;NULL BYTE
SOH        EQU  01H          ;START OF HEADER
.
.
.
US         EQU  1FH          ;UNIT SEPARATOR
DEL        EQU  7FH          ;DELETE
```

How could these definitions be made known in another file—a source program—without actually inserting the equates into that file?

13. In order for a printout of a listing file to look nice, it is desirable to have each subroutine begin at the top of a page. How is this accomplished?
14. Write the definition for a macro that could be used to fill a block of external data memory with a data constant. Pass the starting address, length, and data constant to the macro as parameters.
15. Write the definition for the following macros:

JGE—jump to LABEL if accumulator is greater than or equal to VALUE

JLT—jump to LABEL if accumulator is less than VALUE

JLE—jump to LABEL if accumulator is less than or equal to VALUE

JOR—jump to LABEL if accumulator is outside the range LOWER and UPPER

16. Write the definition for a macro called CJNE_DPTR that will jump to LABEL if the data pointer does not contain VALUE. Define the macro so that the contents of all registers and memory locations are left intact.

8

PROGRAM STRUCTURE AND DESIGN

8.1 INTRODUCTION

What makes one program better than the next? Beyond simple views such as "it works," the answer to this question is complex and depends on many factors: maintenance requirements, computer language, quality of documentation, development time, program length, execution time, reliability, security, and so on. In this chapter we introduce the characteristics of good programs and some techniques for developing good programs. We begin with an introduction to structured programming techniques.

Structured programming is a technique for organizing and coding programs that reduces complexity, improves clarity, and facilitates debugging and modifying. The idea of properly structuring programs is emphasized in most programming tasks, and we advance the idea here as well. The power of this approach can be appreciated by considering the following statement: All programs may be written using only three structures. This seems too good to be true, but it's not. "Statements," "loops," and "choices" form a complete set of structures, and all programs can be realized using only these three structures. Program control is passed through the structures without unconditional branches to other structures. Each structure has one entry point and one exit point. Typically, a structured program contains a hierarchy of subroutines, each with a single entry point and a single exit point.[1]

The purpose of this chapter is to introduce structured programming as applied to assembly language programming. Although high-level languages (such as Pascal or C) promote structured programming through their statements (WHILE, FOR, etc.) and notational conventions (indentation), assembly language lacks such inherent properties. Nevertheless, assembly language programming can benefit tremendously through the use of structured techniques.

---

[1]In high-level languages, programs are composed of functions or procedures.

Progressing toward our goal—producing good assembly language programs, the example problems are solved using three methods:

- □ Flowcharts
- □ Pseudo code
- □ Assembly language

Solving programming problems in assembly language is, of course, our terminal objective; however, flowcharts and pseudo code are useful tools for the initial stages. Both of these are "visual" tools, facilitating the formulation and understanding of the problem. They allow a problem to be described in terms of "what must be done" rather than "how it is to be done." The solution can often be expressed in flowcharts or pseudo code in machine-independent terms, without considering the intricacies of the target machine's instruction set.

It is not likely that both pseudo code and flowcharts are used. The preferred choice is largely a matter of personal style. The most common flowcharting symbols are shown in Figure 8–1.

Pseudo code is just what the name suggests: "sort of" a computer language. The idea has been used informally in the past as a convenient way to sketch out solutions to

FIGURE 8–1
Common symbols for flowcharting

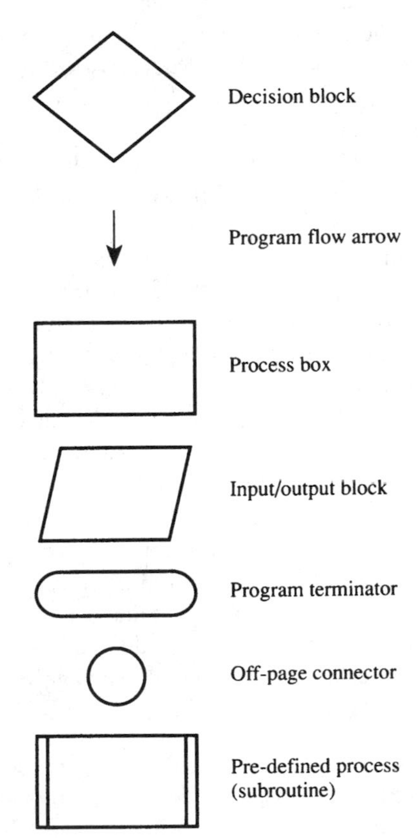

programming problems. As applied here, pseudo code mimics the syntax of Pascal or C in its notation of structure, yet at the same time it encourages the use of natural language in describing actions. Thus, statements such as

```
[get a character from the keyboard]
```

may appear in pseudo code. The benefit in using pseudo code lies in the strict adherence to structure in combination with informal language. Thus we may see

```
IF [condition is true]
   THEN [do statement 1]
   ELSE BEGIN
      [do statement 2]
      [do statement 3]
   END
```

or

```
IF [the temperature is less than 20 degrees Celsius]
   THEN [wear a jacket]
   ELSE BEGIN
      [wear a short sleeve shirt]
      [bring sunglasses]
   END
```

The use of keywords, indentation, and order are essential for the effective use of pseudo code. Our goal is to clearly demonstrate the solution to a programming problem using flowcharts and/or pseudo code, such that the translation to assembly language is easier than direct coding in assembly language. The final product is also easier to read, debug, and maintain. Pseudo code will be defined formally in a later section.

8.2 ADVANTAGES AND DISADVANTAGES OF STRUCTURED PROGRAMMING

The advantages of adopting a structured approach to programming are numerous. These include the following:

- The sequence of operations is simple to trace, thus facilitating debugging.
- There are a finite number of structures with standardized terminology.
- Structures lend themselves easily to building subroutines.
- The set of structures is complete; that is, all programs can be written using three structures.
- Structures are self-documenting and, therefore, easy to read.
- Structures are easy to describe in flowcharts, syntax diagrams, pseudo code, and so on.
- Structured programming results in increased programmer productivity—programs can be written faster.

However, some tradeoffs occur. Structured programming has disadvantages, such as the following:

- Only a few high-level languages (Pascal, C, PL/M) accept the structures directly; others require an extra stage of translation.
- Structured programs may execute slower and require more memory than the unstructured equivalent.
- Some problems (a minority) are more difficult to solve using only the three structures rather than using a brute-force "spaghetti" approach.
- Nested structures can be difficult to follow.

8.3 THE THREE STRUCTURES

All programming problems can be solved using three structures:

- Statements
- Loops
- Choice

The completeness of the three structures seems unlikely; but, with the addition of nesting (structures within structures), it is easily demonstrated that any programming problem can be solved using only three structures. Let's examine each in detail.

8.3.1 Statements

Statements provide the basic mechanism to do something. Possibilities include the simple assigning of a value to a variable, such as

```
[count = 0]
```

or a call to a subroutine, such as

```
PRINT_STRING("Select Option:")
```

Anywhere a single statement can be used, a group of statements, or a **statement block,** can be used. This is accomplished in pseudo code by enclosing the statements between the keywords BEGIN and END as follows:

```
BEGIN
        [statement 1]
        [statement 2]
        [statement 3]
END
```

Note that the statements within the statement block are indented from the BEGIN and END keywords. This is an important feature of structured programming.

8.3.2 The Loop Structure

The second of the basic structures is the "loop," used to repeatedly perform an operation. Adding a series of numbers or searching a list for a value are two examples of program-

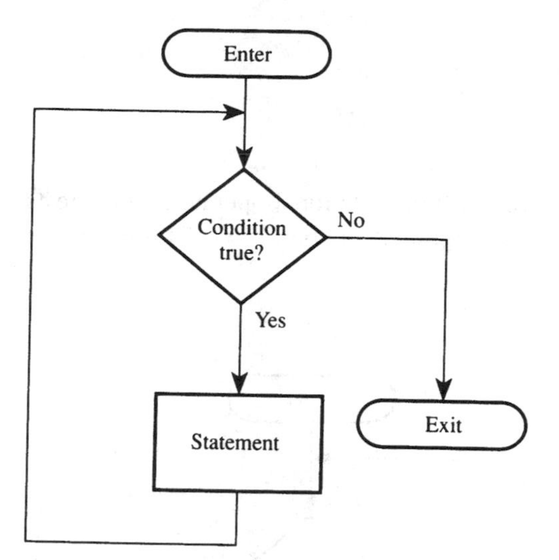

FIGURE 8–2
Flowchart for the WHILE/DO structure

ming problems that require loops. The term "iteration" is also used in this context. Although there are several possible forms of loops, only two are necessary, WHILE/DO and REPEAT/UNTIL.

8.3.2.1 The WHILE/DO Statement

The WHILE/DO statement provides the easiest means for implementing loops. It is called a "statement," since it is treated like a statement—as a single unit with a single entry point (the beginning) and a single exit point (the ending). The pseudo code format is

```
WHILE [condition] DO
      [statement]
```

The "condition" is a "relational expression" which evaluates to "true" or "false." If the condition is true, the statement (or statement block) following the "DO" is executed and then the condition is reevaluated. This is repeated until the relational expression yields a false response; this causes "statement" to be skipped with program execution continuing at the next statement. The WHILE/DO structure is shown in the flowchart in Figure 8–2.

Example 8–1: WHILE/DO Structure

Illustrate a WHILE/DO structure such that while the 8051 carry flag is set, a statement is executed.

Solution.
Pseudo Code:

```
WHILE [c == 1] DO
       [statement]
```

(*Note:* The double equal sign is used to distinguish the relational operator, which tests for equality, from the assignment operator, the single equal sign (see 8.4 Pseudo Code Syntax).

Flowchart:

FIGURE 8–3
Flowchart for example 8–1

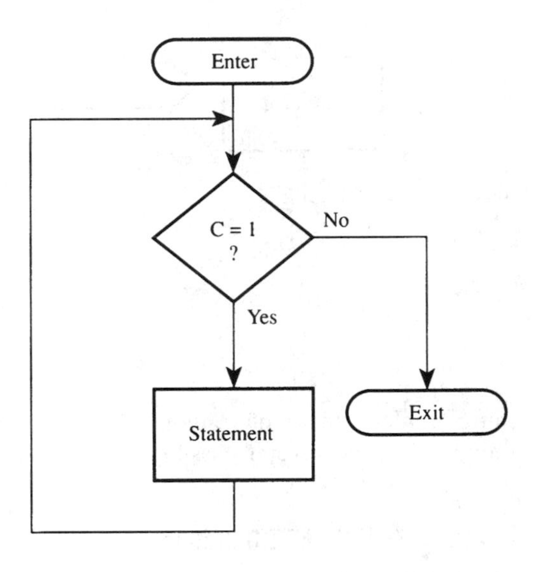

8051 Code:

```
ENTER:        JNC EXIT
STATEMENT:    (statement)
              JMP ENTER
EXIT:         (continue)
```

As a general rule, the actions within the statement block must affect at least one variable in the relational expression, otherwise a bug in the form of an infinite loop results.

Example 8–2: SUM Subroutine

Write an 8051 subroutine called SUM to calculate the sum of a series of numbers. Parameters passed to the subroutine include the length of the series in R7 and the starting address of the series in R0. (Assume the series is in 8051 internal memory.) Return with the sum in the accumulator.

Solution.
Pseudo Code:

```
[sum = 0]
WHILE [length > 0] DO BEGIN
            [sum = sum + @pointer]
            [increment pointer]
            [decrement length]
END
```

Flowchart:

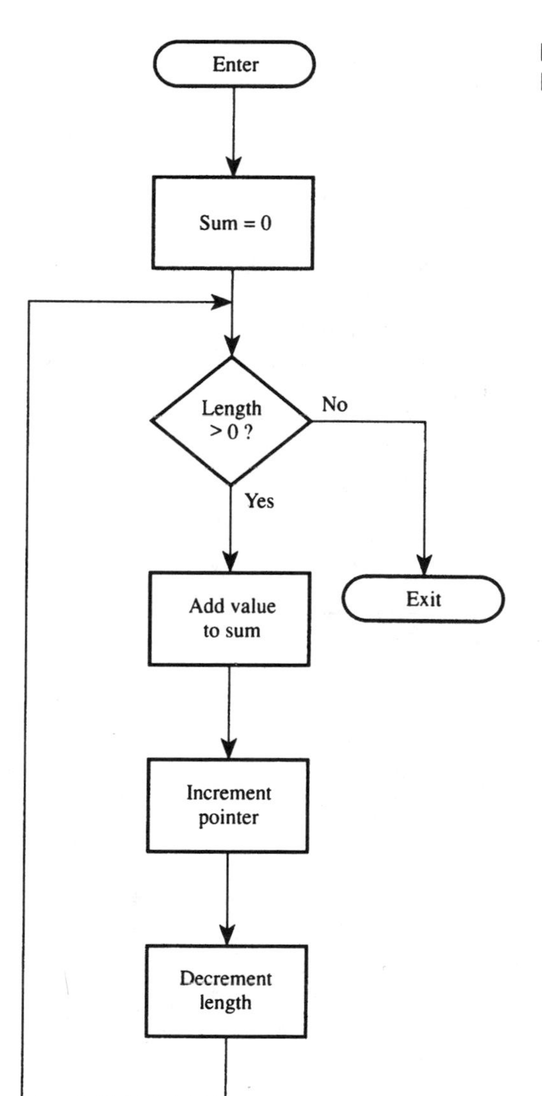

FIGURE 8–4
Flowchart for example 8–2

8051 Code: (closely structured; 13 bytes)

```
SUM:        CLR  A
LOOP:       CJNE R7,#0,STATEMENT
            JMP  EXIT
STATEMENT:  ADD  A,@R0
            INC  R0
            DEC  R7
            JMP  LOOP
EXIT:       RET
```

(loosely structured; 9 bytes)

```
SUM:        CLR  A
            INC  R7
MORE:       DJNZ R7,SKIP
            RET
SKIP:       ADD  A,@R0
            INC  R0
            SJMP MORE
```

Notice above that the loosely structured solution is shorter (and faster) than the closely structured solution. Experienced programmers will, no doubt, code simple examples such as this intuitively and follow a loose structure. Novice programmers, however, can benefit by solving problems clearly in pseudo code first, and then progress to assembly language while following the pseudo code structure.

Example 8–3: WHILE/DO Structure

Illustrate a WHILE/DO structure using the following compound condition: the accumulator not equal to carriage return (0DH) and R7 not equal to 0.

Solution.
Pseudo Code:

```
WHILE [ACC != CR AND R7 != 0] DO
        [statement]
```

Flowchart:

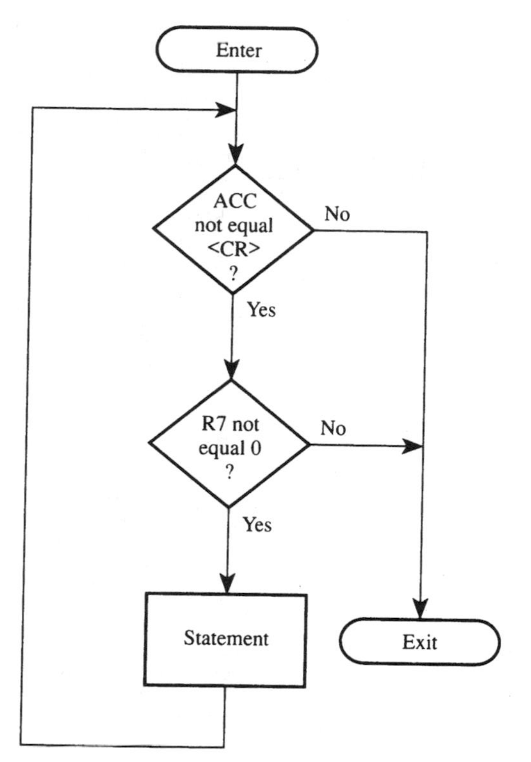

FIGURE 8–5
Flowchart for example 8–3

8051 Code:

```
ENTER:       CJNE A,#0DH,SKIP
             JMP  EXIT
SKIP:        CJNE R7,#0,STATEMENT
             JMP  EXIT
STATEMENT:   (one or more statements)
             .
             .
             .
             JMP  ENTER
EXIT:        (continue)
```

8.3.2.2 The REPEAT/UNTIL Statement

Similar to the WHILE/DO statement is the REPEAT/UNTIL statement, which is useful when the "repeat statement" must be performed at least once. WHILE/DO statements test the condition first, thus the statement might not execute at all.

Pseudo Code:

```
REPEAT [statement]
UNTIL [condition]
```

Flowchart:

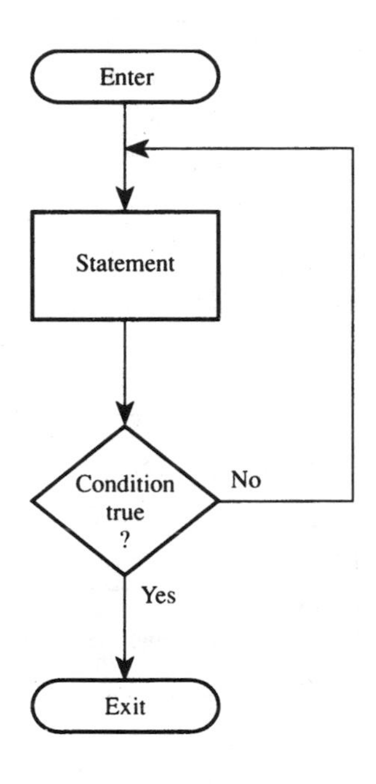

FIGURE 8–6
Flowchart for REPEAT/UNTIL structure

Example 8–4: Search Subroutine

Write an 8051 subroutine to search a null-terminated string pointed at by R0 and determine if the letter Z is in the string. Return with ACC = Z if it is in the string, or ACC = 0 otherwise.

Solution.
Pseudo Code:

```
REPEAT
       [ACC = @pointer]
       [increment pointer]
Until [ACC == 'Z' or ACC == 0]
```

Flowchart:

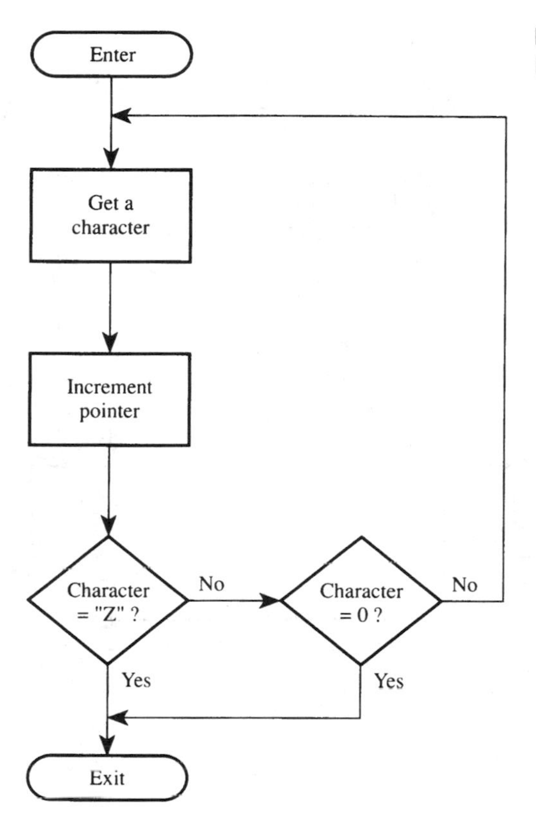

FIGURE 8–7
Flowchart for example 8–4

8051 Code:

```
STATEMENT:  MOV  A,@R0
            INC  R0
            JZ   EXIT
            CJNE A,#'Z',STATEMENT
EXIT:       RET
```

8.3.3 The Choice Structure

The third basic structure is that of "choice"—the programmer's "fork in the road." The two most common arrangements are the IF/THEN/ELSE statement and the CASE statement.

8.3.3.1 The IF/THEN/ELSE Statement

The IF/THEN/ELSE statement is used when one of two statements (or statement blocks) must be chosen, depending on a condition. The ELSE part of the statement is optional.

Pseudo Code:

```
IF [condition]
            THEN [statement 1]
            ELSE [statement 2]
```

Flowchart:

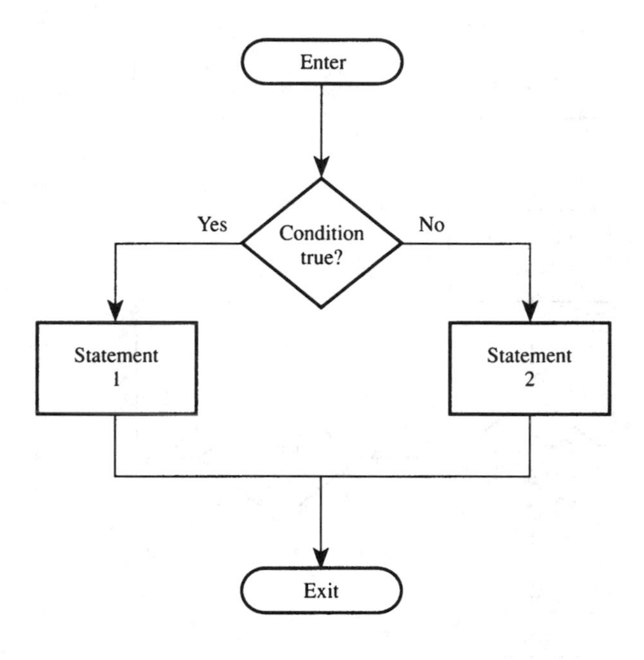

FIGURE 8–8
Flowchart for the IF/THEN/ELSE structure

Example 8–5: Character Test

Write a sequence of instructions to input and test a character from the serial port. If the character is displayable ASCII (i.e., in the range 20H to 7EH), echo it as is, otherwise echo a period (.).

Solution.
Pseudo Code:

```
[input character]
IF [character == graphic]
            THEN [echo character]
            ELSE [echo '.']
```

Flowchart:

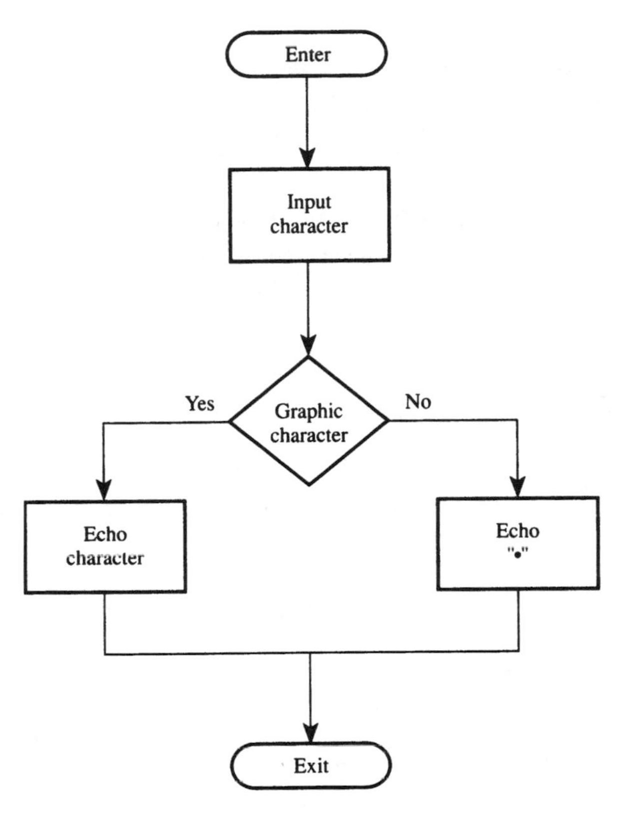

FIGURE 8–9
Flowchart for example 8–5

8051 Code: (closely structured; 14 bytes)

```
ENTER:      ACALL   INCH
            ACALL   ISGRPH
            JNC     STMENT2
STMENT1:    ACALL   OUTCHR
            JMP     EXIT
STMENT2:    MOV     A,#'.'
            ACALL   OUTCHR
EXIT:       (continue)
```

(loosely structured; 10 bytes)

```
            ACALL   INCH
            ACALL   ISGRPH
            JC      SKIP
            MOV     A,#'.'
SKIP:       ACALL   OUTCHR
            (continue)
```

Modify the structure to repeat indefinitely:

```
WHILE [1] DO BEGIN
            [input character]
            IF [character == graphic]
                 THEN [echo character]
                 ELSE [echo '.']
END
```

8.3.3.2 The CASE Statement

The CASE statement is a handy variation of the IF/THEN/ELSE statement. It is used when one statement from many must be chosen as determined by a value.
Pseudo Code:

```
CASE [expression] OF
            0: [statement 0]
            1: [statement 1]
            2: [statement 2]
            .
            .
            .
            n: [statement n]
            [default statement]
END_CASE
```

Flowchart:

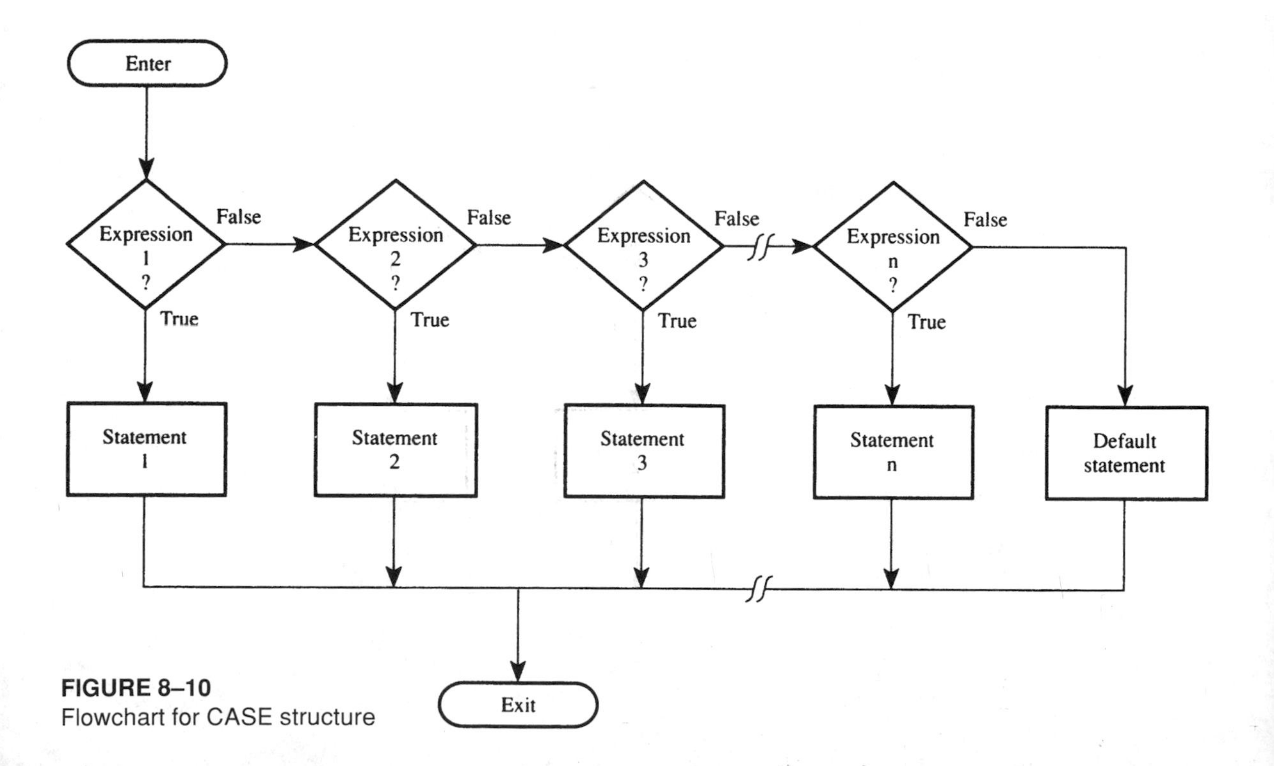

FIGURE 8–10
Flowchart for CASE structure

Example 8–6: User Response

A menu-driven program requires a user response of 0, 1, 2, or 3 to select one of four possible actions. Write a sequence of instructions to input a character from the keyboard and jump to ACT0, ACT1, ACT2, or ACT3, depending on the user response. Omit error checking.

Solution.
Pseudo Code:

```
[input a character]
CASE [character] OF
            '0': [statement 0]
            '1': [statement 1]
            '2': [statement 2]
            '3': [statement 3]
END_CASE
```

Flowchart:

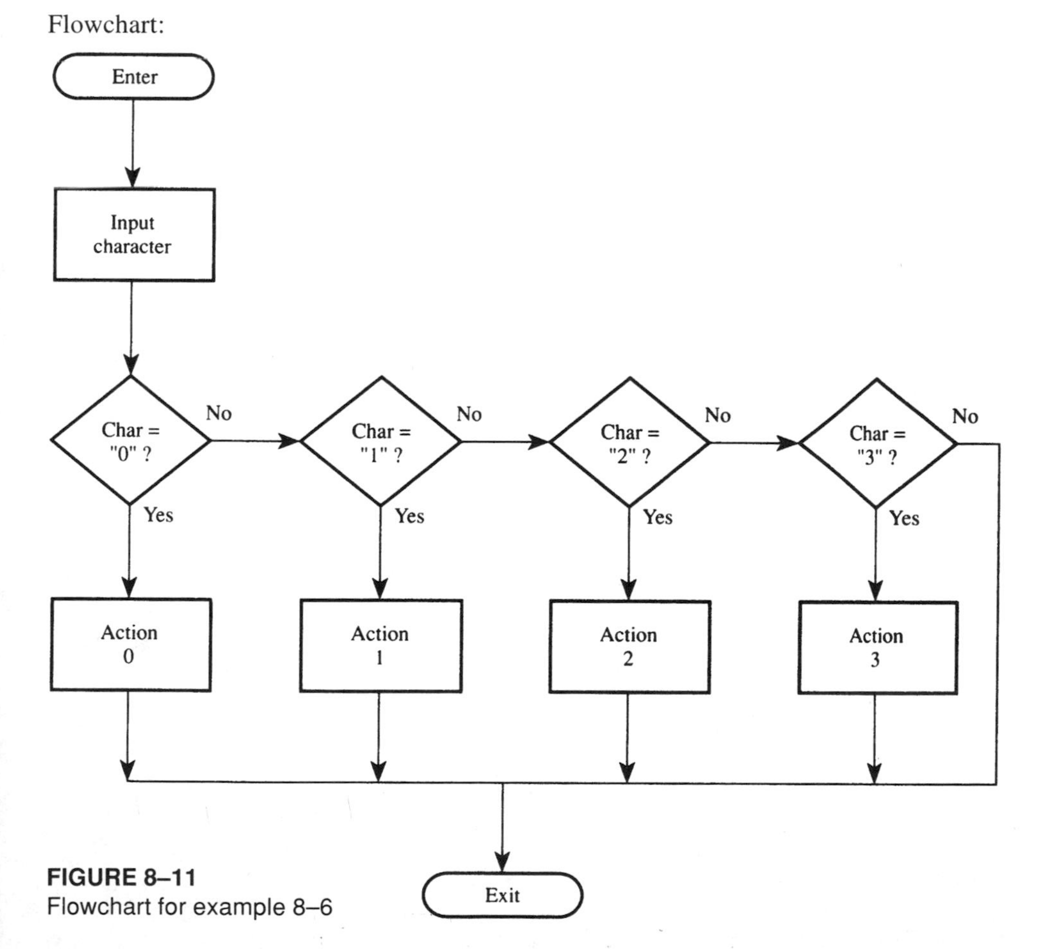

FIGURE 8–11
Flowchart for example 8–6

8051 Code: (closely structured)

```
              CALL INCH
              CJNE A,#'0',SKIP1
ACT0:         .
              .
              .
              JMP  EXIT
SKIP1:        CJNE A,#'1',SKIP2
ACT1:         .
              .
              .
              JMP  EXIT
SKIP2:        CJNE A,#'2',SKIP3
ACT2:         .
              .
              .
SKIP3:        CJNE A,#'3',EXIT
ACT3:         .
              .
              .
EXIT:         (continue)
```

(loosely structured)

```
              CALL INCH                    ;REDUCE to 2 BITS
              ANL  A,#3                    ;WORD OFFSET
              RL   A
              MOV  DPTR,#TABLE
              JMP  @A+DPTR
TABLE:        AJMP ACT0
              AJMP ACT1
              AJMP ACT2
ACT3:         .
              .
              .
              JMP  EXIT
ACT0:         .
              .
              .
              JMP  EXIT
ACT1:         .
              .
              .
              JMP  EXIT
ACT2:         .
              .
              .
EXIT:         (continue)
```

8.3.3.3 The GOTO Statement

GOTO statements can always be avoided using the structures just presented. Sometimes a GOTO statement provides an easy method of terminating a structure when errors occur; however, exercise extreme caution. When the program is coded in assembly language, GOTO statements usually become unconditional jump instructions. A problem will arise, for example, if a subroutine is entered in the usual way (a CALL subroutine instruction), but is exited using a jump instruction rather than a return from subroutine instruction. The return address will be left on the stack, and eventually a stack overflow will occur.

8.4 PSEUDO CODE SYNTAX

Since pseudo code is similar to a high-level language such as Pascal or C, it is worthwhile defining it somewhat more formally, so that, for example, a pseudo code program can be written by one programmer and converted to assembly language by another programmer.

We should acknowledge too that pseudo code is not always the best approach for designing programs. While it offers the advantage of easy construction on a word processor (with subsequent modifications), it suffers from a disadvantage common to other programming languages: pseudo code programs are written line-by-line, so parallel operations are not immediately obvious. Using flowcharts, on the other hand, parallel operations can be placed physically adjacent to one another, thus improving the conceptual model (see Figure 8–10).

Before presenting a formal syntax, the following tips are offered to enhance the power of solving programming problems using pseudo code.

- Use descriptive language for statements.
- Avoid machine dependencies in statements.
- Enclose conditions and statements in brackets: [].
- Begin all subroutines with their names followed by a set of parentheses: (). Parameters passed to subroutines are entered (by name or by value) within the parentheses.
- End all subroutines with RETURN followed by parentheses. Return values are entered within the parentheses.

  Examples of subroutines:

```
INCHAR()          OUTCHR(char)      STRLEN(pointer)
  [statement]      [statement         [statement]
  . . .            . . .              . . .
RETURN(char)      RETURN()          RETURN(length)
```

- Use lowercase text except for reserved words and subroutine names.
- Indent all statements from the structure entry and exit points. When a LOOP or CHOOSE structure is started, the statements within the structure appear at the next level of indentation.
- Use the commercial at sign (@) for indirect addressing.

The following is a suggested syntax for pseudo code.

Reserved Words:

```
BEGIN      END
REPEAT     UNTIL
WHILE      DO
IF         THEN      ELSE
CASE       OF
RETURN
```

Arithmetic Operators:

```
+       addition
−       subtraction
*       multiplication
/       division
%       modulus (remainder after division)
```

Related Operators: (result is true or false)

```
==      true if values equal to each other
!=      true if values not equal
<       true if first value less than second
<=      true if first value <= second value
>       true if first value > second value
>=      true if first value >= second value
&&      true if both values are true
||      true if either value is true
```

Bitwise Logical Operators:

```
&       logical AND
|       logical OR
∧       logical exclusive OR
~       logical NOT (one's complement)
>>      logical shift right
<<      logical shift left
```

Assignment Operators:

```
=       set equal to
op =    assignment operation shorthand where "op" is one
        of
        + − * / % << >> & ∧|
        e.g.: j + = 4 is equivalent to j = j + 4
```

Precedence Operation:

```
( )
```

Indirect Address:

```
@
```

Operator Precedence:

```
()
~       @
*       /       %
+       -
<<      >>
<       <=      >       >=
==      !=
&
^
|
&&
||
=       +=      -=      *=      etc.
```

Note 1. Do not confuse relational operators with bitwise logical operators. Bitwise logical operators are generally used in assignment statements such as

```
[lower_nibble = byte & 0FH]
```

whereas relational operators are generally used in conditional expressions such as

```
IF (char != 'Q' && char != 0DH) THEN . . .
```

Note 2. Do not confuse the relational operator "==" with the assignment operator "=." For example, the Boolean expression

```
j == 9
```

is either true or false depending on whether or not j is equal to the value 9, whereas the assignment statement

```
j = 9
```

assigns the value 9 to the variable j.

Structures:

Statement:

```
[do something]
```

Statement Block:

```
BEGIN
        [statement]
        [statement]
        . . .
END
```

WHILE/DO:

```
WHILE [condition] DO
        [statement]
```

REPEAT/UNTIL:

```
REPEAT
        [statement]
UNTIL [condition]
```

IF/THEN/ELSE:

```
IF [condition]
        THEN [statement 1]
        (ELSE [statement 2])
```

CASE/OF:

```
CASE [expression] OF
        1: [statement 1]
        2: [statement 2]
        3: [statement 3]
        .
        .
        .
        n: [statement n]
        [default statement]
END
```

8.5 ASSEMBLY LANGUAGE PROGRAMMING STYLE

It is important to adopt a clear and consistent style in assembly language programming. This is particularly important when working as part of a team, since individuals must be able to read and understand each other's programs.

The assembly language solutions to problems to this point have been deliberately sketchy. For larger programming tasks, however, a more critical approach is required. The following tips are offered to help improve assembly language programming style.

8.5.1 Labels

Use labels that are descriptive of the destination they represent. For example, when branching back to repeatedly perform an operation, use a label such as "LOOP," "BACK," "MORE," etc. When skipping over a few instructions in the program, use a label such as "SKIP" or "AHEAD." When repeatedly checking a status bit, use a label such as "WAIT" or "AGAIN."

The choice of labels is restricted somewhat when using a simple memory-resident or absolute assembler. These assemblers treat the entire program as a unit, thus limiting the use of common labels. Several techniques circumvent this problem. Common labels can be sequentially numbered, such as SKIP1, SKIP2, SKIP3, etc.; or perhaps within subroutines all labels can use the name of the subroutine followed by a number, such as SEND, SEND2, SEND3, etc. There is an obvious loss of clarity here, since the labels SEND2 and SEND3 are not likely to reflect the skipping or looping actions taking place.

More sophisticated assemblers, such as ASM51, allow each subroutine (or a common group of subroutines) to exist as a separate file that is assembled independent of the main program. The main program is also assembled on its own and then combined with the subroutines using a linking and locating program that, among other things, resolves external references between the files. This type of assembler, usually called a "relocatable" assembler, allows the same label to appear in different files.

8.5.2 Comments

The use of comments cannot be overemphasized, particularly in assembly language programming, which is inherently abstract. All lines of code, except those with truly obvious actions, should include a comment.

Conditional jump instructions are effectively commented using a question similar to the flowchart question for a similar operation. The "yes" and "no" answers to the question should appear in comments at the lines representing the "jump" and "no jump" actions. For example, in the INLINE subroutine below, note the style of comments used to test for the carriage return <CR>code.

```
;****************************************************************
;
;INLINE      INPUT LINE OF CHARACTERS
;            LINE MUST END WITH <CR>
;            MAXIMUM LENGTH 31 CHARACTERS INCLUDE <CR>
;
;ENTER:      NO CONDITIONS
;EXIT:       ASCII CODES IN INTERNAL DATA RAM
;            0 STORED AT END OF LINE
;USES        INCHAR, OUTCHR
;
;****************************************************************

;
INLINE:      PUSH   00H                ;SAVE R0 ON STACK
             PUSH   07H                ;SAVE R7 ON STACK
             PUSH   ACC                ;SAVE ACCUMULATOR ON STACK
             MOV    R0,#60H            ;SET UP BUFFER AT 60H
             MOV    R7,#31             ;MAXIMUM LENGTH OF LINE
STMENT:      ACALL  INCHAR             ;INPUT A CHARACTER
             ACALL  OUTCHR             ;ECHO TO CONSOLE
             MOV    @R0,A              ;STORE IN BUFFER
             INC    R0                 ;INCREMENT BUFFER POINTER
             DEC    R7                 ;DECREMENT LENGTH COUNTER
             CJNE   A,#0DH,SKIP        ;IS CHARACTER = <CR>?
             SJMP   EXIT               ;YES: EXIT
SKIP:        CJNE   R7,#0,STMENT       ;NO: GET ANOTHER CHARACTER
EXIT:        MOV    @R0,#0
             POP    ACC                ;RETRIEVE REGISTERS FROM
                                       ; STACK
             POP    07H
             POP    00H
             RET
```

8.5.3 Comment Blocks

Comment lines are essential at the beginning of each subroutine. Since subroutines perform well-defined tasks commonly needed throughout a program, they should be general-purpose and well documented. Each subroutine is preceded by a **comment block,** a series of comment lines that explicitly state

- The name of the subroutine
- The operation performed
- Entry conditions
- Exit conditions
- Name of other subroutines used (if any)
- Name of registers affected by the subroutine (if any)

The INLINE subroutine above is a good example of a well-commented subroutine.

8.5.4 Saving Registers on the Stack

As applications grow in size and complexity, new subroutines are often written that build upon and use existing subroutines. Thus, subroutines are calling other subroutines which in turn call other subroutines, and so on. These are called "nested subroutines." There is no danger in nesting subroutines so long as the stack has enough room to hold the return addresses. This is not a problem, since nesting beyond several levels is rare.

A potential problem, however, lies in the use of registers within subroutines. As the hierarchy of subroutines grows, it becomes more and more difficult to keep track of what registers are affected by subroutines. A solid programming practice, then, is to save registers on the stack that are altered by a subroutine, and then restore them at the end of the subroutine. Note that the INLINE subroutine shown above saves and retrieves R0, R7, and the accumulator using the stack. When INLINE returns to the calling subroutine, these registers contain the same value as when INLINE was called.

8.5.5 The Use of Equates

Defining constants with equate statements makes programs easier to read and maintain. Equates appear at the beginning of a program to define constants such as carriage return (<CR>) and line feed (<LF>), or addresses of registers inside peripheral ICs such as STATUS or CONTROL.

The constant can be used throughout the program by substituting the equated symbol for the value. When the program is assembled, the value is substituted for the symbol. A generous use of equates makes a program more maintainable, as well as more readable. If a constant must be changed, only one line needs changing—the line where the symbol is equated. When the program is reassembled, the new value is automatically substituted wherever the symbol is used.

8.5.6 The Use of Subroutines

As programs grow in size, it is important to "divide and conquer;" that is, subdivide large and complex operations into small and simple operations. These small and simple operations are programmed as subroutines. Subroutines are hierarchical in that simple subroutines can be used by more complex subroutines, and so on.

Flowcharts:

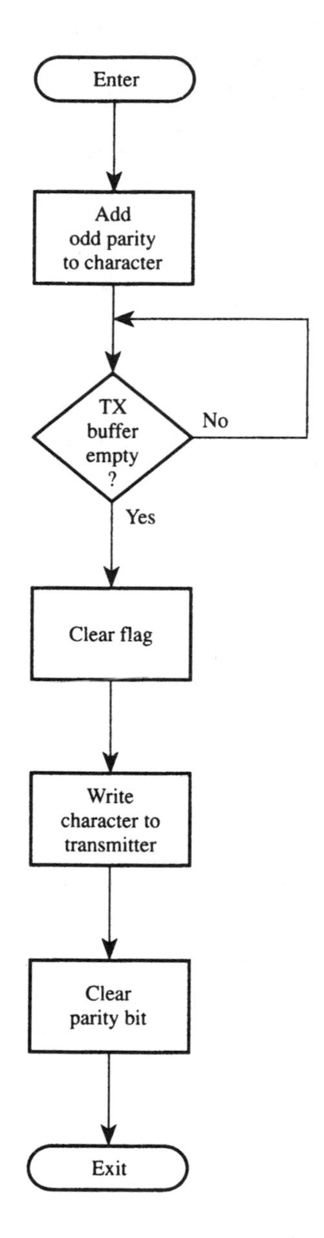

FIGURE 8–12
Flowchart for OUTCHR subroutine

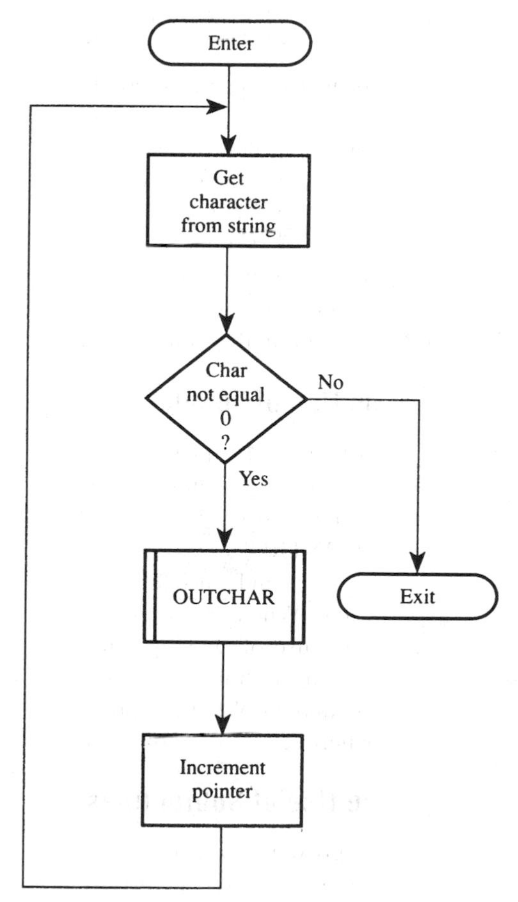

FIGURE 8–13
Flowchart for OUTSTR subroutine

A flowchart references a subroutine using the "predefined process" box. (See Figure 8–1.) The use of this symbol indicates that another flowchart elsewhere describes the details of the operation.

Subroutines are constructed in pseudo code as complete sections of code, beginning with their names and parentheses. Within the parentheses are the names or values of parameters passed to the subroutine (if any). Each subroutine ends with the keyword RETURN followed by parentheses containing the name or value of parameters returned by the subroutine (if any).

Perhaps the simplest example of subroutine hierarchy is the output string (OUTSTR) and output character (OUTCHR) subroutines. The OUTSTR subroutine (a high-level routine) calls the OUTCHR subroutine (a low-level routine).

The flowcharts, pseudo code, and 8051 assembly language solutions are shown below.

Pseudo Code:

```
OUTCHR(char)
   [put odd parity in bit 7]
   REPEAT [test transmit buffer]
        UNTIL [buffer empty]
   [clear transmit buffer empty flag]
   [move char to transmit buffer]
   [clear parity bit]
RETURN()

OUTSTR(pointer)
    WHILE [(char = @pointer) != 0] BEGIN
         OUTCHR(char)
         [increment pointer]
    END
RETURN()
```

Software:

```
OUTCHR:     MOV  C,P           ;PUT PARITY BIT IN C FLAG
            CPL  C             ;CHANGE TO ODD PARITY
            MOV  ACC.7,C       ;ADD TO CHARACTER
AGAIN:      JNB  TI,AGAIN      ;TX EMPTY?
            CLR  TI            ;YES: CLEAR FLAG AND
            MOV  SBUF,A        ; SEND CHARACTER
            CLR  ACC.7         ;STRIP OFF PARITY BIT AND
            RET                ; RETURN

OUTSTR:     MOV  A,@DPTR       ;GET CHARACTER
            JZ   EXIT          ;IF 0, DONE
            CALL OUTCHR        ;OTHERWISE SEND IT
            INC  DPTR          ;INCREMENT POINTER
            SJMP OUTSTR        ; AND GET NEXT CHARACTER
EXIT:       RET
```

8.5.7 Program Organization

Although programs are often written piecemeal (i.e., subroutines are written separately from the main program), all programs should be consistent in their final organization. In general, the sections of a program are ordered as follows:

- Equates
- Initialization instructions
- Main body of program
- Subroutines
- Data constant definitions (DB and DW)
- RAM data locations defined using the DS directive

All but the last item above are called the "code segment" and the RAM data locations are called the "data segment." Code and data segments are traditionally separate, since code is often destined for ROM or EPROM whereas RAM data are always destined for RAM. Note that data constants and strings defined using the DB or DW directives are part of the code segment (not the data segment), since these data are unchanging constants and, therefore, are part of the program.

8.6 SUMMARY

This chapter has introduced structure programming techniques through flowcharting and pseudo code. Suggestions were offered on designing and presenting programs to enhance their readability. In the next chapter, some of the tools and techniques for developing programs are presented.

PROBLEMS

1. Illustrate a WHILE/DO structure such that while the accumulator is less than or equal to 7EH, a statement block is executed.
2. Illustrate a WHILE/DO structure using the following compound condition: (accumulator greater than zero and R7 greater than zero) or the carry flag equal to 1. Treat the values in the accumulator and R7 as unsigned integers.
3. Write a subroutine to find the place of the most significant 1 in the accumulator. Return with "place" in R7. For example, if ACC = 00010000B, return with R7 = 4.
4. Write a subroutine called INLINE to input a line of characters from the console and place them in internal memory starting at location 60H. Maximum line length is 31 characters, including the carriage return. Put 0 at the end-of-line.

9

TOOLS AND TECHNIQUES FOR PROGRAM DEVELOPMENT

9.1 INTRODUCTION

In this chapter, the process of developing microcontroller- or microprocessor-based products is described as it follows a series of steps and utilizes a variety of tools. In progressing from concept to product, numerous steps are involved and numerous tools are used. The most common steps and tools are presented as found in typical design scenarios employing the 8051 microcontroller.

Design is a highly creative activity, and in recognition of this we state at the outset that substantial leeway is required for individuals or development teams. Such autonomy may be difficult to achieve for very large or safety-critical projects, however. Admittedly, in such environments the management of the process and the validation of the results must satisfy a higher order. The present chapter addresses the development of relatively small-scale products, such as controllers for microwave ovens, automobile dashboards, computer peripherals, electronic typewriters, or high-fidelity audio equipment.

The steps required and the tools and techniques available are presented and elaborated on, and examples are given. Developing an understanding of the steps is important, but strict adherence to their sequence is not advocated. It is felt that forcing the development process along ordered, isolated activities is usually overstressed and probably wrong. Later in the chapter we will present an all-in-one development scenario, where the available resources are known and called upon following the instinct of the designer. We begin by examining the steps in the development cycle.

9.2 THE DEVELOPMENT CYCLE

Proceeding from concept to product is usually shown in a flow diagram known as the **development cycle,** similar to that shown in Figure 9–1. The reader may notice that there is nothing particularly "cyclic" about the steps shown. Indeed, the figure shows the ideal and impossible scenario of "no breakdowns." Of course, problems arise. **Debugging** (finding and fixing problems) is needed at every step in the development cycle with cor-

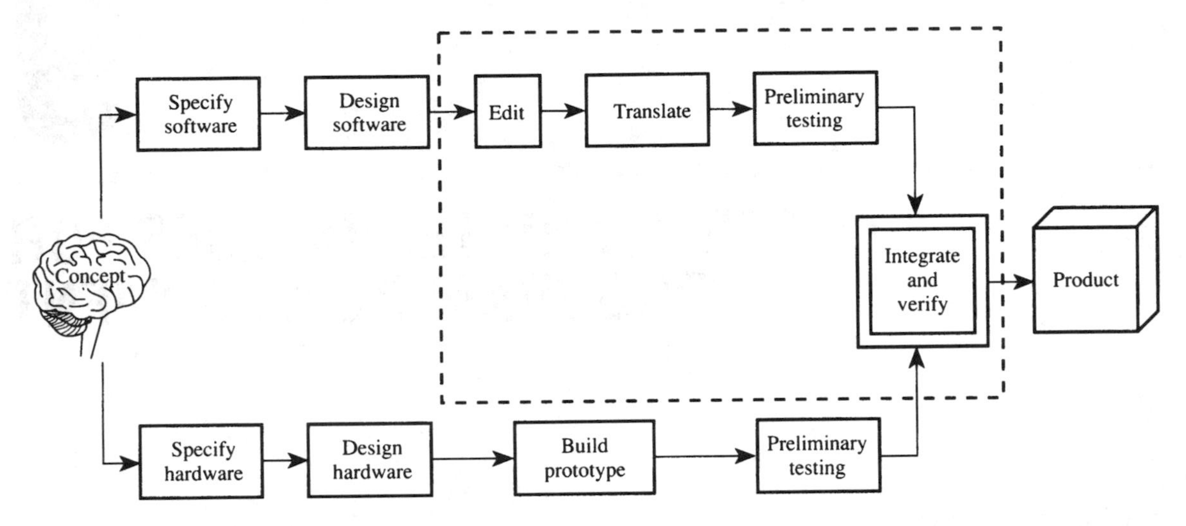

FIGURE 9–1
The development cycle

rections introduced by reengaging in an earlier activity. Depending on the severity of the error, the correction may be trivial or, in the extreme, may return the designer to the concept stage. Thus, there is an implied connection in Figure 9–1 from the output of any step in the development cycle to any earlier step.

The steps along the top path in Figure 9–1 correspond to software development, while those along the bottom correspond to hardware development. The two paths meet at a critical and complicated step called "integration and verification," which leads to acceptance of the design as a "product." Not shown are various steps subsequent to acceptance of the design. These include, for example, manufacturing, testing, distribution, and marketing. The dotted line in Figure 9–1 encompasses the steps of primary concern in this chapter (and book). These will be elaborated in more detail later. But first, we begin by examining the steps in software development.

9.2.1 Software Development

The steps in the top path in Figure 9–1 are discussed in this section, beginning with the specification of the application software.

Specifying Software. Specifying software is the task of explicitly stating what the software will do. This may be approached in several ways. At a superficial level, specifications may first address the user interface; that is, how the user will interact with and control the system. (What effects will result from and be observed for each action taken?) If switches, dials, or audio or visual indicators are employed on the prototype hardware, the explicit purpose and operation of each should be stated.

Formal methods have been devised by computer scientists for specifying software requirements; however, they are not generally used in the design of microcontroller-based applications, which are small in comparison to application software destined for mainframe computers.

Software specifications may also address details of system operation below the user level. For example, a controller for a photocopier may monitor internal conditions necessary for normal or safe operation, such as temperature, current, voltage, or paper movement. These conditions are largely independent of the user interface, but still must be accommodated by software.

Specifications can be modularized by system function with entry and exit conditions defined to allow intermodule communication. The techniques described in the previous chapter for documenting subroutines are a reasonable first step in specifying software.

Interrupt-driven systems require careful planning and have unique characteristics that must be addressed at the specification stage. Activities without time-critical requirements may be placed in the foreground loop or in a round-robin sequence for handling by timed interrupts. Time-critical activities generate high-priority interrupts that take over the system for immediate handling. Software specifications may emphasize execution time on such systems. How long does each subroutine or interrupt service routine (ISR) take to execute? How often is each ISR executed? ISRs that execute asynchronously (in response to an event) may take over the system at any time. It may be necessary to block them in some instances or to preempt (interrupt) them in others. Software specifications for such systems must address priority levels, polling sequences, and the possibility of dynamically reassigning priority levels or polling sequences within ISRs.

Designing Software. Designing the software is a task designers are likely to jump into without a lot of planning. There are two common techniques for designing software prior to coding: flowcharts and pseudo code. These were the topic of Chapter 8.

Editing and Translation. The editing and translation of software occur, at least initially, in a tight cycle. Errors detected by the assembler are quickly corrected by editing the source file and reassembling. Since the assembler has no idea of the purpose of the program and checks only for "grammatical" errors (e.g., missing commas, undefined instructions), the errors detected are **syntax errors.** They are also called **assemble-time errors.**

Preliminary Testing. A **run-time** error will not appear until the program is executed by a simulator or in the target system. These errors may be elusive, requiring careful observation of CPU activity at each stage in the program. A **debugger** is a system program that executes a user program for the purpose of finding run-time errors. The debugger includes features such as executing the program until a certain address (a **breakpoint**) is reached, and **single-stepping** through instructions while displaying CPU registers, status bits, or input/output ports.

9.2.2 Hardware Development

For the most part, this book has not emphasized hardware development. Since the 8051 is a highly integrated device, we have focused on learning the 8051's internal architecture and exploiting its on-chip resources through software. The examples presented thus far have used only simple interfaces to external components.

Specifying Hardware. Specifying the hardware involves assigning quantitative data to system functions. For example, a robotic arm project should be specified in terms of number of articulations, reach, speed, accuracy, torque, power requirements, and so on. Designers are often required to provide a specification sheet analogous to that accompanying an audio amplifier or VCR. Other hardware specifications include physical size and weight, CPU speed, amount and type of memory, memory map assignments, I/O ports, optional features, etc.

Designing Hardware. The conventional method of hardware design, employing a pencil and a logic template, is still widely used, but may be enhanced through computer-aided design (CAD) software. Although many CAD tools are for the mechanical or civil engineering disciplines, some are specifically geared for electronic engineering. The two most common examples are tools for drawing schematic diagrams and tools for laying out printed circuit boards (PCBs). Although these programs have a long learning curve, the results are impressive. Some schematic drawing programs produce files that can be read by PCB programs to automatically generate a layout.

Building the Prototype. There are pathetically few shortcuts for the labors of prototyping. Whether breadboarding a simple interface to a bus or port connector on a single-board computer (SBC), or wire wrapping an entire controller board, the techniques of prototyping are only developed with a great deal of practice. Large companies with large budgets may proceed directly to a printed circuit board format, even for the first iteration of hardware design. Projects undertaken by small companies, students, or hobbyists, however, are more likely to use the traditional wire wrapping method for prototypes.

Preliminary Testing. The first test of hardware is undertaken in the absence of any application software. Step-wise testing is important: there's no point in measuring a clock signal using an oscilloscope before the presence of power-supply voltages has been verified. The following sequence may be followed:

- □ Visual checks
- □ Continuity checks
- □ DC measurements
- □ AC measurements

Visual and continuity checks should occur before power is applied to the board. Continuity checks using an ohmmeter should be conducted from the IC side of the prototype, from IC pin to IC pin. This way, the IC pin-to-socket and socket pin-to-wire connection are both verified. ICs should be removed when power is first applied to the prototype. DC voltages should be verified throughout the board with a voltmeter. Finally, AC measurements are made with the ICs installed to verify clock signals, and so on.

After verifying the connections, voltages, and clock signals, debugging becomes pragmatic: Is the prototype functioning as planned? If not, corrective action may take the designer back to the construction, design, or specification of the hardware.

If the design is a complete system with a CPU, a single wiring error may prevent the CPU from completing its reset sequence: The first instruction after reset may never execute! A powerful debugging trick is to drive the CPU's reset line with a low frequency square wave ($\approx$1 kHz) and observe (with an oscilloscope or logic analyzer) bus activity immediately following reset.

Functional testing of the board may require application software or a monitor program to "work" the board through its motions. It is at this stage that software must assist in completing the development cycle.

9.3 INTEGRATION AND VERIFICATION

The most difficult stage in the development cycle occurs when hardware meets software. Some very subtle bugs that eluded simulation (if undertaken) emerge under real-time execution. The problem is confounded by the need for a full complement of resources: hardware such as the PC development system, target system, power supply, cables, and test equipment; and software such as the monitor program, operating system, terminal emulation program, and so on.

We shall elaborate on the integration and verification step by first expanding the area within the dotted line in Figure 9–1. (See Figure 9–2.)

Figure 9–2 shows utility programs and development tools within circles, user files within squares, and "execution environments" within double-lined squares. The use of an editor to create a source file is straightforward. The translation step (from Figure 9–1) is shown in two stages. An assembler (e.g., ASM51) converts a source file to an object file, and a linker/locator (e.g., RL51) combines one or more relocatable object files into a single absolute object file for execution in a target system or simulator. The assembler and linker/locator also create listing files.

The most common filename suffixes are shown in parentheses for each file type. Although any filename and suffix usually can be provided as an argument, assemblers vary in their choice of default suffixes.

If the program was written originally in a single file following an absolute format, linking and locating are not necessary. In this case, the alternate path in Figure 9–2 shows the assembler generating an absolute object file.

It is also possible (although not emphasized in this book) that high-level languages, such as C or PL/M, are used instead of, or in addition to, assembly language. Translation requires a **cross-compiler** to generate the relocatable object modules for linking and locating.

A **librarian** may also participate, such as Intel's LIB51. Relocatable object modules that are general-purpose and useful for many projects (most likely subroutines) may be stored in "libraries." RL51 receives the library name as an argument and searches the library for the code (subroutines) corresponding to previously declared external symbols that have not been resolved at that point in linking/locating.

9.3.1 Software Simulation

Five execution environments are shown in Figure 9–2. Preliminary testing (see Figure 9–1) proceeds in the absence of the target system. This is shown in Figure 9–2 as software simulation. A **simulator** is a program that executes on the development system and

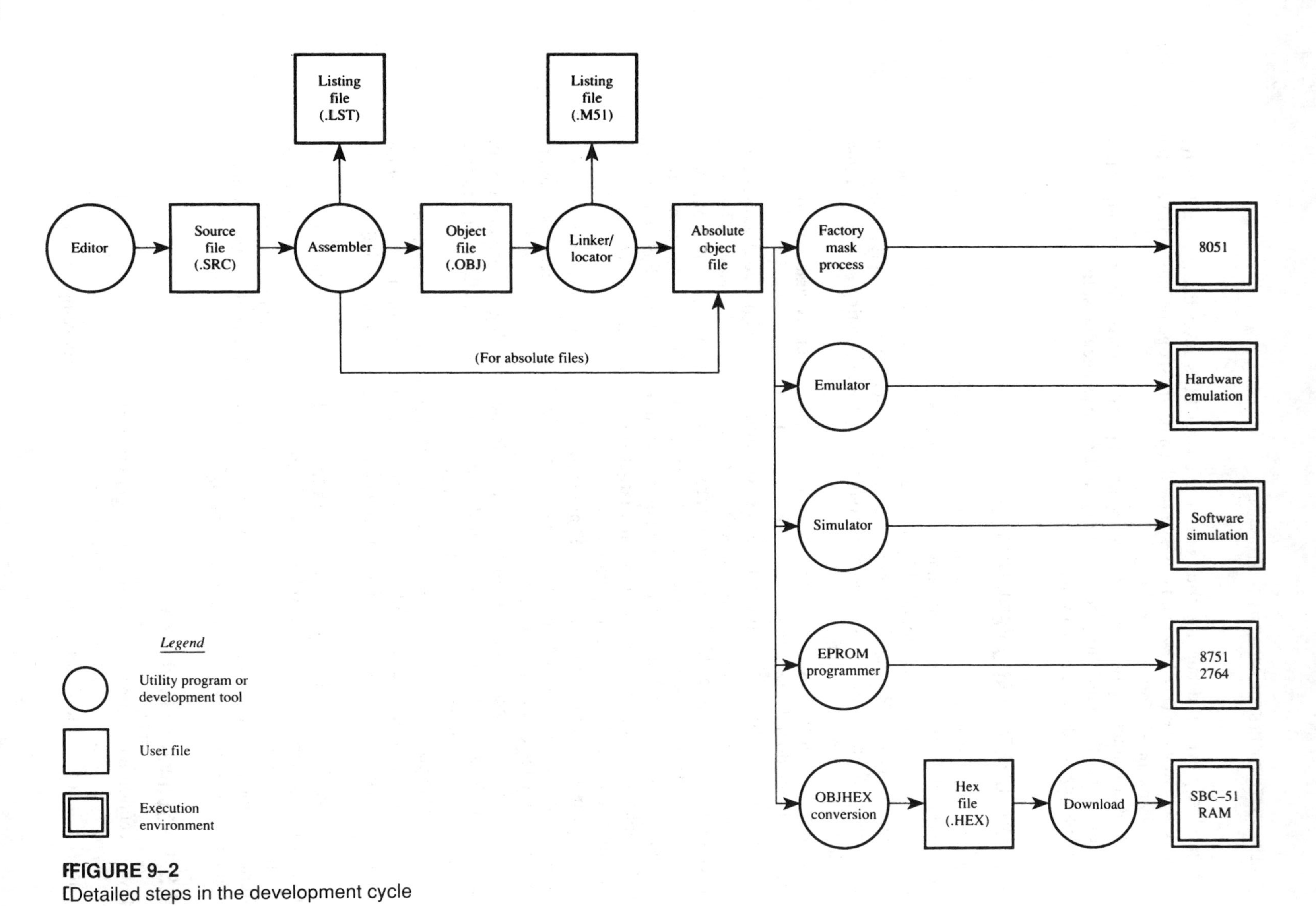

FIGURE 9–2
Detailed steps in the development cycle

imitates the architecture of the target machine. An 8051 simulator, for example, would contain a fictitious (or "simulated") register for each of the special function registers and fictitious memory locations corresponding to the 8051's internal and external memory spaces. Programs are executed in simulation mode with progress presented on the development system's CRT display. Simulators are useful for early testing; however, portions of the application program that directly manipulate hardware must be integrated with the target system for testing.

9.3.2 Hardware Emulation

A direct connection between the development system and the target system is possible through a **hardware emulator** (or **in-circuit emulator**). The emulator contains a processor that replaces the processor IC in the target system. The emulator processor, however, is under the direct control of the development system. This allows software to execute in the environment of the target system without leaving the development system. Commands are available to single-step the software, execute to a breakpoint (or the nth occurrence of a breakpoint), and so on. Furthermore, execution is at full speed, so time-dependent bugs may surface that eluded debugging under simulation.

The main drawback of hardware emulators is cost. PC-hosted units sell in the \$2,000 to \$7,000 (U.S.) range, which is beyond the budget of most hobbyists and stretches the budgets of most colleges or universities (if equipping an entire laboratory, for example). Companies supporting professional development environments, however, will not hesitate to invest in hardware emulators. The benefit in accelerating the product development process easily justifies the cost.

9.3.3 Execution from RAM

An effective and simple scenario for testing software in the target system is possible, even if a hardware emulator is not available. If the target system contains external RAM configured to overlap the external code space (using the method discussed in Chapter 2; see 2.6.4., Overlapping the External Code and Data Spaces), then the absolute object program can be transferred, or "downloaded," from the development system to the target system and executed in the target system.

Intel Hexadecimal Format. As shown in Figure 9–2, a extra stage of translation is required to convert the absolute object file to a standard ASCII format for transmission. Since object files contain binary codes, they cannot be displayed or printed. This weakness is alleviated by splitting each binary byte into two nibbles and converting each nibble to the corresponding hexadecimal ASCII character. For example, the byte 1AH cannot be transmitted to a printer because in ASCII it represents a control character rather than a graphic character. However, the bytes 31H and 41H can be transmitted to a printer because they correspond to graphic or displayable ASCII codes. In fact, these two bytes will print as "1A." (See Appendix F.)

One standard for storing machine language programs in a displayable or printable format is known as "Intel hexadecimal format." An Intel hex file is a series of lines or "hex records" containing the following fields:

Field	Bytes	Description
Record mark	1	":" indicates start-of-record
Record length	2	number of data bytes in record
Load address	4	starting address for data bytes
Record type	2	00 = data record; 01 = end record
Data bytes	0-16	data
Checksum	2	sum of all bytes in record + checksum = 0

These fields are shown in the Intel hexadecimal file in Figure 9–3. Conversion programs are available that receive an absolute object program as input, convert the machine language bytes to Intel hexadecimal format, and generate a hex file as output. Intel's conversion utility is called OH.

9.3.4 Execution from EPROM

Once a satisfactory degree of performance is obtained through execution in RAM (or through in-circuit emulation), the software is burned into EPROM and installed in the system as **firmware.** Two types of EPROMs are identified in Figure 9–2 as examples. The 8751 is the EPROM version of the 8051, and the 2764 is a common, general-purpose EPROM used in many microprocessor- or microcontroller-based products. Systems designed using an 8751 benefit in that Ports 0 and 2 are available for I/O, rather than functioning as the address and data buses. However, 8751s are relatively expensive compared to 2764s ($30 versus $5, for example).

9.3.5 The Factory Mask Process

If a final design is destined for mass production, then a cost-effective alternative to EPROM is a factory mask ROM, such as the 8051. An 8051 is functionally identical to

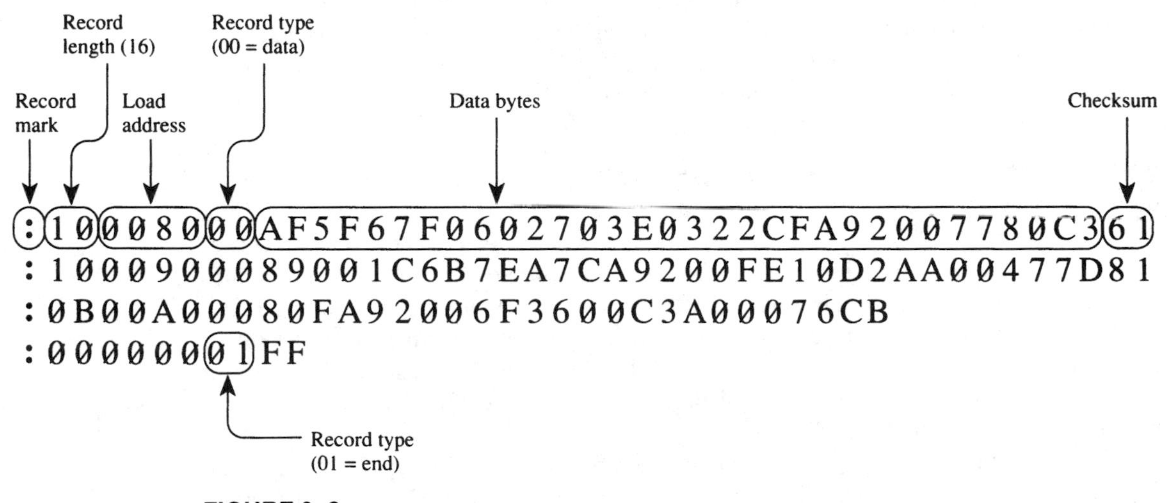

FIGURE 9–3
Intel hexadecimal format

an 8751; however, code memory cannot be changed on an 8051. The data are permanently entered during the IC manufacturing cycle using a "mask"—essentially a photographic plate that passes or masks (i.e., blocks) light during a stage of manufacturing. Connections to memory cells in the 8051 are either made or blocked, thus programming each cell as a 1 or 0.

The choice of using an 8751 versus an 8051 is largely economic. A factory mask device is considerably cheaper than the EPROM device; however, there is a large setup fee to produce the mask and initiate a custom manufacturing cycle. A tradeoff point can be identified to determine the feasibility of each approach. For example, if 8751s sell for $25 and 8051s sell for $5 plus a $5,000 setup fee, then the break-even point is

$$
\begin{aligned}
25\,n &= 5\,n + 5000 \\
20\,n &= 5000 \\
n &= 250 \text{ units}
\end{aligned}
$$

A production run of 250 units or more would justify the use of the 8051 over the 8751.

The situation is more complicated when comparing designs using an 8051 versus an 8031 + 2764, for example. In the latter case, the 8031 + 2764 alternative is much cheaper than an 8751 with on-chip EPROM, so the tradeoff point occurs at much greater quantities. If an 8031 + 2764 sells for, say, $7, then the break-even point is

$$
\begin{aligned}
7\,n &= 5\,n + 5000 \\
2\,n &= 5000 \\
n &= 2500 \text{ units}
\end{aligned}
$$

A production run of 1000 units would not justify use of the 8051—or so it seems. The use of external EPROM means that Ports 0 and 2 are unavailable for I/O. This may be a critical point that prevents the 8031 + 2764 approach. Even if the loss of on-chip I/O is not a concern, other factors enter. The 8031 + 2764 approach requires two ICs instead of one. This complicates manufacturing, testing, maintenance, reliability, procurement, and a host of other seemingly innocent, but nevertheless real, dimensions of product design. Furthermore, the 8031 + 2764 design will be physically larger than the 8051 design. If the final product necessitates a small form factor, then the 8051 may have to be used, regardless of the additional cost.

9.4 COMMANDS AND ENVIRONMENTS

In this section the overall development environment is considered. We present the notion that at any time the designer is working within an "environment" with commands doing the work. The central environment is the operating system on the host system, which is most likely MS-DOS running on a member of the PC family of microcomputers. As suggested in Figure 9–4, some commands return to MS-DOS upon completion, while others evoke a new environment.

Invoking Commands. Commands are either **resident** (e.g., DIR) or **transient** (e.g., FORMAT, DISKCOPY). A resident command is in memory at all times, ready for execution (e.g., DIR). A transient command is an executable disk file that is loaded into memory for execution (e.g., FORMAT).

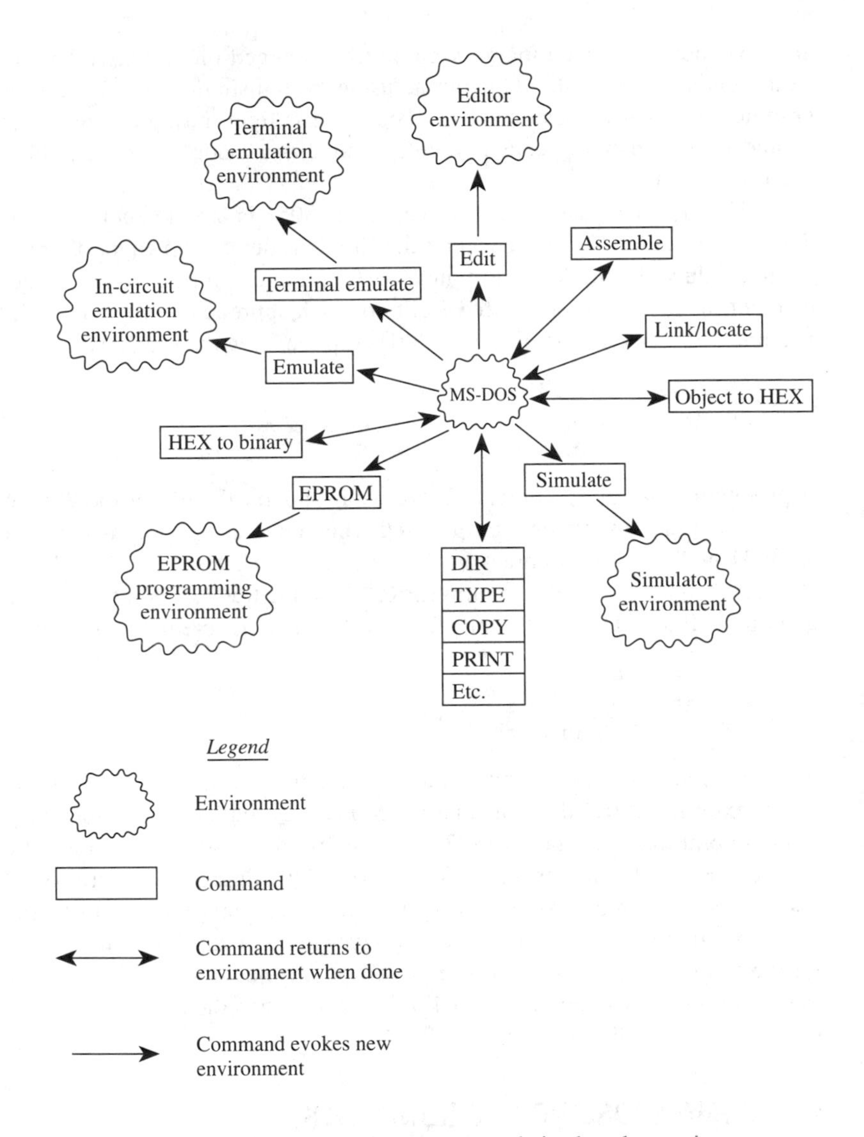

FIGURE 9–4
The development environment

Application programs are similar to transient commands in that they exist as an executable disk file and are invoked from the MS-DOS prompt. However, there are still many possibilities. Commands or applications may be invoked as part of a batch file, by a function key, or from a menu-driven user interface acting as a front-end for MS-DOS.

If command arguments are needed, there are many possibilities again. Although arguments are typically entered on the invocation line following the command, some commands have default values for arguments, or prompt the user for arguments. Unfortunately, there is no standard mechanism, such as the "dialogue box" used in the *Macintosh* interface, to retrieve extra information needed for a command or application.

Some applications, such as editors, "take over" the system and bring the user into a new environment for subsequent activities.

Environments. As evident in Figure 9–4, some software tools such as the simulator, in-circuit emulator, or EPROM programmer evoke their own environment. Learning the nuances of each takes time, due to the great variety of techniques for directing the activities of the environment: cursor keys, function keys, first-letter commands, menu highlighting, default paths, and so on. It is often possible to switch among environments while leaving them active. For example, terminal emulators and editors usually allow switching to DOS momentarily to execute commands. The MS-DOS command EXIT immediately brings the user back the suspended environment.

Methodology. As research in artificial intelligence and cognitive science has discovered, modeling human "problem solving" is a slippery business. Humans appear to approach the elements of a situation in parallel, simultaneously weighing possible actions and proceeding by intuition. The methodology suggested here recognizes this human quality. The steps in the development cycle and the tools and techniques afforded by the development environment should be clearly understood, but the overall process should support substantial freedom.

The basic operation of commands is to "translate," "view," or "evoke" (a new environment). The results of translation should be viewed to verify results. We can take the attitude of not believing the outcome of any translation (assembling, EPROM programming, etc.) and verify everything by viewing results. Tools for viewing are commands such as DIR (Were the expected output files created?), TYPE (What's in the output file?), EDIT, PRINT, and so on.

9.5 SUMMARY

The tools and techniques available for designing microcontroller-based products have been introduced in this chapter. There is no substitute for experience, however. Success in design requires considerable intuition, a valuable commodity that cannot be delivered in a textbook. The age-old expression "trial and error" still rings true as the main technique employed by designers for turning ideas into real products.

PROBLEMS

1. If 8751 EPROMs sell for $30 in any quantity and a mask-programmed 8051 sells for $3 plus a $10,000 setup fee, how many units are necessary to justify use of the 8051 device? What is the savings for projected sales of 3,000 units of the final product if the 8051 is used instead of the 8751?
2. Below is an 8051 program in Intel hex format.

```
:10080000758911 7F007E0575A88AD28FD28D80FEF2
:10081000C28C758C3C758AB0DE087E050FBF09025C
:1008200007F00D28C32048322FB90FC0CFC7AFCAD5E
:0A083000FD0AFD5CFDA6FDC8FDC831
:00000001FF
```

(a) What is the starting address of the program?

(b) What is the length of the program?

(c) What is the last address of the program?

3. The following is a single line from an Intel hex file with an error in the checksum. The incorrect checksum appears in the last two characters as "00". What is the correct checksum?

```
:1008000007589117C007F0575A8FFD28FD28D80FE00
```

4. The contents of an Intel hex file are shown below.

```
:0901000007820765508B880FA2237
:00000001FF
```

Recreate the original source program that this file represents.

10

DESIGN AND INTERFACE EXAMPLES

10.1 INTRODUCTION

Many of the 8051's hardware and software features are brought together in this chapter through several design and interface examples. The first is an 8051 single-board computer—the SBC-51—suitable for learning about the 8051 or developing 8051-based products. The SBC-51 uses a substantial monitor program offering basic commands for system operation and user interaction. The monitor program (MON51) is described in detail in Appendix G.

The interface examples are advanced in comparison to those presented in previous chapters. Each example includes a hardware schematic, a statement of the design objective, a software listing of a program that achieves the design objective, and a general description of the operation of the hardware and software. The software listings are extensively commented and should be consulted for specific details.

10.2 THE SBC-51

Several companies offer 8051 single-board computers similar to that described in this section. Surprisingly, the basic design of an 8051 single-board computer does not vary substantially among the various products offered. Since many features are "on-chip," designing an 8051 single-board computer is straightforward. For the most part, only the basic connections to external memory and the interface to a host computer are required.

A monitor program in EPROM is also required. The most basic system requirements, such as examining and changing memory locations or downloading application programs from a host computer, are needed to get "up and running." The SBC-51 described here works together with a simple monitor program to provide these basic functions.

Figure 10–1 contains the schematic diagram for the SBC-51. The entire design includes only 10 ICs, yet is powerful and flexible enough to support the development of sophisticated 8051-based products. Central to the operation of the SBC-51 is a monitor program that resides in EPROM and communicates with a video display terminal (VDT) connected to the 8051. The monitor program is described in detail in Appendix G.

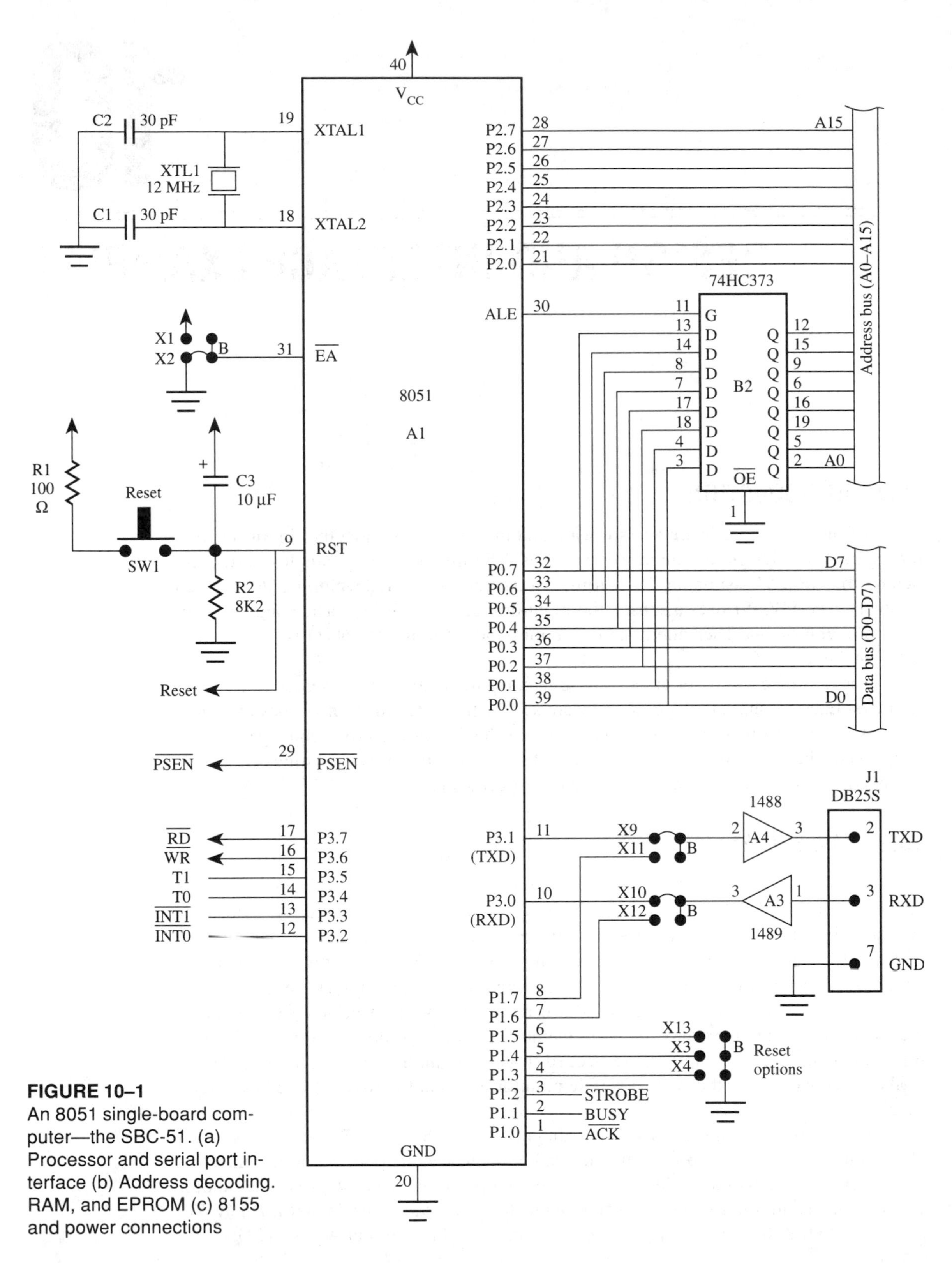

FIGURE 10–1
An 8051 single-board computer—the SBC-51. (a) Processor and serial port interface (b) Address decoding. RAM, and EPROM (c) 8155 and power connections

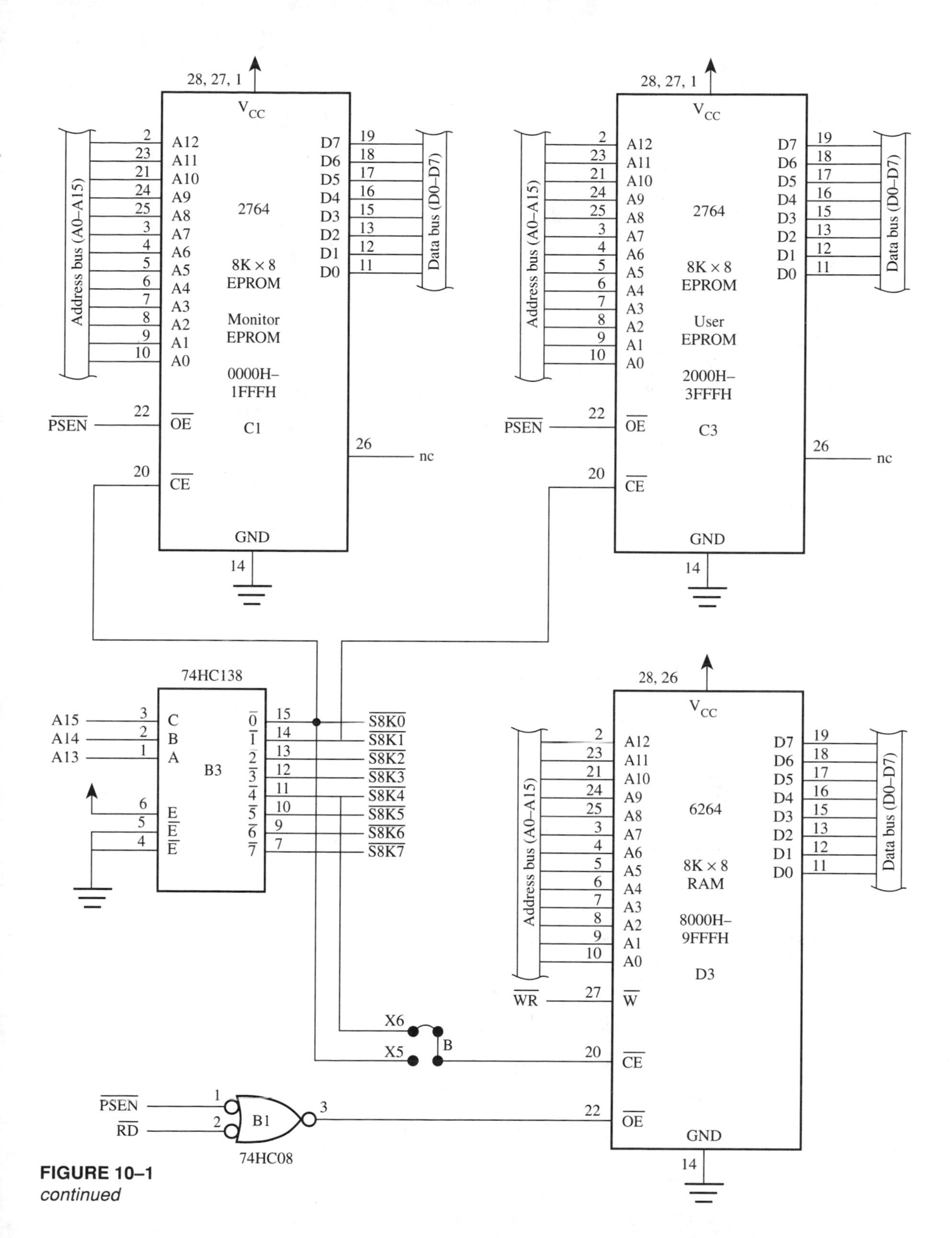

FIGURE 10–1
continued

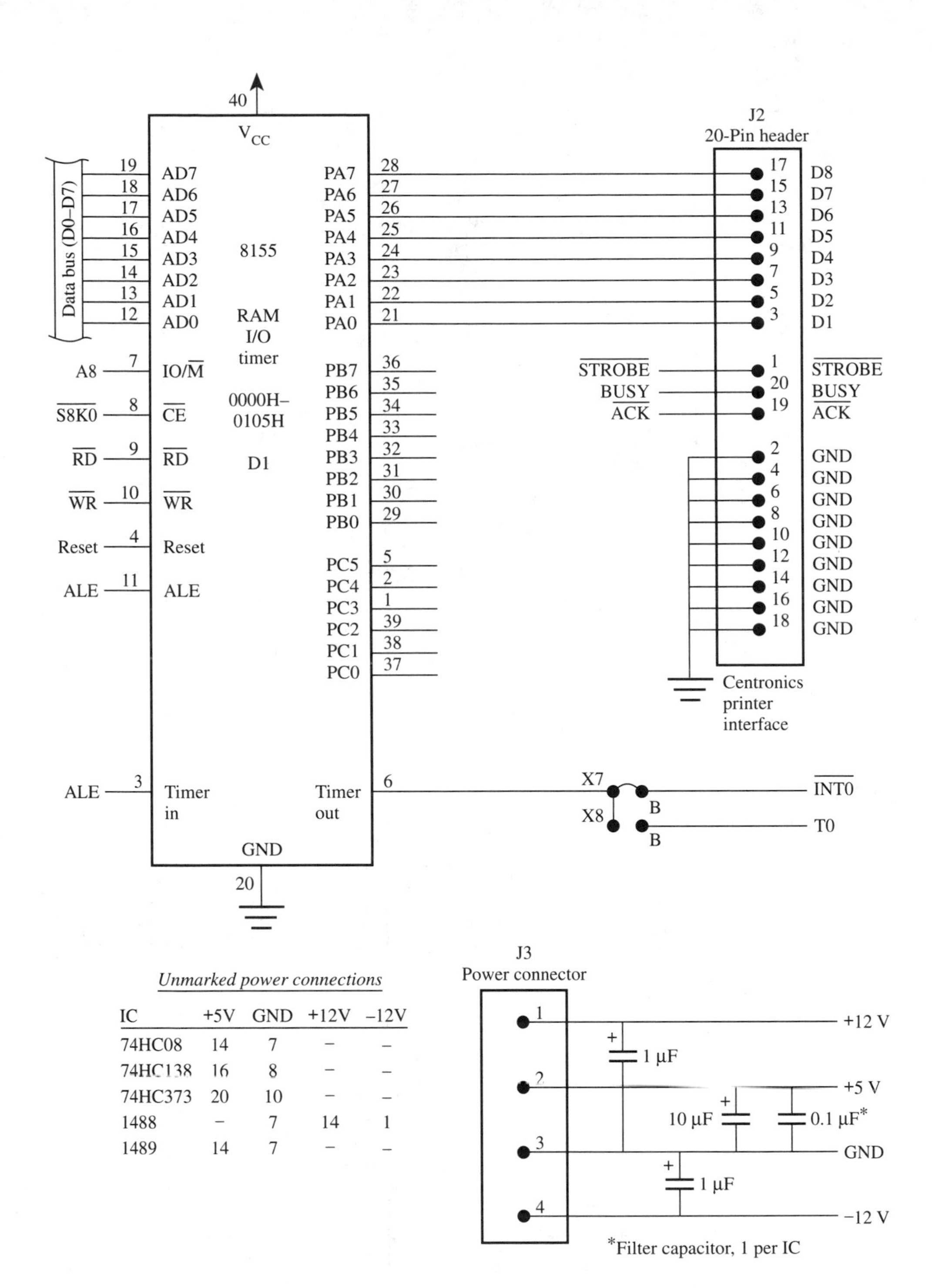

Unmarked power connections

IC	+5V	GND	+12V	–12V
74HC08	14	7	–	–
74HC138	16	8	–	–
74HC373	20	10	–	–
1488	–	7	14	1
1489	14	7	–	–

FIGURE 10–1
continued

The SBC-51 includes, in addition to the standard 80C31 features, 16K bytes of external EPROM, 8.25K bytes of external RAM, an extra 14-bit timer, and 22 extra input/output lines. The configuration shown in Figure 10–1 includes the following components and parts:

- 10 integrated circuits
- 15 capacitors
- 2 resistors
- 1 crystal
- 1 push-button switch
- 3 connectors
- 13 configuration jumpers

Since external memory is used, Port 0 and Port 2 are unavailable for input/output. Although Ports 1 and 3 are partially utilized for special features, some Port 0 and Port 3 lines may be used for input/output purposes, depending on the configuration.

The 80C31 clock source is a 12 MHz crystal connected in the usual way. (See Figure 2–2.) The RST (reset) line is driven by an R-C network for power-on reset and by a push-button switch for manual reset. Port 0 doubles as the data bus (D0 to D7) and the low-byte of the address bus (A0 to A7), as discussed earlier. (See 2.6 External Memory.) A 74HC373 octal latch is clocked by ALE to hold the low-byte of the address bus for the duration of a memory cycle. Since the 80C31 does not include on-chip ROM, execution is from external EPROM, and so $\overline{\text{EA}}$ (external access) is connected to ground through configuration jumper X2.

The connection to the host computer or VDT uses a serial RS232C interface. The DB25S connector is wired as a DTE (data terminal equipment) with transmit data (TXD) on pin 2, receive data (RXD) on pin 3, and ground on pin 7. A 1488 RS232 line driver connects to TXD and a 1489 RS232 line receiver connects to RXD. The default connection to the 80C31 is through jumpers X9 and X10 with P3.1 as TXD and P3.0 as RXD. Optionally, through jumpers X11 and X12, the TXD and RXD functions can be provided through software using P1.7 and P1.6.

Port 1 lines 3, 4, and 5 are read by the monitor program upon reset to evoke special features. After reset, however, these lines are available for general-purpose I/O. If the printer interface is used, Port 1 lines 0, 1, and 2 are the handshake signals. If the printer interface is not used, these lines are available for general-purpose I/O.

The 74HC138 decodes the upper three bits on the address bus (A15 to A13) and generates eight select lines, one for each 8K block of memory. These are called $\overline{\text{S8K0}}$ (for "select 8K block 0") through to $\overline{\text{S8K7}}$. Four ICs are selected by these lines: two 2764 EPROMs, a 6264 RAM, and an 8155 RAM/IO/TIMER.

Two 2764 8K by 8 EPROMs are shown in Figure 10–1. The first (labeled "MONITOR EPROM") is selected by $\overline{\text{S8K0}}$ and resides in the external code space from address 0000H to 1FFFH. Since the SBC-51 will begin execution from address 0000H immediately after a system reset, the monitor program must reside in this IC. The second 2764 is labeled "USER EPROM" and is selected by $\overline{\text{S8K1}}$ for execution at addresses 2000H to 3FFFH. This IC is intended for user applications and is not needed for basic system operation. Note that both EPROMs are selected only if $\overline{\text{CE}}$ (chip enable; pin 20) is active (or low) and $\overline{\text{OE}}$ is also active (or low). $\overline{\text{OE}}$ is driven by the 80C31's $\overline{\text{PSEN}}$ line; thus selection is in the external code space, as expected.

The 6264 8K by 8 RAM IC is selected by $\overline{\text{S8K4}}$ (if jumper X6 is installed, as shown), so it resides at addresses 8000H to 9FFFH. The RAM is selected to occupy both the external data space and the external code space using the method described earlier. (See Section 2.6.4, Overlapping the External Code and Data Spaces.) This allows user programs to be loaded (or written) to the RAM as "data memory" and then executed as "code memory."

The 8155 RAM/IO/TIMER is a peripheral interface IC that was added to demonstrate the expansion capabilities of the SBC-51. It is easy to add other peripheral interface ICs in a similar way. The 8155 is selected by $\overline{\text{S8K0}}$, placing it at the bottom of memory. No conflict occurs with the monitor EPROM (which also resides at the bottom of memory, but in the external code space) because the 8155 is further selected for read and write operations using $\overline{\text{RD}}$ and $\overline{\text{WR}}$.

The 8155 contains the following features:

- 256 bytes of RAM
- 22 input/output lines
- 14-bit timer

Address line A8 connects to the 8155's IO/$\overline{\text{M}}$ line (pin 7) and selects the RAM when low and the I/O lines or timer when high. The I/O lines and timer are accessed from six addresses, so the total address range of the 8155 is 0000H to 0105H (256 + 6 addresses). These are summarized below.

Address	Purpose
0000H	first RAM address
. . .	Other RAM addresses
00FFH	last RAM address
0100H	Interval/command register
0101H	Port A
0102H	Port B
0103H	Port C
0104H	Low-order 8 bits of timer count
0105H	High-order 6 bits of timer count & 2 bits of timer mode

Although the manufacturer's data sheet should be consulted for details of the 8155's operation, configuring the I/O ports is extremely easy. By default all port lines are inputs after a system reset; therefore no "initialize" operation is needed to read input devices connected to the 8155. To read Port A into the accumulator, for example, the following instruction sequence is used:

```
MOV  DPTR,#0101H          ;DPTR points to 8155 Port A
MOVX A,@DPTR              ;read Port A into Acc
```

To program Port A and Port B as outputs, 1s must first be written into the command register bits 0 and 1, respectively. For example, to configure Port B as an output port and leave Port A and Port C as input, the following instruction sequence is used:

```
MOV  DPTR,#0100H          ;8155 command register
MOV  A,#00000010B         ;Port B = output
MOVX @DPTR,A              ;initialize 8155
```

Port C is configured as an output by writing 1s to the command register bits 2 and 3. All three ports would be configured as output as follows:

```
MOV  DPTR,#0100H          ;8155 command register
MOV  A,#00001111B         ;all ports = output
MOVX @DPTR,A              ;initialize 8155
```

Port A of the 8155 is shown connected to a 20-pin header labeled "Centronics printer interface." This interface is for demonstration purposes only. MON51 includes a PCHAR (print character) subroutine and directs output to the VDT *and* a parallel printer if CONTROL-Z is entered on the keyboard. (See Appendix G.) Of course, Port A can be used for other purposes if desired.

Power-supply connections are also shown in Figure 10–1. The filter capacitors are particularly important for the +5 volt supply to avoid glitches due to inductive effects when digital devices switch. If the SBC-51 is constructed on a prototype board (for example, by wire wrapping), these capacitors should be considered critical. Place a 10 μF electrolytic capacitor where power enters the prototype board, and 0.01 μF ceramic capacitors beside the socket for each IC, wired between the +5 volt pin and the ground pin.

Since the SBC-51 is small and inexpensive, it is easy to construct a prototype and gain hands-on experience through the monitor program and the interfacing examples in this chapter. Wire wrapping is the most practical method of construction. The SBC-51 is also available assembled and tested on a printed-circuit board (see Figure 10-2).[1]

This concludes our description of the SBC-51. The following sections contain examples of interfaces to peripheral devices that have been developed to connect to the SBC-51 (or a similar 8051 single-board computer).

10.3 HEXADECIMAL KEYPAD INTERFACE

Interfaces to keypads are common for microcontroller-based designs. Keypad input and LED output are an economical choice for a user interface and are often adequate for complex applications. Examples include the user interface to microwave ovens or automated banking machines. Figure 10–3 shows an interface between Port 1 and a hexadecimal keypad. The keypad contains 16 keys arranged in four rows and four columns. The row lines are connected to Port 1 bits 4–7, the column lines to Port 1 bits 0–3.

[1]The printed-circuit board version of the SBC-51 is available from URDA, Inc., 1811 Jancey St., Suite #200, Pittsburgh, PA, USA, 15206.

FIGURE 10–2
The printed-circuit board version of the SBC-51. (Courtesy URDA, Inc.)

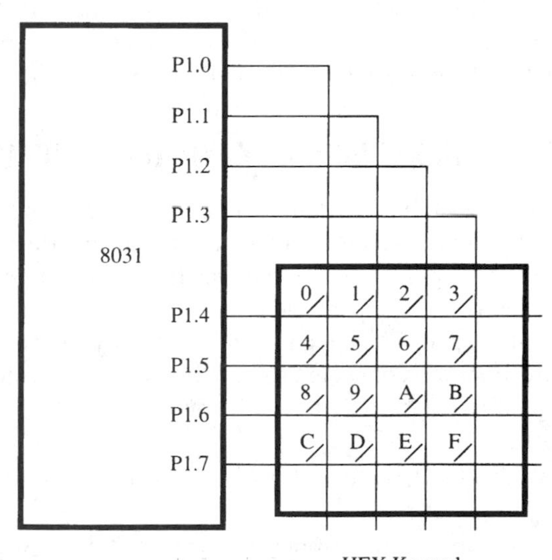

FIGURE 10–3
Interface to hexadecimal keypad

Design Objective

Write a program that continually reads hexadecimal characters from the keypad and echoes the corresponding ASCII code to the console.

On the surface, this example seems quite simple. The software can be divided into the following steps:

1. Get a hexadecimal character from the keypad.
2. Convert the hexadecimal code to ASCII.
3. Send the ASCII code to the VDT.
4. Go to step 1.

In fact, the software solution shown in Figure 10–4 follows this exact pattern (see lines 16–19). Of course, the work is done in the subroutines. Note that steps 2 and 3 above are implemented by calling subroutines in MON51. Of course, the code could have been extracted from MON51 and placed in the listing in Figure 10–4, but that's wasteful. Instead, the MON51 entry points for these subroutines are defined near the top of the listing (in lines 12–13) using the symbols HTOA and OUTCHR, and then the subroutines are called in the MAIN program loop in the usual way. Incidentally, the entry points for MON51 subroutines can be found in the symbol table created by RL51 when MON51 was linked and located. The entry points for HTOA and OUTCHR, for example, are found in Appendix G.

The real challenge for this example is writing the subroutines IN_HEX and GET_KEY. GET_KEY does the work of scanning the row and column lines of the keypad to determine if a key is pressed. If no key is pressed, it returns with C = 0. If a key is pressed, it returns with C = 1 and the hexadecimal code for the key in the accumulator bits 0–3.

IN_HEX performs software debouncing. Since the keypad is a series of mechanical switches, contact closure and release include bounce—the rapid but brief make-and-break of the switch contacts. Debouncing is performed by calling GET_HEX repeatedly until 50 consecutive calls return with C = 1. Any call to GET_HEX returning with C = 0 is interpreted as noise (i.e., bounce) and the counter is reset. After detecting a legitimate key closure, IN_HEX then waits for 50 consecutive calls to GET_HEX returning with C = 0. This ensures a clean key release before the next call to GET_HEX.

The software in Figure 10–4 works, but it is not particularly elegant. Since interrupts are not used, the program's utility within a larger application is limited. A reasonable improvement, therefore, is to redesign the software using interrupts. An interrupt-driven interface is illustrated in the next example.

10.4 INTERFACE TO MULTIPLE 7-SEGMENT LEDS

An interface to a 7-segment LED display was presented in a problem at the end of Chapter 3. (See Figure 3–5.) Unfortunately, the interface used seven lines on Port 1, so it represents a poor allocation of the 8051's on-chip resources. In this section, we demonstrate

```
LOC   OBJ            LINE   SOURCE
                       1    $DEBUG
                       2    $NOPAGING
                       3    $NOSYMBOLS
                       4    ;FILE: KEYPAD.SRC
                       5    ;*****************************************************
                       6    ;                KEYPAD INTERFACE EXAMPLE            *
                       7    ;                                                    *
                       8    ; This program reads hexadecimal characters from a  *
                       9    ; keypad attached to Port 1 and echos keys pressed  *
                      10    ; to the console.                                    *
                      11    ;*****************************************************
  033C                12    HTOA     EQU     033CH    ;MON51 subroutines (V12)
  01DE                13    OUTCHR   EQU     01DEH
                      14
8000                  15             ORG     8000H
8000 12800B           16    MAIN:    CALL    IN_HEX   ;get code from keypad
8003 12033C           17             CALL    HTOA     ;convert to ASCII
8006 1201DE           18             CALL    OUTCHR   ;echo to console
8009 80F5             19             SJMP    MAIN     ;repeat
                      20
                      21    ;*****************************************************
                      22    ; IN_HEX - input hex code from keypad with debouncing *
                      23    ;          for key press and key release (50 repeat  *
                      24    ;          operations for each)                      *
                      25    ;*****************************************************
800B 7B32             26    IN_HEX:  MOV     R3,#50   ;debounce count
800D 128022           27    BACK:    CALL    GET_KEY  ;key pressed?
8010 50F9             28             JNC     IN_HEX   ;no:  check again
8012 DBF9             29             DJNZ    R3,BACK  ;yes: repeat 50 times
8014 C0E0             30             PUSH    ACC      ;save hex code
8016 7B32             31    BACK2:   MOV     R3,#50   ;wait for key up
8018 128022           32    BACK3:   CALL    GET_KEY  ;key pressed?
801B 40F9             33             JC      BACK2    ;yes: keep checking
801D DBF9             34             DJNZ    R3,BACK3 ;no:  repeat 50 times
801F D0E0             35             POP     ACC      ;recover hex code and
8021 22               36             RET              ; return
                      37
                      38    ;*****************************************************
                      39    ; GET_KEY - get keypad status                        *
                      40    ;         - return with C = 0 if no key pressed      *
                      41    ;         - return with C = 1 and hex code in ACC if *
                      42    ;           a key is pressed                         *
                      43    ;*****************************************************
8022 74FE             44    GET_KEY: MOV     A,#0FEH          ;start with column 0
8024 7E04             45             MOV     R6,#4            ;use R6 as counter
8026 F590             46    TEST:    MOV     P1,A             ;activate colmn line
8028 FF               47             MOV     R7,A             ;save ACC
8029 E590             48             MOV     A,P1             ;read back Port 0
802B 54F0             49             ANL     A,#0F0H          ;isolate row lines
802D B4F007           50             CJNE    A,#0F0H,KEY_HIT ;row line active?
8030 EF               51             MOV     A,R7             ;no: move to next
8031 23               52             RL      A                ;    column line
8032 DEF2             53             DJNZ    R6,TEST
8034 C3               54             CLR     C                ;no key pressed
8035 8015             55             SJMP    EXIT             ;return with C = 0
8037 FF               56    KEY_HIT: MOV     R7,A             ;save in R6
8038 7404             57             MOV     A,#4             ;prepare to caculate
803A C3               58             CLR     C                ; column weighting
803B 9E               59             SUBB    A,R6             ;4 - R6 = weighting
803C FE               60             MOV     R6,A             ;save in R6
803D EF               61             MOV     A,R7             ;restore scan code
803E C4               62             SWAP    A                ;put in low nibble
803F 7D04             63             MOV     R5,#4            ;use R5 as counter
8041 13               64    AGAIN:   RRC     A                ;rotate until 0
```

FIGURE 10–4
Software for keypad interface

```
8042 5006        65                JNC       DONE           ;done when C = 0
8044 0E          66                INC       R6             ;add 4 until active
8045 0E          67                INC       R6             ; row found
8046 0E          68                INC       R6
8047 0E          69                INC       R6
8048 DDF7        70                DJNZ      R5,AGAIN
804A D3          71       DONE:    SETB      C              ;C = 1 (key pressed)
804B EE          72                MOV       A,R6           ;code in A (whew!!!)
804C 22          73       EXIT:    RET
                 74                END
```

FIGURE 10–4
continued

an interface to four 7-segment LEDs using only three of the 8051's I/O lines. This, obviously, is a much-improved design, particularly if multiple segments must be connected.

Central to the design is the Motorola MC14499 7-segment decoder/driver, which includes much of the circuitry necessary to drive four displays. The only additional components are a 0.015 μF timing capacitor, seven 47 Ω current-limiting resistors and four 2N3904 transistors. Figure 10–5 shows the connections between the 80C51, the MC14499, and the four 7-segment LEDs.

Design Objective

Assume BCD digits are stored in internal RAM locations 70H and 71H. Copy the BCD digits to the LED display 10 times per second using interrupts.

The software to accomplish the above objective is shown in Figure 10–6. The listing illustrates a number of concepts discussed earlier. The low-level details of sending data to the MC14499 are found in the subroutines UPDATE and OUT8. At a higher level, this example illustrates the design of interrupt-driven applications with a significant amount of foreground *and* background activity (unlike the examples in Chapter 6, which operated only in the background). The interrupts for this example coexist with MON51, which does not itself use interrupts. The monitor program executes in the foreground while the program in Figure 10–6 executes at interrupt-level in the background. When the program is started (e.g., by entering the MON51 command GO8000; see Appendix G), conditions are initialized for the necessary interrupt-initiated updating of the LED displays, and then control quickly passes back to the monitor program. Monitor commands can be executed in the usual way; meanwhile, interrupts are occurring in the background. If, for example, the monitor SET command is used to change internal RAM locations 70H and 71H, the changes are seen immediately (within 0.1 s) on the 7-segment LED displays.

Note the overall structure of the program. The following sections appear in order:

- Assembler controls (lines 1–3)
- Comment block (lines 4–30)
- Definition of symbols (31–38)
- Define storage declarations (lines 40–42)
- Jump table for program and interrupt entry points (lines 44–51)

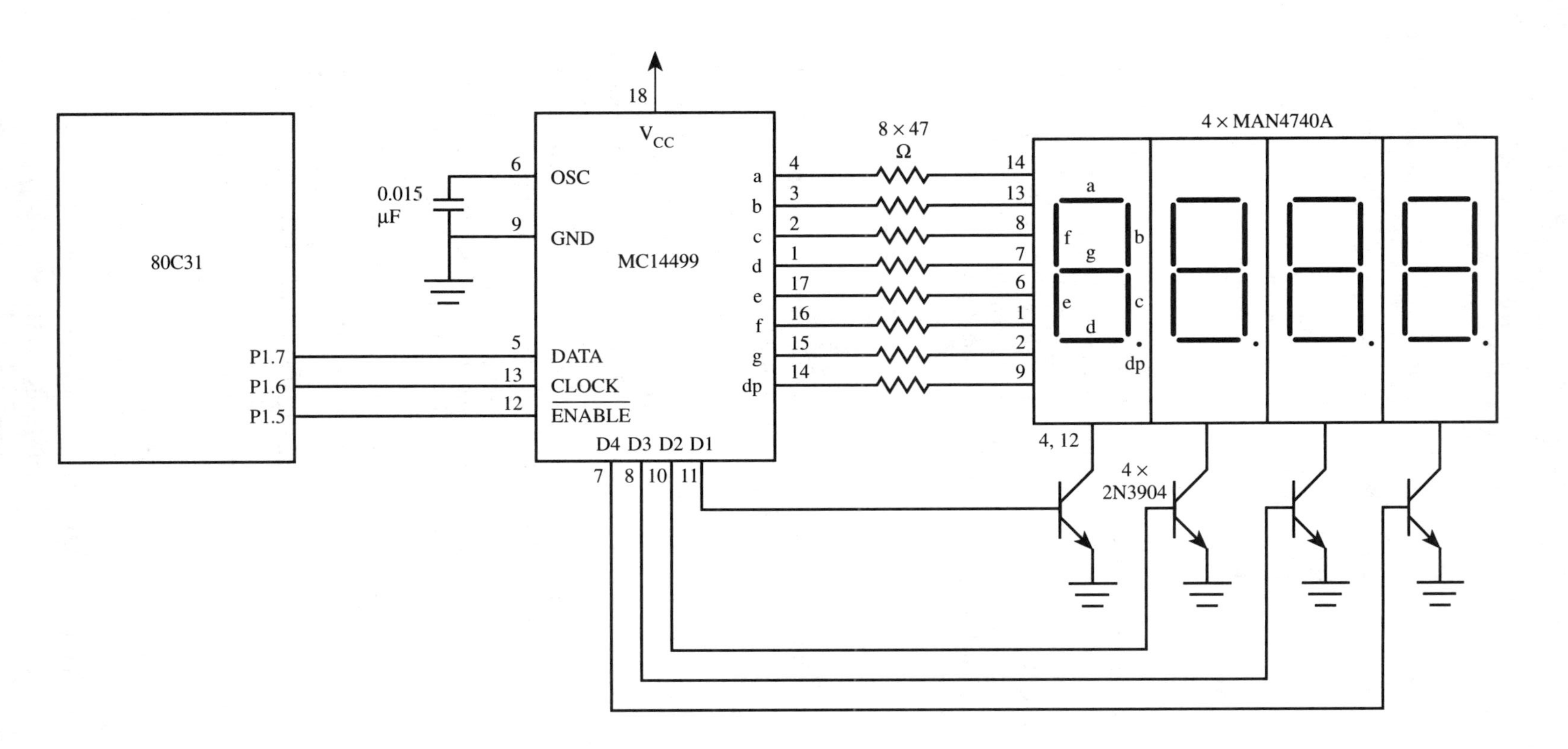

FIGURE 10–5
Interface to MC14499 and four 7-segment LEDs

- □ Main section (MAIN; lines 56–69)
- □ External interrupt service routine (EXT0ISR; lines 74–77)
- □ Update LED display subroutine (UPDATE; lines 89–97)
- □ Output byte subroutine (OUT8; lines 103–113)
- □ Code to handle unimplemented interrupts (lines 118–123)

The program is written for execution at address 8000H in the SBC-51's 6264 RAM IC. Since interrupts vector through locations at the bottom of memory, the monitor program includes a jump table redirecting interrupts to addresses starting at address 8000H. (See Appendix G.) The program entry point is conveniently 8000H; however, an LJMP instruction (line 45; see Figure 10–6) passes control to the label MAIN. All the initialize instructions are contained in lines 56–68. The MAIN section terminates by jumping back to the monitor program.

10.5 LOUDSPEAKER INTERFACE

Figure 10–7 shows an interface between an 8031 and a loudspeaker. Small loudspeakers, such as those found in personal computers or children's toys, can be driven from a single logic gate, as shown. One side of the loudspeaker's coil connects to +5 volts, the other to the output of a 74LS04 logic inverter. The inverter is required because it has a higher drive capability than the port lines on the 8031.

Design Objective

Write an interrupt-driven program that continually plays an A-major musical scale.

Musical melodies are easy to generate from an 8051 using a simple loudspeaker interface. We begin with some music theory. The frequency for each note in an A-major musical scale is given in the comment block at the top of the software listing in Figure 10–8 (lines 14–21). The first frequency is 440 Hz (called "A above middle C"), which is the international reference frequency for musical instruments using the equal-tempered scale (e.g., the piano). The frequency of all other notes can be determined by multiplying this frequency by $2^{n/12}$, where n is the number of steps (or "semitones") to the note being calculated. The easiest example is A′, one octave, or 12 steps, above A, which has a frequency of $440 \times 2^{12/12} = 880$ Hz. This is the last note in our musical scale. (See Figure 10–8, line 21.) With reference to the bottom note (or "root") in any major scale, the scale in steps is 2, 4, 5, 7, 9, 11, and 12. For example, the note "E" in Figure 10–8 (line 18) is seven steps above the root; thus its frequency is $440 \times 2^{7/12} = 659.26$ Hz.

To create a musical scale, two timings are required: the timing from one note to the next, and the timing for toggling the port bit that drives the loudspeaker. These two timings are vastly different. To play the melody at a rate of four notes/second, for example, a timeout (or interrupt) is needed every 250 ms. To create the frequency for the first note in the scale, a timeout is needed every 1.136 ms. (See Figure 10–8, line 14.)

```
 LOC  OBJ            LINE   SOURCE
                        1   $DEBUG
                        2   $NOPAGING
                        3   $NOSYMBOLS
                        4   ;FILE: MC14499.SRC
                        5   ;*******************************************************
                        6   ;                 MC14499 INTERFACE EXAMPLE            *
                        7   ;                                                      *
                        8   ; This program updates a 4-digit display 10 times per *
                        9   ; second using interrupts.  The digits are 7-segment   *
                       10   ; LEDs driven by an MC14499 decoder/driver connected   *
                       11   ; to P1.5 (-ENABLE), P1.6 (CLOCK), and P1.7 (DATA      *
                       12   ; IN).  Interrupts are generated by the 8155's TIMER   *
                       13   ; OUT line connected to -INT0.  TIMER OUT oscillates   *
                       14   ; at 500 Hz and generates an interrupt on each 1-to-0 *
                       15   ; transition.  An interrupt counter is used to update *
                       16   ; the display every 50 interrupts, for an update       *
                       17   ; frequency of 10 Hz.                                  *
                       18   ;                                                      *
                       19   ; The example illustrates the foreground/background    *
                       20   ; concept for interrupt-driven systems.  Once the      *
                       21   ; 8155 is intitialized and External 0 interrupts are   *
                       22   ; enabled, the program returns to the monitor program.*
                       23   ; MON51 itself does not use interrupts; however, it    *
                       24   ; executes as usual in the foreground while            *
                       25   ; interrupts take place in the background.  If the     *
                       26   ; MON51 command SI (set internal memory) is used to    *
                       27   ; change locations DIGITS or DIGITS+1, then the value *
                       28   ; written is immediately seen (within 0.1 s) on the    *
                       29   ; LED display.                                         *
                       30   ;*******************************************************
  00BC                 31   MON51    CODE    00BCH       ;MON51 (V12) entry
  0100                 32   X8155    XDATA   0100H       ;8155 address
  0104                 33   TIMER    XDATA   X8155 + 4   ;timer registers
  0FA0                 34   COUNT    EQU     4000        ;interrupts @ 2000 us
  0040                 35   MODE     EQU     01000000B   ;timer mode bits
  0097                 36   DIN      BIT     P1.7        ;MC14499 interface lines
  0096                 37   CLOCK    BIT     P1.6
  0095                 38   ENABLE   BIT     P1.5
                       39
 ----                  40            DSEG    AT 70H      ;absolute internal segment
 0070                  41   DIGITS:  DS      2           ; (no conflict with MON51)
 0072                  42   ICOUNT:  DS      1
                       43
 ----                  44            CSEG    AT 8000H
 8000 028015           45            LJMP    MAIN        ; program entry point
 8003 028031           46            LJMP    EX0ISR      ; 8155 interrupt
 8006 02805D           47            LJMP    T0ISR       ; Timer 0 interrupt
 8009 02805D           48            LJMP    EX1ISR      ; External 1 interrupt
 800C 02805D           49            LJMP    T1ISR       ; Timer 1 interrupt
 800F 02805D           50            LJMP    SPISR       ; Serial Port interrupt
 8012 02805D           51            LJMP    T2ISR       ; Timer 2 interrupt
                       52
                       53   ;*******************************************************
                       54   ; MAIN PROGRAM BEGINS (INIT 8155 & ENABLE INTERRUPTS) *
                       55   ;*******************************************************
 8015 900104           56   MAIN:    MOV     DPTR,#TIMER ;initialize 8155 timer
 8018 74A0             57            MOV     A,#LOW(COUNT)
 801A F0               58            MOVX    @DPTR,A
 801B A3               59            INC     DPTR        ;initialize high register
 801C 744F             60            MOV     A,#HIGH(COUNT) OR MODE
 801E F0               61            MOVX    @DPTR,A
 801F 900100           62            MOV     DPTR,#X8155 ;8155 command register
 8022 74C0             63            MOV     A,#0C0H     ;start timer command
 8024 F0               64            MOVX    @DPTR,A     ;500 Hz square wave
```

FIGURE 10–6
Software for MC14499 interface

```
8025 757232      65          MOV     ICOUNT,#50     ;initialize int. counter
8028 D2AF        66          SETB    EA             ;enable interrupts
802A D2A8        67          SETB    EX0            ;enable External 0 int.
802C D288        68          SETB    IT0            ;negative-edge triggered
802E 0200BC      69          LJMP    MON51          ;return to MON51
                 70
                 71  ;*****************************************************
                 72  ; EXTERNAL 0 INTERRUPT SERVICE ROUTINE                *
                 73  ;*****************************************************
8031 D57205      74  EX0ISR: DJNZ    ICOUNT,EXIT    ;on 50th interrupt,
8034 757232      75          MOV     ICOUNT,#50     ; reset counter and
8037 113A        76          ACALL   UPDATE         ; refresh LED display
8039 32          77  EXIT:   RETI
                 78
                 79  ;*****************************************************
                 80  ; UDATE 4-DIGIT LED DISPLAY (EXECUTION TIME = 84 us)  *
                 81  ;                                                     *
                 82  ; ENTER:        Four BCD digits in internal memory    *
                 83  ;               locations DIGITS and DIGITS+1 (MSD in *
                 84  ;               high nibble of DIGITS)                *
                 85  ; EXIT:         MC14499 display updated               *
                 86  ; USES:         P1.5, P1.6, P1.7                      *
                 87  ;               All memory locations and regs intact  *
                 88  ;*****************************************************
803A C0E0        89  UPDATE: PUSH    ACC           ;save Accumulator on stack
803C C295        90          CLR     ENABLE        ;prepare MC14499
803E E570        91          MOV     A,DIGITS      ;get first two digits
8040 114B        92          ACALL   OUT8          ;send two digits
8042 E571        93          MOV     A,DIGITS + 1  ;get second byte
8044 114B        94          ACALL   OUT8          ;send last two digits
8046 D295        95          SETB    ENABLE        ;disable MC14499
8048 D0E0        96          POP     ACC           ;restore ACC from stack
804A 22          97          RET
                 98
                 99  ;*****************************************************
                100  ; SEND 8 BITS IN ACCUMULATOR TO MC14499 (MSB FIRST)   *
                101  ;*****************************************************
                102          USING   0          ;assume reg. bank 0 enabled
804B C007       103  OUT8:   PUSH    AR7        ;save R7 on stack
804D 7F08       104          MOV     R7,#8      ;use R7 as bit counter
804F 33         105  AGAIN:  RLC     A          ;put bit in C flag
8050 9297       106          MOV     DIN,C      ;send it to MC14499
8052 C296       107          CLR     CLOCK      ;3 us low pulse on clock line
8054 00         108          NOP                ;NOPs needed to stretch pulse
8055 00         109          NOP                ; (minimum pulse width is
8056 D296       110          SETB    CLOCK      ; is 2 us)
8058 DFF5       111          DJNZ    R7,AGAIN   ;repeat until all 8 bits sent
805A D007       112          POP     AR7        ;restore R7 from stack
805C 22         113          RET
                114
                115  ;*****************************************************
                116  ; UNUSED INTERRUPTS (ERROR; RETURN TO MONITOR PROGRAM)*
                117  ;*****************************************************
                118  T0ISR:
                119  EX1ISR:
                120  T1ISR:
                121  SPISR:
805D C2AF       122  T2ISR:  CLR     EA         ;shut off interrupts &
805F 0200BC     123          LJMP    MON51      ; return to MON51
                124          END
```

FIGURE 10–6
continued

FIGURE 10–7
Interface to a loudspeaker

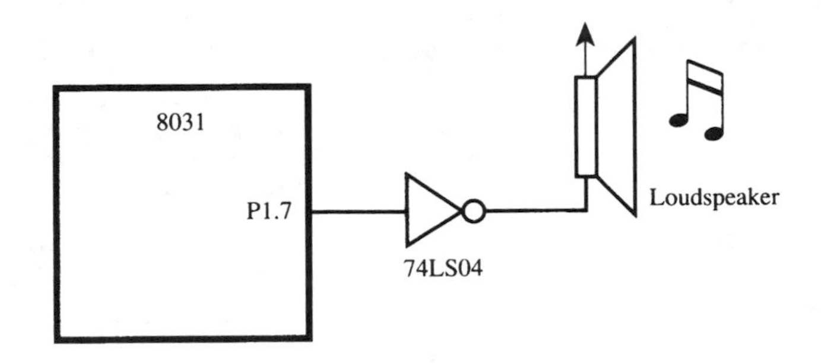

The software in Figure 10–8 initializes both timers for 16-bit timer mode (line 43) and uses Timer 0 interrupts for the note changes, and Timer 1 interrupts for the frequency of notes. The reload values for the note frequencies are read from a look-up table (lines 90–104). Consult the listing in Figure 10–8 for further details.

10.6 NON-VOLATILE RAM INTERFACE

Nonvolatile RAMs (NVRAMs) are semiconductor memories that maintain their contents in the absence of power. NVRAMs incorporate both standard static RAM cells and electrically erasable programmable ROM (EEPROM) cells. Each bit of the static RAM is overlaid with a bit of EEPROM. Data can be transferred back and forth between the two memories.

NVRAMs occupy an important niche in microprocessor- and microcontroller-based applications. They are used to store setup data or parameters that are changed occasionally by the user but must be retained when power is lost.

As an example, many VDT designs avoid the use of DIP switches (which are prone to failure) and use NVRAMs to store setup information such as baud rate, parity on/off, parity odd/even, and so on. Each time the VDT is turned on, these parameters are recalled from NVRAM and the system is initialized accordingly. When a parameter is changed by the user (via the keyboard) the new value is stored in NVRAM.

Modems with an auto-dial feature usually hold phone numbers in internal memory. These phone numbers are often stored in a NVRAM so they will be retained in the event of a power outage. Ten phone numbers with seven digits each can be stored in 35 bytes (by encoding each digit in BCD notation).

The NVRAM used for this interface example is an X2444 manufactured by Xicor,[1] a company that specializes in NVRAMs and EEPROMs. The X2444 contains 256 bits of static RAM overlaid by 256 bits of EEPROM. Data can be transferred back and forth between the two memories either by instructions sent from the processor over the serial interface or by toggling the external $\overline{\text{STORE}}$ and $\overline{\text{RECALL}}$ inputs. Nonvolatile data are retained in the EEPROM, while independent data are accessed and updated in the RAM. The X2444 features are summarized in the first page of its data sheet, reproduced in Figure 10–9.

In this interface example, the $\overline{\text{STORE}}$ and $\overline{\text{RECALL}}$ lines are not used. The various modes of operation are entered by sending the X2444 serial instructions through 8051 port pins.

---

[1]XICOR, Inc., 851 Buckeye Court, Milpitas, CA 95035

```
                1    $debug
                2    $nopaging
                3    $nosymbols
                4    ;FILE: SCALE.SRC
                5    ;*******************************************************
                6    ;            LOUDSPEAKER INTERFACE EXAMPLE              *
                7    ;                                                       *
                8    ; This program plays an A major musical scale using     *
                9    ; a loudspeaker driven by a inverter through P1.7       *
               10    ;*******************************************************
               11    ;                                                       *
               12    ;  Note  Frequency (Hz)  Period (us)  Period/2 (us)     *
               13    ;  ----  --------------  -----------  -------------     *
               14    ;   A        440.00         2273          1136          *
               15    ;   B        493.88         2025          1012          *
               16    ;   C#       554.37         1804           902          *
               17    ;   D        587.33         1703           851          *
               18    ;   E        659.26         1517           758          *
               19    ;   F#       739.99         1351           676          *
               20    ;   G#       830.61         1204           602          *
               21    ;   A'       880.00         1136           568          *
               22    ;*******************************************************
  00BC         23    MONITOR  CODE  00BCH       ;MON51 (V12) entry point
  3CB0         24    COUNT    EQU   -50000      ;0.05 seconds per timeout
  0005         25    REPEAT   EQU   5           ;5 x 0.05 = 0.25 seconds/note
               26
               27    ;*******************************************************
               28    ; Note: X3 not installed on SBC-51, therefore           *
               29    ; interrupts directed to the following jump table       *
               30    ; beginning at 8000H                                    *
               31    ;*******************************************************
8000           32             ORG   8000H       ;RAM entry points for...
8000 028015    33             LJMP  MAIN        ; main program
8003 02806B    34             LJMP  EXT0ISR     ; External 0 interrupt
8006 028025    35             LJMP  T0ISR       ; Timer 0 interrupt
8009 02806B    36             LJMP  EXT1ISR     ; External 1 interrupt
800C 02803A    37             LJMP  T1ISR       ; Timer 1 interrupt
800F 02806B    38             LJMP  SPISR       ; Serial Port interrupt
8012 02806B    39             LJMP  T2ISR       ; Timer 2 interrupt
               40
               41    ;*******************************************************
               42    ; MAIN PROGRAM BEGINS                                   *
               43    ;*******************************************************
8015 758911    44    MAIN:    MOV   TMOD,#11H   ;both timers 16-bit mode
8018 7F00      45             MOV   R7,#0       ;use R7 as note counter
801A 7E05      46             MOV   R6,#REPEAT  ;use R6 as timeout counter
801C 75A88A    47             MOV   IE,#8AH     ;Timer 0 & 1 interrupts on
801F D28F      48             SETB  TF1         ;force Timer 1 interrupt
8021 D28D      49             SETB  TF0         ;force Timer 0 interrupt
8023 80FE      50             SJMP  $           ;ZzZzZzZz time for a nap
               51
               52    ;*******************************************************
               53    ; TIMER 0 INTERRUPT SERVICE ROUTINE (EVERY 0.05 SEC.) *
               54    ;*******************************************************
8025 C28C      55    T0ISR:   CLR   TR0                   ;stop timer
8027 758C3C    56             MOV   TH0,#HIGH (COUNT)     ;reload
802A 758AB0    57             MOV   TL0,#LOW  (COUNT)
802D DE08      58             DJNZ  R6,EXIT               ;if not 5th int, exit
802F 7E05      59             MOV   R6,#REPEAT            ;if 5th, reset
8031 0F        60             INC   R7                    ;increment note
8032 BF0C02    61             CJNE  R7,#LENGTH,EXIT       ;beyond last note?
8035 7F00      62             MOV   R7,#0                 ;yes: reset, A=440 Hz
8037 D28C      63    EXIT:    SETB  TR0                   ;no:  start timer, go
8039 32        64             RETI                        ;     back to ZzZzZzZ
```

FIGURE 10–8
Software for loudspeaker interface

```
                      65
                      66   ;*****************************************************
                      67   ; TIMER 1 INTERRUPT SERVICE ROUTINE (PITCH OF NOTES)  *
                      68   ;                                                    *
                      69   ; Note: The output frequencies are slightly off due  *
                      70   ; to the length of this ISR.  Timer reload values    *
                      71   ; need adjusting.                                    *
                      72   ;*****************************************************
803A B297             73   T1ISR:   CPL   P1.7      ;music maestro!
803C C28E             74            CLR   TR1       ;stop timer
803E EF               75            MOV   A,R7      ;get note counter
803F 23               76            RL    A         ;multiply (2 bytes/note)
8040 128050           77            CALL  GETBYTE   ;get high-byte of count
8043 F58D             78            MOV   TH1,A     ;put in timer high register
8045 EF               79            MOV   A,R7      ;get note counter again
8046 23               80            RL    A         ;align on word boundary
8047 04               81            INC   A         ;past high-byte (whew!)
8048 128050           82            CALL  GETBYTE   ;get low-byte of count
804B F58B             83            MOV   TL1,A     ;put in timer low register
804D D28E             84            SETB  TR1       ;start timer
804F 32               85            RETI            ;time for a rest
                      86
                      87   ;*****************************************************
                      88   ; GET A BYTE FROM LOOK-UP OF NOTES IN A MAJOR SCALE   *
                      89   ;*****************************************************
8050 04               90   GETBYTE: INC   A         ;table look-up subroutine
8051 83               91            MOVC  A,@A+PC
8052 22               92            RET
8053 FB90             93   TABLE:   DW    -1136     ;A
8055 FB90             94            DW    -1136     ;A (play again; half note)
8057 FC0C             95            DW    -1012     ;B (quarter note, etc.)
8059 FC7A             96            DW    -902      ;C# - major third
805B FCAD             97            DW    -851      ;D
805D FD0A             98            DW    -758      ;E  - perfect fifth
805F FD5C             99            DW    -676      ;F#
8061 FDA6            100            DW    -602      ;G#
8063 FDC8            101            DW    -568      ;A'
8065 FDC8            102            DW    -568      ;A' (play 4 times; whole note)
8067 FDC8            103            DW    -568
8069 FDC8            104            DW    -568
  000C               105   LENGTH   EQU   ($ - TABLE) / 2 ;LENGTH = # of notes
                     106
                     107   ;*****************************************************
                     108   ; UNUSED INTERRUPTS - BACK TO MONITOR PROGRAM (ERROR) *
                     109   ;*****************************************************
                     110   EXT0ISR:
                     111   EXT1ISR:
                     112   SPISR:
806B C2AF            113   T2ISR:   CLR   EA        ;shut off interrupts and
806D 0200BC          114            LJMP  MONITOR ; return to MON51
                     115            END
```

FIGURE 10–8
continued

256 Bit | **Commercial Industrial** | **X2444 X2444I** | **16 x 16 Bit**

Nonvolatile Static RAM

FEATURES

- **Ideal for use with Single Chip Microcomputers**
  - **—Static Timing**
  - **—Minimum I/O Interface**
  - **—Serial Port Compatible (COPS™, 8051)**
  - **—Easily Interfaces to Microcontroller Ports**
  - **—Minimum Support Circuits**
- **Software and Hardware Control of Nonvolatile Functions**
  - **—Maximum Store Protection**
- **TTL Compatible**
- **16 x 16 Organization**
- **Low Power Dissipation**
  - **—Active Current: 15 mA Typical**
  - **—Store Current: 8 mA Typical**
  - **—Standby Current: 6 mA Typical**
  - **—Sleep Current: 5 mA Typical**
- **8 Pin Mini-DIP Package**

DESCRIPTION

The Xicor X2444 is a serial 256 bit NOVRAM featuring a static RAM configured 16 x 16, overlaid bit for bit with a nonvolatile E²PROM array. The X2444 is fabricated with the same reliable N-channel floating gate MOS technology used in all Xicor 5V nonvolatile memories.

The Xicor NOVRAM design allows data to be transferred between the two memory arrays by means of software commands or external hardware inputs. A store operation (RAM data to E²PROM) is completed in 10 ms or less and a recall operation (E²PROM data to RAM) is completed in 2.5 μs or less.

Xicor NOVRAMs are designed for unlimited write operations to RAM, either from the host or recalls from E²PROM and a minimum 100,000 store operations. Data retention is specified to be greater than 100 years.

COPS™ is a trademark of National Semiconductor Corp.

PIN CONFIGURATION

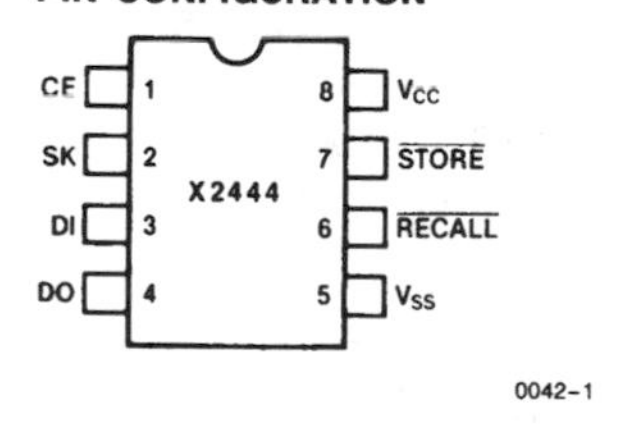

PIN NAMES

CE	Chip Enable
SK	Serial Clock
DI	Serial Data In
DO	Serial Data Out
$\overline{RECALL}$	Recall
$\overline{STORE}$	Store
V_{CC}	+5V
V_{SS}	Ground

FUNCTIONAL DIAGRAM

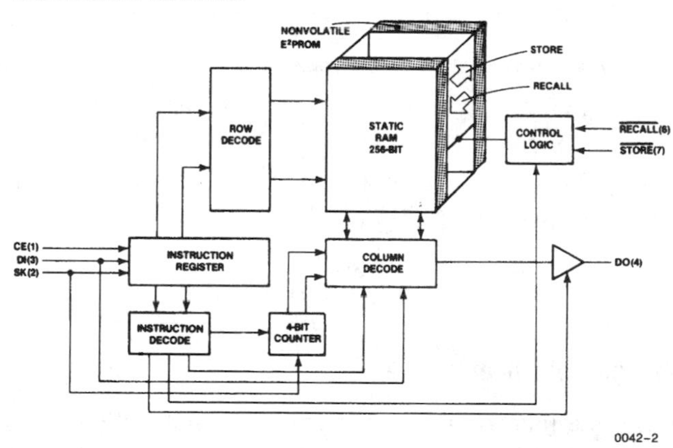

May 1987

FIGURE 10–9
Cover page for the X2444 non-volatile RAM data sheet

FIGURE 10–10
Interface to X2444 non-volatile RAM

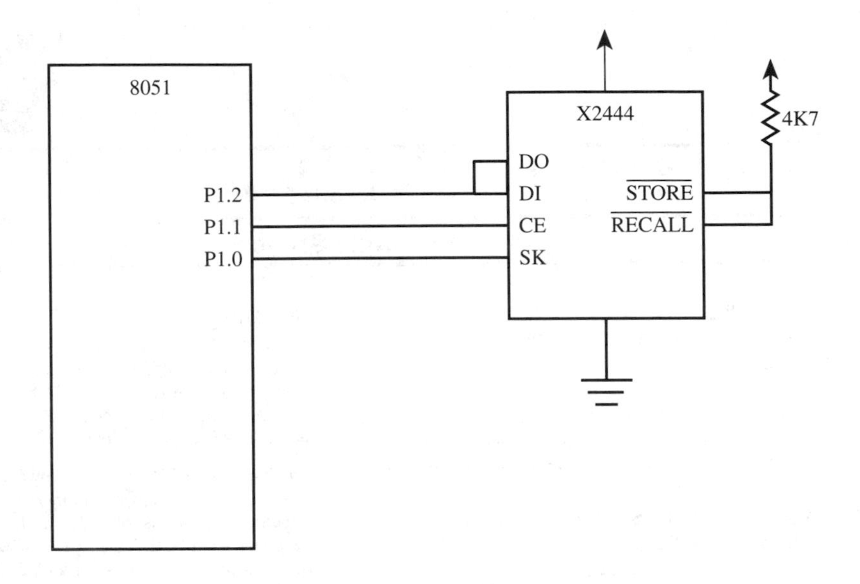

The interface to the 8051 is shown in Figure 10–10. Only 3 lines are used:

- P1.0—SK (serial clock)
- P1.1—CE (chip enable)
- P1.2—DI/DO (data input/output)

Instructions are sent to the X2444 by bringing CE high and then clocking an 8-bit opcode into the X2444 via the SK and DI/DO lines. The following opcodes are required for this example:

Instruction	Opcode	Operation
RCL	85H	Recall EEPROM data into RAM
WREN	84H	Set write enable latch
STORE	81H	Store RAM data into EEPROM
WRITE	1AAAA011B	Write data into RAM address AAAA
READ	1AAAA111B	Read data from RAM address AAAA

Design Objective

Write the following two programs. The first, called SAVE, copies the contents of 8051 internal locations 60H–7FH to the X2444 EEPROM. The second, called RECOVER, reads previously saved data from the X2444 EEPROM and restores it to locations 60H–7FH.

These are two distinct programs. Typically, the SAVE program is executed whenever nonvolatile information is changed (for example, by a user altering a configuration parameter). The RECOVER program is executed each time the system is powered up or re-

```
                 1       $DEBUG
                 2       $NOPAGING
                 3       $NOSYMBOLS
                 4       ;FILE: NVRAM.SRC
                 5       ;*****************************************************
                 6       ;                X2444 INTERFACE EXAMPLE             *
                 7       ;                                                    *
                 8       ; Two subroutines are shown below that SAVE or       *
                 9       ; RECOVER data between a X2444 non-volatile RAM and  *
                10       ; 32 bytes of the 8051's internal RAM.               *
                11       ;*****************************************************
  0085          12       RECALL  EQU     85H     ;X2444 recall instruction
  0084          13       WRITE   EQU     84H     ;X22444 write enable instruction
  0081          14       STORE   EQU     81H     ;X2444 store instruction
  0082          15       SLEEP   EQU     82H     ;X2444 sleep istruction
  0083          16       W_DATA  EQU     83H     ;X2444 write data instruction
  0087          17       R_DATA  EQU     87H     ;X2444 read data instruction
  00BC          18       MON51   EQU     00BCH   ;MON51 entry point (V12)
  0020          19       LENGTH  EQU     32      ;32 bytes saved/restored
  0092          20       DIN     BIT     P1.2    ;X2444 interface lines
  0091          21       ENABLE  BIT     P1.1
  0090          22       CLOCK   BIT     P1.0
                23
----            24               DSEG    AT 60H
0060            25       NVRAM:  DS      LENGTH   ;60H-7FH saved/recovered
                26
----            27               CSEG    AT 8000H
8000 110A       28       WX2444: ACALL   SAVE     ;8000H entry point for write
8002 0200BC     29               LJMP    MON51
8005 1149       30       RX2444: ACALL   RECOVER  ;8005H entry point for read
8007 0200BC     31               LJMP    MON51
                32
                33       ;*****************************************************
                34       ; SAVE 8031 RAM LOCATIONS 60H-7FH IN X2444 NVRAM     *
                35       ;*****************************************************
800A 7860       36       SAVE:   MOV     R0,#NVRAM       ;R0 -> locations to save
800C C291       37               CLR     ENABLE          ;disable X2444
800E 7485       38               MOV     A,#RECALL       ;recall instruction
8010 D291       39               SETB    ENABLE
8012 1184       40               ACALL   W_BYTE
8014 C291       41               CLR     ENABLE
8016 7484       42               MOV     A,#WRITE        ;write enable prepares
8018 D291       43               SETB    ENABLE          ; X2444 to be written to
801A 1184       44               ACALL   W_BYTE
801C C291       45               CLR     ENABLE
801E 7F00       46               MOV     R7,#0           ;R7 = X2444 address
8020 EF         47       AGAIN:  MOV     A,R7            ;put address in ACC
8021 23         48               RL      A               ;put in bits 3,4,5,6
8022 23         49               RL      A
8023 23         50               RL      A
8024 4483       51               ORL     A,#W_DATA       ;build write instruction
8026 D291       52               SETB    ENABLE
8028 1184       53               ACALL   W_BYTE
802A 7D02       54               MOV     R5,#2
802C E6         55       LOOP:   MOV     A,@R0           ;get 8051 data
802D 08         56               INC     R0              ;point to next byte
802E 1184       57               ACALL   W_BYTE          ;sent byte to X2444
8030 DDFA       58               DJNZ    R5,LOOP         ;repeat (send 2nd byte)
8032 C291       59               CLR     ENABLE
8034 0F         60               INC     R7              ;increment X2444 address
8035 BF10E8     61               CJNE    R7,#16,AGAIN    ;if not finished, again
8038 7481       62               MOV     A,#STORE        ;if finished, copy to EEPROM
803A D291       63               SETB    ENABLE
803C 1184       64               ACALL   W_BYTE
```

FIGURE 10–11
Software for X2444 interface

```
803E C291         65              CLR     ENABLE
8040 7482         66              MOV     A,#SLEEP       ;put X2444 to sleep
8042 D291         67              SETB    ENABLE
8044 1184         68              ACALL   W_BYTE
8046 C291         69              CLR     ENABLE
8048 22           70              RET                    ;DONE!
                  71
                  72     ;*****************************************************
                  73     ; RECOVER 8051 RAM LOCATIONS 60H-7FH FROM X2444 NVRAM *
                  74     ;*****************************************************
8049 7860         75     RECOVER: MOV     R0,#NVRAM
804B C291         76              CLR     ENABLE
804D 7485         77              MOV     A,#RECALL      ;recall instruction
804F D291         78              SETB    ENABLE
8051 1184         79              ACALL   W_BYTE
8053 C291         80              CLR     ENABLE
8055 7F00         81              MOV     R7,#0          ;R7 = X2444 address
8057 EF           82     AGAIN2:  MOV     A,R7           ;put address in ACC
8058 23           83              RL      A              ;build read instruction
8059 23           84              RL      A
805A 23           85              RL      A
805B 4487         86              ORL     A,#R_DATA
805D D291         87              SETB    ENABLE
805F 1184         88              ACALL   W_BYTE         ;send read instruction
8061 7D02         89              MOV     R5,#2          ; (+ address)
8063 1178         90     LOOP2:   ACALL   R_BYTE         ;read byte of data
8065 F6           91              MOV     @R0,A          ;put in 8051 RAM
8066 08           92              INC     R0             ;point to next location
8067 DDFA         93              DJNZ    R5,LOOP2
8069 C291         94              CLR     ENABLE
806B 0F           95              INC     R7             ;increment X2444 address
806C BF10E8       96              CJNE    R7,#16,AGAIN2  ;repeat until last
806F 7482         97              MOV     A,#SLEEP       ;put X2444 to sleep
8071 D291         98              SETB    ENABLE
8073 1184         99              ACALL   W_BYTE
8075 C291        100              CLR     ENABLE
8077 22          101              RET                    ;DONE!
                 102
                 103     ;*****************************************************
                 104     ; READ A BYTE OF DATA FROM X2444                      *
                 105     ;*****************************************************
8078 7E08        106     R_BYTE:  MOV     R6,#8      ;use R6 as bit counter
807A A292        107     AGAIN3:  MOV     C,DIN      ;put X2444 data bit in C
807C 33          108              RLC     A          ;build byte in Accumulator
807D D290        109              SETB    CLOCK      ;toggle clock line (1 us)
807F C290        110              CLR     CLOCK
8081 DEF7        111              DJNZ    R6,AGAIN3  ;if not last bit, do again
8083 22          112              RET
                 113
                 114     ;*****************************************************
                 115     ; WRITE A BYTE OF DATA TO X2444                       *
                 116     ;*****************************************************
8084 7E08        117     W_BYTE:  MOV     R6,#8      ;use R6 as bit counter
8086 33          118     AGAIN4:  RLC     A          ;put bit to write in C
8087 9292        119              MOV     DIN,C      ;put in X2444 DATA IN line
8089 D290        120              SETB    CLOCK      ;clock bit into X2444
808B C290        121              CLR     CLOCK
808D DEF7        122              DJNZ    R6,AGAIN4  ;if not last bit, do again
808F 22          123              RET
                 124              END
```

FIGURE 10–11
continued

set. For this example, the nonvolatile information is kept in the 8051 internal locations 60H–7FH (presumably for access by a control program executing in firmware). The software listing is shown in Figure 10–11.

The operations of saving and recovering data involve the following steps:

Write Data into the X2444

```
1. Execute RCL (recall) instruction.
2. Execute WREN (set write enable latch) instruction.
3. Write data into X2444 RAM.
4. Execute STO (store RAM into EEPROM) instruction.
5. Execute SLEEP instruction.
```

Read Data From X2444

```
1. Execute RCL (recall) instruction.
2. Read data from X2444 RAM.
3. Execute SLEEP instruction.
```

As an example of what the software drivers must do, Figure 10–12 illustrates the timing diagram to send the RCL instruction to the X2444. Several of the bits are actually "don't cares" (as specified in the data sheet); however, they are shown as 0s in the figure.

The timing for the WRITE data and READ data instructions is slightly different. For these, the 8-bit opcode is followed immediately by 16 bits of data, and chip enable remains high for all 24 bits. For the read instruction, the eight bits (the opcode) are written to the X2444, then 16 data bits are read from the X2444. Separate subroutines are used for reading eight bits (R_BYTE; lines 106–112) and writing eight bits (W_BYTE; lines 117–123). For specific details, consult the software listing.

10.7 INPUT/OUTPUT EXPANSION

Our next example illustrates a simple way to increase the number of input lines on the 8051. Three port lines are used to interface to multiple (in this example, 2) 74HC165 parallel-in serial-out shift registers. (See Figure 10–13.) The additional inputs are sampled periodically by pulsing the SHIFT/$\overline{\text{LOAD}}$ line low. The data are then read into the 8051 by reading the DATA IN line and pulsing the CLOCK line. Each pulse on the clock line shifts the data ("down," as shown in Figure 10–13), so the next read to DATA IN reads the next bit, and so on.

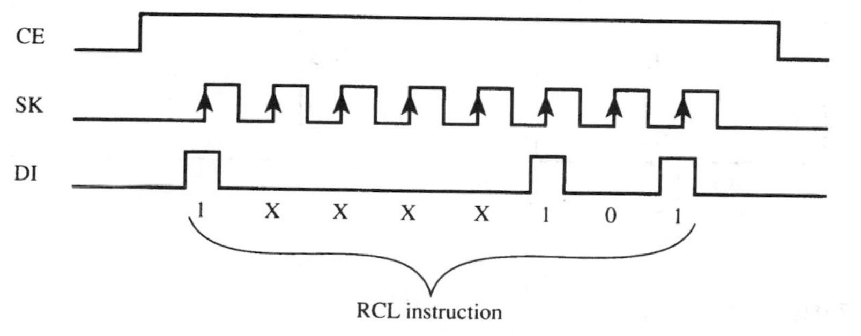

FIGURE 10–12
Timing for the X2444 recall instruction

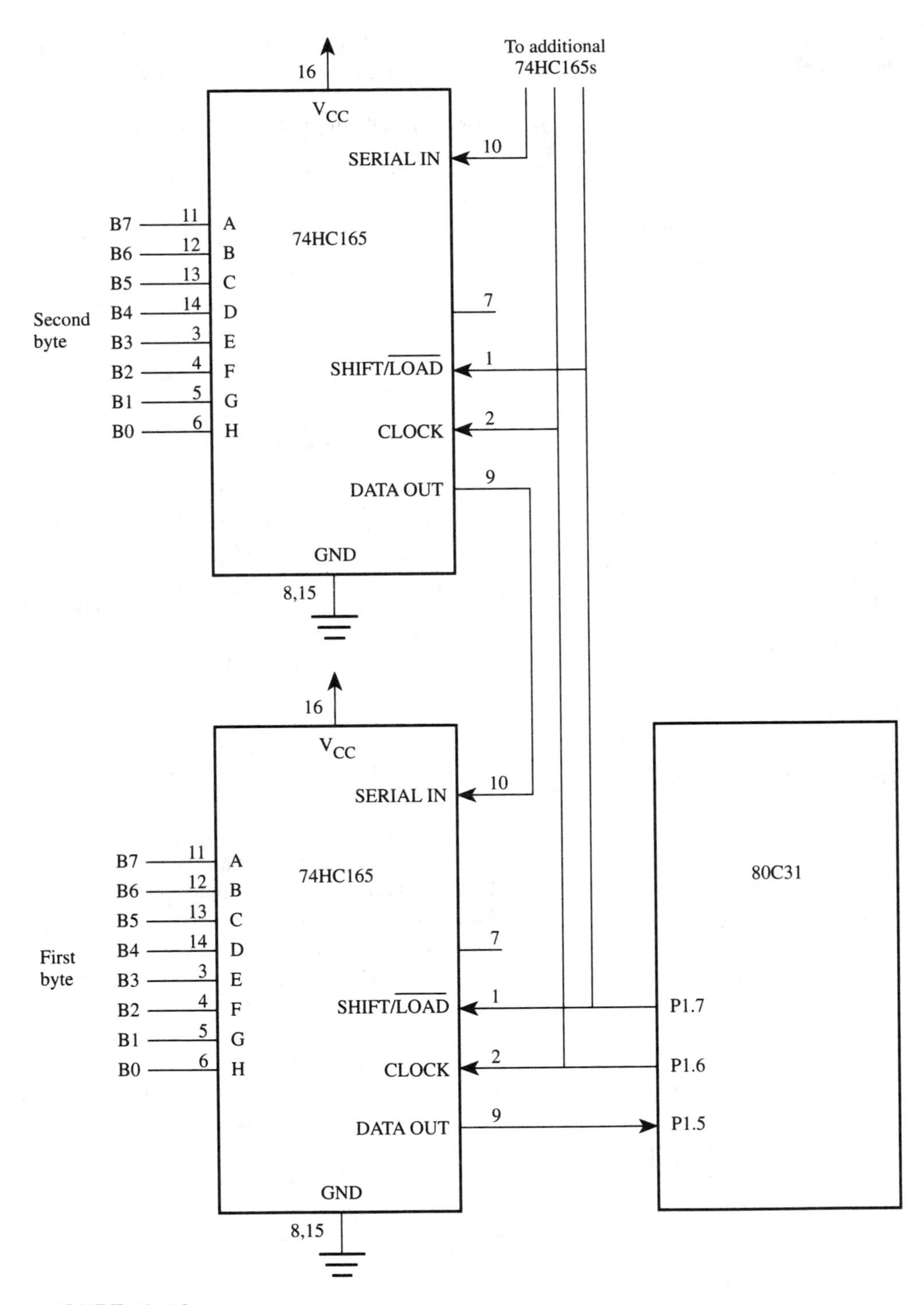

FIGURE 10–13
Interface to two 74HC165s

Design Objective

Write a subroutine that copies the state of the 16 inputs in Figure 10–13 to 8051 internal RAM locations 25H and 26H.

The software to accomplish this is shown in Figure 10–14. Note that the main program loop consists of calls to two subroutines: GET_BYTES and DISPLAY_RESULTS (lines 34–35). The latter subroutine is included to illustrate a useful technique for debugging when resources are limited. DISPLAY_RESULTS (lines 72–83) reads the data from internal locations 25H and 26H and sends each nibble to the console as a hexadecimal character. This provides a simple visual interface to verify if the program and interface are working. As input lines are toggled high and low, changes will immediately appear on the console (if the interface and program are working properly).

The GET_BYTES subroutine (lines 44–58) takes 112 μs to execute when two 74HC165s are used and the system operates from a 12 MHz crystal. If the inputs were sampled, for example, 20 times per second, GET_BYTES would consume 112 ÷ 50,000 = 0.2% of the CPU's execution time. This is minimal; however, increasing the number of input lines and/or the sampling rate may start to impact overall system performance. Consult the software listing for further details.

10.8 ANALOG OUTPUT

Interfacing to the real world often requires generating or sensing analog conditions. Generating and controlling an analog output signal from a microcontroller is easy. This design example uses two resistors, two capacitors, a potentiometer, an LM301 op amp, and MC1408L8 8-bit digital-to-analog converter (DAC). Both ICs are inexpensive and readily available. The eight data inputs to the DAC are driven from port 1 on the 8031 (see Figure 10–15). After building the circuit and connecting it to the SBC-51, it should be tested using monitor commands. Measure the output voltage at pin 6 of the LM301 (Vo) while writing different values to port 1 and adjusting the 1K potentiometer. The output should vary from 0 volts (P1 = 00H) to about 10 volts (P1 = FFH).

Once the circuit is operating correctly, we are ready to have fun with the interface software. The usual test program is a sawtooth waveform generator that sends a value to the DAC, increments the value, sends it again, and so on (see question 3 at the end of this chapter). However, we will embark on a much more ambitious design—a digitally controlled sine wave generator.

Design Objective

Write a program to generate a sine wave using the DAC interface in Figure 10–15. Use a constant call STEP to set the frequency of the sine wave. Make the program interrupt-driven with an update rate of 10 kHz.

```
                     1    $DEBUG
                     2    $NOSYMBOLS
                     3    $NOPAGING
                     4    ;FILE: HC165.SRC
                     5    ;*****************************************************
                     6    ;               74HC165 INTERFACE EXAMPLE            *
                     7    ;                                                    *
                     8    ; The subroutine GET_BYTES below reads multiple (in  *
                     9    ; this case two) 74HC165 parallel-in serial-out shift *
                    10    ; registers attached to P1.7 (SHIFT/LOAD), P1.6      *
                    11    ; (CLOCK), and P1.5 (DATA OUT).  The bytes read are  *
                    12    ; placed in bit-addressable locations starting at the *
                    13    ; byte address BUFFER.                               *
                    14    ;*****************************************************
  000D              15    CR      EQU     0DH
  0002              16    COUNT   EQU     2       ;number of 74HC165s
  0097              17    SHIFT   BIT     P1.7    ;74HC165 SHIFT/LOAD input
                    18                            ; 1 = shift, 0 = load
  0096              19    CLOCK   BIT     P1.6    ;74HC165 CLOCK input
  0095              20    DOUT    BIT     P1.5    ;74HC165 DATA OUT output
  0282              21    OUTSTR  CODE    0282H   ;subroutines in MON51(V12)
  028D              22    OUT2HEX CODE    028DH   ;output byte as two hex char.
  01DE              23    OUTCHR  CODE    01DEH
                    24
8000                25            ORG     8000H   ;begin code segment at 8000H
8000 D296           26            SETB    CLOCK   ;set interface lines initially in
8002 D297           27            SETB    SHIFT   ; case not already
8004 D295           28            SETB    DOUT    ;DOUT must be set (input)
                    29
                    30    ;*****************************************************
                    31    ; MAIN LOOP (KEPT SMALL FOR THIS EXAMPLE)            *
                    32    ;*****************************************************
8006 128029         33            CALL    SEND_HELLO_MESSAGE  ;banner message
8009 128011         34    REPEAT: CALL    GET_BYTES           ;read 74HC165s
800C 128050         35            CALL    DISPLAY_RESULTS     ;show results
800F 80F8           36            JMP     REPEAT              ;loop
                    37
                    38    ;*****************************************************
                    39    ; GET BYTES FROM 74HC165s & PLACE IN INTERNAL RAM    *
                    40    ;                                                    *
                    41    ; Execution time = 112 microseconds (@ 12 MHz).      *
                    42    ; Execution time for N 74HC165s = 6 + (N x 53) us    *
                    43    ;*****************************************************
                    44    GET_BYTES:
8011 7E02           45            MOV     R6,#COUNT   ;use R6 as byte counter
8013 7825           46            MOV     R0,#BUFFER  ;use R0 as pointer to buffer
8015 C297           47            CLR     SHIFT       ;load into 74HC165s by
8017 D297           48            SETB    SHIFT       ; pulsing SHIFT/LOAD low
8019 7F08           49    AGAIN:  MOV     R7,#8       ;use R7 as bit counter
801B A295           50    LOOP:   MOV     C,DOUT      ;get a bit (put it in C)
801D 13             51            RRC     A           ;put in ACC.0 (LSB 1st)
801E C296           52            CLR     CLOCK       ;pulse CLOCK line (shifts
8020 D296           53            SETB    CLOCK       ; bits toward DATA OUT)
8022 DFF7           54            DJNZ    R7,LOOP     ;if not 8th shift, repeat
8024 F6             55            MOV     @R0,A       ;if 8th shift, put in buf.
8025 08             56            INC     R0          ;increment pointer to buf.
8026 DEF1           57            DJNZ    R6,AGAIN    ;get two bytes
8028 22             58            RET
                    59
                    60    ;*****************************************************
                    61    ; SEND HELLO MESSAGE TO CONSOLE (DEBUGGING AID)      *
                    62    ;*****************************************************
                    63    SEND_HELLO_MESSAGE:
8029 908030         64            MOV     DPTR,#BANNER    ;point to hello message
```

FIGURE 10–14
Software for 74HC165 interface

```
802C 120282      65              CALL    OUTSTR          ;send it to console
802F 22          66              RET
8030 2A2A2A20    67      BANNER: DB      '*** TEST 74HC165 INTERFACE ***',CR,0
8034 54455354
8038 20373448
803C 43313635
8040 20494E54
8044 45524641
8048 4345202A
804C 2A2A
804E 0D
804F 00
                 68
                 69      ;****************************************************
                 70      ; DISPLAY RESULTS ON CONSOLE (DEBUGGING AID)        *
                 71      ;****************************************************
                 72      DISPLAY_RESULTS:                ;display bytes
8050 7825        73              MOV     R0,#BUFFER      ;R0 points to bytes
8052 7E02        74              MOV     R6,#COUNT       ;R6 is # of bytes read
8054 E6          75      LOOP2:  MOV     A,@R0           ;get byte
8055 08          76              INC     R0              ;increment pointer
8056 12028D      77              CALL    OUT2HEX         ;output as 2 hex char.
8059 7420        78              MOV     A,#' '          ;separate bytes
805B 1201DE      79              CALL    OUTCHR
805E DEF4        80              DJNZ    R6,LOOP2        ;repeat for each byte
8060 740D        81              MOV     A,#CR           ;begin a new line
8062 1201DE      82              CALL    OUTCHR          ;send CR (LF too!)
8065 22          83              RET
                 84
                 85      ;****************************************************
                 86      ; CREATE BUFFER IN BIT-ADDRESSABLE INTERNAL RAM     *
                 87      ;****************************************************
----             88              DSEG    AT 25H  ;on-chip data segment in
0025             89      BUFFER: DS      COUNT   ; in bit-addressable space
                 90              END
```

FIGURE 10–14
continued

Since the number-crunching capabilities of the 8031 are very limited, the only reasonable approach to this problem is to use a look-up table. We need a table with 8-bit values corresponding to one period of a sine wave. The values should start around 127, increase to 255, decrease through 127 to 0, and rise back up to 127, following the pattern of a sine wave.

A reasonable rendition of a sine wave requires a relatively large table; so the question arises, how do we generate the table? Manual methods are impractical. The easiest approach is to write a program in some other high-level language to create the table and save the entries in a file. The table is then imported into our 8031 source program and off we go. Figure 10–16 is a simple C program called table51.c that will do the job for us. The program generates a 1024-entry sine wave table with values constrained between 0 and 255. The output is written to an output file called sine51.src. Each entry is preceded with " DB " for compatibility with 8031 source code.

The 8031 sine wave program is shown in Figure 10–17. The main loop (lines 36–40) does three things: initialize timer 0 to interrupt every 100 μs, turn on interrupts, and sit in an infinite loop. The timer 0 interrupt service routine (lines 41–51) does all the work. Every 100 μs a value is read from the look-up table using the DPTR and then written to port 1. A constant called STEP is used as the increment through the table. STEP is

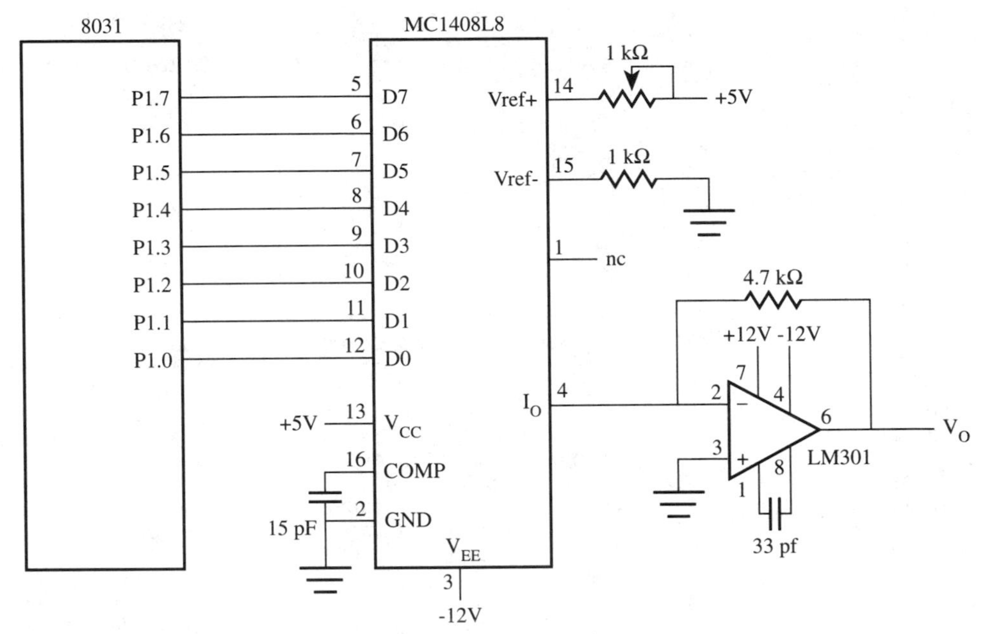

FIGURE 10–15

```
/*****************************************************/
/* table51.c - program to generate a sinewave table */
/*                                                   */
/* The table consists of 1024 entries between 0      */
/* and 255.  Each entry is preceded by " DB " for    */
/* inclusion in a 8051 source program.  The table    */
/* is written to the file sine51.src                 */
/*****************************************************/

#include <stdio.h>
#include <math.h>

#define PI   3.1415927
#define MAX  1024
#define BYTE 255

main()
{
   FILE *fp, *fopen();
   double x, y;

   fp = fopen("sine51.src", "w");
   for(x = 0; x < MAX; ++x) {
      y = ((sin((x / MAX) * (2 * PI)) + 1) / 2) * BYTE;
      fprintf(fp," DB %3d\n", (int)y);
   }
}
```

FIGURE 10–16

```
                         1      $debug
                         2      $nopaging
                         3      $nosymbols
                         4      ;FILE: DAC.SRC
                         5      ;*****************************************************
                         6      ;               MC1408L8 INTERFACE EXAMPLE           *
                         7      ;                                                    *
                         8      ; This program generates a sine wave using a sine    *
                         9      ; wave look-up table and an interface to a MC1408L8  *
                        10      ; 8-bit digital-to-analog converter.  The program is *
                        11      ; interrupt-driven.                                  *
                        12      ;                                                    *
                        13      ; Data are read from a 1024-entry sine wave table and *
                        14      ; sent to the DAC every 100 us.  Each value sent is  *
                        15      ; STEP locations past the previous value sent (with  *
                        16      ; wrap around once the end is reached).  The period  *
                        17      ; of the sine wave is 100 x (1024 / STEP) us.  For   *
                        18      ; example, if STEP is 20H, the sine wave has a period *
                        19      ; of 100 x (1024 / 32) = 3.2 ms and a frequency of   *
                        20      ; 313 Hz.                                            *
                        21      ;                                                    *
                        22      ; Note:  Initialize STEP in internal location 50H    *
                        23      ; before running the program.                        *
                        24      ;*****************************************************
  00BC                  25      MONITOR  CODE    00BCH       ;MON51 entry (V12)
  0050                  26      STEP     DATA    50H         ;put STEP in internal RAM
                        27
8000                    28               ORG     8000H       ;start at 8000H
8000 028015             29               LJMP    MAIN        ;initialize timer
8003 028037             30               LJMP    EXT0ISR     ;unsused
8006 028022             31               LJMP    T0ISR       ;every 100 us, update DAC
8009 028037             32               LJMP    EXT1ISR     ;unused
800C 028037             33               LJMP    T1ISR       ;unused
800F 028037             34               LJMP    SPISR       ;unused
8012 028037             35               LJMP    T2ISR       ;unused
8015 758902             36      MAIN:    MOV     TMOD,#02H   ;8-bit, auto reload
8018 758C9C             37               MOV     TH0,#-100   ;100 us delay
801B D28C               38               SETB    TR0         ;start timer
801D 75A882             39               MOV     IE,#82H     ;enable timer 0 interrupts
8020 80FE               40               SJMP    $           ;main loop does nothing!
8022 E550               41      T0ISR:   MOV     A,STEP      ;add STEP to DPTR
8024 2582               42               ADD     A,DPL
8026 F582               43               MOV     DPL,A
8028 5002               44               JNC     SKIP
802A 0583               45               INC     DPH
802C 538303             46      SKIP:    ANL     DPH,#03H    ;wrap around, if necessary
802F 438384             47               ORL     DPH,#HIGH(TABLE)
8032 E4                 48               CLR     A
8033 93                 49               MOVC    A,@A+DPTR   ;get entry
8034 F590               50               MOV     P1,A        ;send it
8036 32                 51               RETI
                        52
                        53      EXT0ISR:                     ;unused interrupts
                        54      EXT1ISR:
                        55      T1ISR:
                        56      T2ISR:
8037 C2AF               57      SPISR:   CLR     EA          ;turn off interrupts and
8039 0200BC             58               LJMP    MONITOR     ; return to MON51
                        59
                        60      ;*****************************************************
                        61      ; The following is a sine wave look-up table.  The   *
                        62      ; table contains 1024 entries and is ORGed to begin  *
                        63      ; at 8400H to allow easy wrap-around of the DPTR     *
                        64      ; once the end of the table is reached.  The entries *
```

FIGURE 10–17

```
                       65        ; are 8-bits each (0 to 255) for output to an 8-bit   *
                       66        ; DAC.  The table was generate from a C program and   *
                       67        ; read into this 8051 program.                         *
                       68        ;*****************************************************
8400                   69              ORG        8400H
8400 7F                70        TABLE: DB        127
8401 80                71              DB         128
8402 81                72              DB         129
8403 81                73              DB         129
8404 82                74              DB         130
                       75        ; Listing turned off after first five entries
                       76        ;----------------------------------------------------
                       77 +1     $NOLIST
                     1093        ; Listing turned back on for last five entries
87FB 7B              1094              DB         123
87FC 7C              1095              DB         124
87FD 7D              1096              DB         125
87FE 7D              1097              DB         125
87FF 7E              1098              DB         126
                     1099              END
```

FIGURE 10–17
continued

defined in line 26 as a byte in internal RAM. It must be initialized using a monitor command. Within each ISR, STEP is added to DPTR to get the address of the next sample. The table is ORGed at 8400H (line 69) so it starts on an even 1K boundary. If the DPTR is incremented past 87FFH (the end of the table), it is adjusted to wrap around through the beginning of the table. Since the table is so big, a $NOLIST assembler directive was used after the first five entries (line 77) to shut off output to the listing file. A $LIST directive was used in line 1092 (not shown) to turn the listing back on for the last five entries. The frequency of the sine wave is controlled by three parameters: STEP, the size of the table, and the timer interrupt period, as explained in lines 16–20 of the listing.

10.9 ANALOG INPUT

Our final design example is an analog input channel. The circuit in Figure 10–18 uses one resistor, one capacitor, a trimpot, and an ADC0804 analog-to-digital converter (ADC). The ADC0804 is an inexpensive ADC (National Semiconductor Corp.) that converts an input voltage to an 8-bit digital word in about 100 μs.

The ADC0804 is controlled by a write input ($\overline{\text{WR}}$) and an interrupt output ($\overline{\text{INTR}}$). A conversion is started by pulsing $\overline{\text{WR}}$ low. When the conversion is complete (100 μs later), the ADC0804 asserts $\overline{\text{INTR}}$, making it low. $\overline{\text{INTR}}$ is de-asserted (high) on the next 1-to-0 transition of $\overline{\text{WR}}$, that initiates the next conversion. $\overline{\text{INTR}}$ and $\overline{\text{WR}}$ connect to the 8031 lines P1.1 and P1.0, respectively. For this example, we use Port A of the 8155 for the data transfer, as shown in the figure.

The ADC0804 operates from an internal clock created by the RC network connecting to pins 19 and 4. The analog input voltage is a differential signal applied to the Vin(+) and Vin(−) inputs on pins 6 and 7. For this example Vin(−) is grounded and Vin(+) is driven from the center tap of the trimpot. Vin(+) will range from 0 to +5 volts, as controlled by the trimpot. Consult the data sheet for a detailed description of the operation of the ADC0804.

FIGURE 10–18

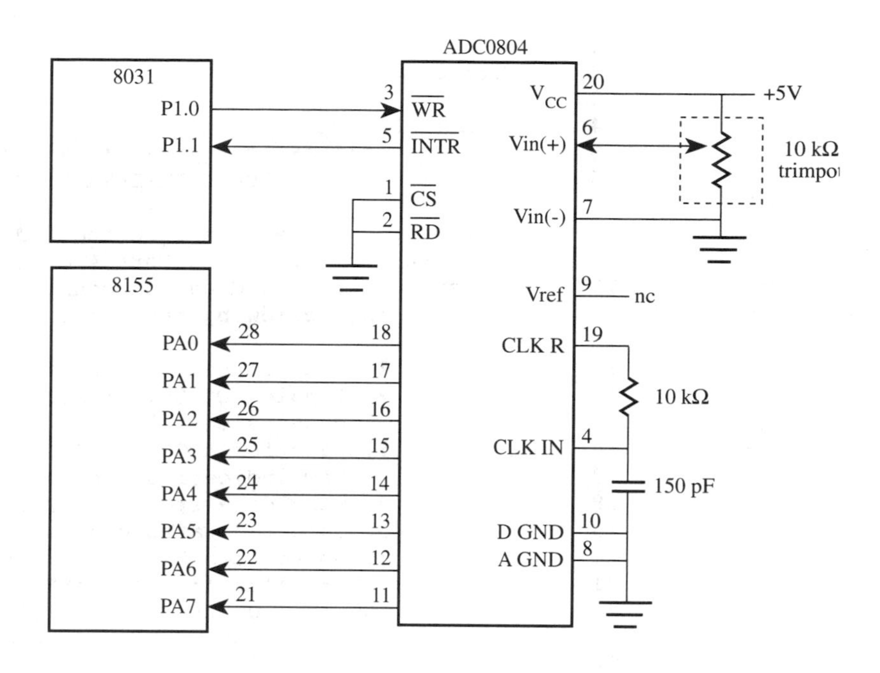

Design Objective

Write a program to continually sense the voltage at the trimpot's center tap (as converted by the ADC0804). Report the result on thc console as an ASCII bytc.

The program in Figure 10–19 achieves the objective stated above. Since the 8155 ports default to input upon reset, an initialize sequence is not necessary. Port A is at address 0101H in external memory and is easily read using a MOVX instruction. A conversion is started by clearing and setting P1.0 (lines 34–35), the ADC0804's $\overline{\text{WR}}$ input. Then, the program sits in a loop waiting for the ADC0804 to finish the conversion and assert $\overline{\text{INTR}}$ at P1.1 (line 36). The data are read in lines 37 and 38 and then sent to the console using MON51's OUT2HX subroutine (line 39). As the program runs, a byte is displayed on the console. It will range from 00H to FFH as the trimpot is adjusted.

The program in Figure 10–19 is a rough first approximation of the potential for analog input. It is possible to replace the trimpot with other analog inputs. Temperature sensing is achieved using a thermistor—a device with a resistance that varies with temperature. Speech input is possible using a microphone. The ADC80804's conversion period of 100 μs translates into a sampling frequency of 10 kHz. This is sufficient to capture signals with up to 5 kHz bandwidth, roughly equivalent to a voice-grade telephone line. Additional circuitry is required to boost the low-level signals provided by typical microphones to the 0–5 volt range expected by the ADC0804. Additionally, a sample-and-hold circuit is needed to maintain a constant voltage for the duration of each conversion. We'll leave it to the reader to explore these possibilities.

```
1       $debug
2       $nopaging
3       $nosymbols
4       ;FILE: ADC.SRC
5       ;*****************************************************
6       ;              ADC0804 INTERFACE EXAMPLE              *
7       ;                                                     *
8       ; This program reads analog input data from an        *
9       ; ADC0804 interfaced to Port A of the 8155.  The      *
10      ; result is reported on the console as a hexadecimal  *
11      ; byte.  The following steps occur:                   *
12      ;                                                     *
13      ;       1. Send banner message to console             *
14      ;       2. Toggle ADC0804 -WR line (P1.0) to begin    *
15      ;          conversion                                 *
16      ;       3. Wait for ADC8084 -INTR line (P1.1) to go   *
17      ;          low indicating "conversion complete"       *
18      ;       4. Read data from 8155 Port A                 *
19      ;       5. Output data to console in hexadecimal      *
20      ;       6. Go to step 2                               *
21      ;*****************************************************
22      PORTA   EQU     0101H          ;8155 Port A
23      CR      EQU     0DH            ;ASCII carriage return
24      LF      EQU     0AH            ;ASCII line feed
25      ESC     EQU     1BH            ;ASCII escape
26      OUT2HX  EQU     028DH          ;MON51 subroutine
27      OUTSTR  EQU     0282H          ;MON51 subroutine
28      WRITE   BIT     P1.0           ;ADC0804 -WR line
29      INTR    BIT     P1.2           ;ADC0804 -INTR line
30
31              ORG     8000H
32      ADC:    MOV     DPTR,#BANNER   ;send message
33              CALL    OUTSTR
34      LOOP:   CLR     WRITE          ;toggle -WR line
35              SETB    WRITE
36              JB      INTR,$         ;wait for -INTR = 0
37              MOV     DPTR,#PORTA    ;init DPTR --> Port A
38              MOVX    A,@DPTR        ;read ADC0804 data
39              CALL    OUT2HX         ;send data to console
40              MOV     DPTR,#LEFT2    ;back-up cursor by 2
41              CALL    OUTSTR
42              SJMP    LOOP           ;repeat
43
44      BANNER: DB      '*** TEST ADC0804 ***',CR,0
45      LEFT2:  DB      ESC,'[2D',0    ;VT100 escape sequence
46              END
```

FIGURE 10–19

10.10 SUMMARY

This concludes our examination of interface examples. The designs presented illustrate many of the concepts required to implement sophisticated interfaces using an 8051 microcontroller.

There is no substitute for hands-on experience, however. The examples in this chapter, and those presented earlier, are best understood through a trial-and-error process. Implementing the examples on a real system, such as the SBC-51, is the best way to develop the concepts presented in this book. This book has provided a basis for students to explore further the possibilities of using a microcontroller such as the 8051 in minimum component designs.

PROBLEMS

1. Use the loudspeaker interface in Figure 10–7 to repeatedly play the musical melody shown in Figure 10–20.
2. Reconfigure the 74HC165 interface in Figure 10–13 and rewrite the accompanying software in Figure 10–14 to use the 8051's serial port (in mode 0) for the clock and data lines.
3. If the following program is used with the digital-to-analog output circuit in Figure 10–15, a sawtooth wave results. (Assume 12 MHz operation.)

```
STEP EQU     1
MAIN MOV     P1,A
     ADD     A,#STEP
     SJMP    MAIN
```

   (a) What is the frequency of the sawtooth wave?

   (b) What value for STEP will achieve an output frequency of 10 kHz (approximately)?

   (c) Derive the equation for frequency, given STEP.

4. Several of the programs in this chapter used MON51 subroutines. If a program is to use the subroutine ISDIG (to check if a byte is an ASCII digit), what address from MON51 must be equated to the symbol ISDIG?
5. Write a program to create a 1 kHz square wave on P1.7 using the 8155 timer and external 0 interrupts. (Hint: consult an 8155 data sheet.)
6. Assume the 74HC165 interface in Figure 10–13 is expanded to 6 ICs, providing 48 additional input lines.

   (a) How should the program in Figure 10–14 be modified to read the 48 input lines?

   (b) Where will the data be stored in the 8031's internal RAM?

   (c) What is the new duration of the GET_BYTES subroutine in Figure 10–14?

   (d) If GET_BYTES is placed in an interrupt service routine that executes every second, what percentage of the CPU's execution time is spent reading the 48 inputs from the 74HC165s?

FIGURE 10–20
Musical melody for problem 1

7. In Figure 10–17, the constant STEP was defined as a byte of internal data at location 50H using the following assembler directive:

```
STEP DATA     50H
```

This is the correct way to define STEP, however the following would also work:

```
STEP EQU      50H
```

In the latter case, type-checking is not performed by ASM51 when the program is assembled. Give an example of an incorrect use of the label STEP that *would not* generate an assemble-time error if STEP were defined with EQU, but *would* generate an error if STEP were defined properly, using DATA.

APPENDIX A QUICK REFERENCE CHART

MNEMONIC		DESCRIPTION
Arithmetic Operations		
ADD	A,source	add source to A
ADD	A,#data	
ADDC	A,#source	add with carry
ADDC	A,#data	
SUBB	A,source	subtract from A
SUBB	A,#data	with borrow
INC	A	increment
INC	source	
DEC	A	decrement
DEC	source	
INC	DPTR	increment DPTR
MUL	AB	multiply A & B
DIV	AB	divide A by B
DA	A	decimal adjust A
Logical Operations		
ANL	A,source	logical AND
ANL	A,#data	
ANL	direct,A	
ANL	direct,#data	
ORL	A,source	logical OR
ORL	A,#data	
ORL	direct,A	
ORL	direct,#data	
XRL	A,source	logical XOR
XRL	A,#data	
XRL	direct,A	
XRL	direct,#data	
CLR	A	clear A
CPL	A	complement A
RL	A	rotate A left
RLC	A	(through C)
RR	A	rotate A right
RRC	A	(through C)
SWAP	A	swap nibbles

LEGEND

Rn	register addressing using R0–R7
direct	8-bit internal address (00H-FFH)
@Ri	indirect addressing using R0 or R1
source	any of [Rn,direct,@ri]
dest	any of [Rn,direct,@ri]
#data	8-bit constant included in instr.
#data 16	16-bit constant
bit	8-bit direct address of bit
rel	signed 8-bit offset
addr11	11-bit address in current 2k page
addr16	16-bit address

FIGURE A–1
Quick reference chart

MNEMONIC		DESCRIPTION	MNEMONIC		DESCRIPTION
Data Transfer			ORL	C,bit	OR bit with C
Operations			ORL	C,/bit	OR NOT bit with C
MOV	A,source	move source	MOV	C,bit	move bit to bit
MOV	A,#data	to destination	MOV	bit,C	
MOV	dest,A		JC	rel	jump if C set
MOV	dest,source		JNC	rel	if C not set
MOV	dest,#data		JB	bit,rel	jump if bit set
MOV	DPTR,#data16		JNB	bit,rel	if bit not set
MOVC	A,@A+DPTR	move from code	JBC	bit,rel	if set then clear
MOVC	A,@A+PC	memory			
MOVX	A,@Ri	move from data	**Program Branching**		
MOVX	A,@DPTR	memory	ACALL	addr11	call subroutine
MOVX	@Ri,A		LCALL	addr16	
MOVX	@DPTR,A		RET		return from sub.
PUSH	direct	push onto stack	RETI		from interrupt
POP	direct	pop from stack	AJMP	addr11	jump
XCH	A,source	exchange bytes	LJMP	addr16	
XCHD	A,@Ri	exchange low order	SJMP	rel	
		digits	JMP	@A+DPTR	
			JZ	rel	jump if A = 0
Boolean Variable Manipulation			JNZ	rel	if A not = 0
CLR	C	clear bit	CJNE	A,direct,rel	compare and jump
CLR	bit		CJNE	A,#data,rel	if not equal
SETB	C	set bit	CJNE	Rn,#data,rel	
SETB	bit		CJNE	@Ri,#data,rel	
CPL	C	complement bit	DJNZ	Rn,rel	decrement and
CPL	bit				jump
ANL	C,bit	AND bit with C	DJNZ	direct,rel	if not zero
ANL	C,/bit	AND NOT bit with C	NOP		no operation

FIGURE A–1
continued

B

APPENDIX B
OPCODE MAP

Instruction Code Summary

L \ H	0	1	2	3	4	5	6	7	8	9	A	B	C	D	E	F
0	NOP	JBC bit, rel	JB bit, rel	JNB bit, rel	JC rel	JNC rel	JZ rel	JNZ rel	SJMP rel	MOV DPTR, # data 16	ORL C, /bit	ANL C. /bit	PUSH dir	POP dir	MOVX A, @DPTR	MOVX @DPTR, A
1	AJMP (P0)	ACALL (P0)	AJMP (P1)	ACALL (P1)	AJMP (P2)	ACALL (P2)	AJMP (P3)	ACALL (P3)	AJMP (P4)	ACALL (P4)	AJMP (P5)	ACALL (P5)	AJMP (P6)	ACALL (P6)	AJMP (P7)	ACALL (P7)
2	LJMP addr16	LCALL addr16	RET	RETI	ORL dir, A	ANL dir, A	XRL dir, A	ORL C, bit	ANL C, bit	MOV C, bit	MOV C, bit	CPL bit	CLR bit	SETB bit	MOVX A, @R0	MOVX @R0, A
3	RR A	RRC A	RL A	RLC A	ORL dir, # data	ANL dir, # data	XRL dir, # data	JMP @A+DPTR	MOVC A, @A+PC	MOVC A, @A+DPTR	INC DPTR	CPL C	CLR C	SETB C	MOVX A, @R1	MOVX @R1, A
4	INC A	DEC A	ADD A, # data	ADDC A, # data	ORL A, # data	ANL A, # data	XRL A, # data	MOV A, # data	DIV AB	SUBB A, # data	MUL AB	CJNE A. # data, rel	SWAP A	DA A	CLR A	CPL A
5	INC dir	DEC dir	ADD A, dir	ADDC A, dir	ORL A, dir	ANL A, dir	XRL A, dir	MOV dir, # data	MOV dir, dir	SUBB A, dir		CJNE A, dir, rel	XCH A, dir	DJNZ dir, rel	MOV A, dir	MOV dir. A
6	INC @R0	DEC @R0	ADD A, @R0	ADDC A, @R0	ORL A, @R0	ANL A, @R0	XRL A, @R0	MOV @R0, # data	MOV dir, @R0	SUBB A, @R0	MOV @R0, dir	CJNE @R0, # data, rel	XCH A, @R0	XCHD A, @R0	MOV A, @R0	MOV @ R0, A
7	INC @R1	DEC @R1	ADD A, @R1	ADDC A, @R1	ORL A, @R1	ANL A, @R1	XRL A, @R1	MOV @R1, # data	MOV dir, @R1	SUBB A, @R1	MOV @R1, dir	CJNE @R1, # data, rel	XCH A, @R1	XCHD A, @R1	MOV A, @R1	MOV @ R1, A
8	INC R0	DEC R0	ADD A, R0	ADDC A, R0	ORL A, R0	ANL A, R0	XRL A, R0	MOV R0, # data	MOV dir, R0	SUBB A, R0	MOV R0, dir	CJNE R0, # data, rel	XCH A, R0	DJNZ R0, rel	MOV A, R0	MOV R0, A
9	INC R1	DEC R1	ADD A, R1	ADDC A, R1	ORL A, R1	ANL A, R1	XRL A, R1	MOV R1, # data	MOV dir, R1	SUBB A, R1	MOV R1, dir	CJNE R1, # data, rel	XCH A, R1	DJNZ R1, rel	MOV A, R1	MOV R1, A
A	INC R2	DEC R2	ADD A, R2	ADDC A, R2	ORL A, R2	ANL A, R2	XRL A, R2	MOV R2, # data	MOV dir, R2	SUBB A, R2	MOV R2, dir	CJNE R2, # data, rel	XCH A, R2	DJNZ R2, rel	MOV A, R2	MOV R2, A
B	INC R3	DEC R3	ADD A, R3	ADDC A, R3	ORL A, R3	ANL A, R3	XRL A, R3	MOV R3, # data	MOV dir, R3	SUBB A, R3	MOV R3, dir	CJNE R3, # data, rel	XCH A, R3	DJNZ R3, rel	MOV A, R3	MOV R3, A
C	INC R4	DEC R4	ADD A, R4	ADDC A, R4	ORL A, R4	ANL A, R4	XRL A, R4	MOV R4, # data	MOV dir, R4	SUBB A, R4	MOV R4, dir	CJNE R4, # data, rel	XCH A, R4	DJNZ R4, rel	MOV A, R4	MOV R4, A
D	INC R5	DEC R5	ADD A, R5	ADDC A, R5	ORL A, R5	ANL A, R5	XRL A, R5	MOV R5, # data	MOV dir, R5	SUBB A, R5	MOV R5, dir	CJNE R5, # data, rel	XCH A, R5	DJNZ R5, rel	MOV A, R5	MOV R5, A
E	INC R6	DEC R6	ADD A, R6	ADDC A, R6	ORL A, R6	ANL A, R6	XRL A, R6	MOV R6, # data	MOV dir, R6	SUBB A, R6	MOV R6, dir	CJNE R6, # data, rel	XCH A, R6	DJNZ R6, rel	MOV A, R6	MOV R6, A
F	INC R7	DEC R7	ADD A, R7	ADDC A, R7	ORL A, R7	ANL A, R7	XRL A, R7	MOV R7, # data	MOV dir, R7	SUBB A, R7	MOV R7, dir	CJNE R7, # data, rel	XCH A, R7	DJNZ R7, rel	MOV A, R7	MOV R7, A

2Byte	3Byte
2Cycle	4Cycle

FIGURE B–1
Opcode map

C

APPENDIX C INSTRUCTION DEFINITIONS[1]

LEGEND

Symbol	Interpretation
←	is replaced by . . .
( )	the contents of . . .
(( ))	the data pointed at by . . .
rrr	one of 8 registers; 000 = R0, 001 = R1, etc.
dddddddd	data bits
aaaaaaaa	address bits
bbbbbbbb	address of a bit
i	indirect addressing using R0 (i = 0) or R1 (i = 1)
eeeeeeee	8-bit relative address

ACALL addr11

Function: Absolute Call

Description: ACALL unconditionally calls a subroutine located at the indicated address. The instruction increments the PC twice to obtain the address of the following instruction, then pushes the 16-bit result onto the stack (low-order byte first) and increments the stack pointer twice. The destination address is obtained by successively concatenating the 5 high-order bits of the incremented PC, opcode bits 7–5, and the second byte of the instruction. The subroutine called must therefore start within the same 2K block of the program

[1]Adapted from *8-Bit Embedded Controllers* (270645). Santa Clara, CA: Intel Corporation, 1991, by permission of Intel Corporation.

memory as the first byte of the instruction following ACALL. No flags are affected.

Example: Initially SP equals 07H. The label "SUBRTN" is a program memory location 0345H. After executing the instruction,

```
ACALL    SUBRTN
```

at location 0123H, the SP contains 09H, internal RAM locations 08H and 09H contain 25H and 01H, respectively, and the PC contains 0345H.

Bytes: 2

Cycles: 2

Encoding: aaa10001 aaaaaaaa

Note: aaa = A10–A8 and aaaaaaaa = A7–A0 of the destination address.

Operation: (PC) ←(PC) + 2
(SP) ←(SP) + 1
((SP)) ←(PC7–PC0)
(SP) ←(SP) + 1
((SP)) ←(PC15–PC8)
(PC10–PC0) ←page address

ADD A, <src-byte>

Function: Add

Description: ADD adds the byte variable indicated to the accumulator, leaving the result in the accumulator. The carry and auxiliary-carry flags are set, respectively, if there is a carry-out from bit 7 or bit 3, and cleared otherwise. When adding unsigned integers, the carry flag indicates an overflow occurred.

OV is set if there is a carry-out of bit 6 but not out of bit 7, or a carry-out of bit 7 but not out of bit 6; otherwise OV is cleared. When adding signed integers, OV indicates a negative number is produced as the sum of 2 positive operands, or a positive sum from 2 negative operands.

Four source-operand addressing modes are allowed: register, direct, register-indirect, or immediate.

Example: The accumulator holds 0C3H (00001100B) and register 0 holds 0AAH (10101010B). The instruction,

```
ACC  A,R0
```

leaves 6DH (01101101B) in the accumulator with the AC flag cleared and both the carry flag and OV set to 1.

ADD A,Rn

Bytes: 1
Cycles: 1
Encoding: 00101rrr
Operation: (A) ←(A) + (Rn)

ADD A,direct

Bytes: 2
Cycles: 1
Encoding: 00100101 aaaaaaaa
Operation: (A) ←(A) + (direct)

ADD A,@Ri

Bytes: 1
Cycles: 1
Encoding: 0010011i
Operation: (A) ←(A) + ((Ri))

ADD A,#data

Bytes: 2
Cycles: 1
Encoding: 00100100 dddddddd
Operation: (A) ←(A) + #data

ADDC A, <src-byte>

Function: Add with Carry

Description: ADDC simultaneously adds the byte variable indicated, the carry flag, and the accumulator contents, leaving the result in the accumulator. The carry and auxiliary-carry flags are set, respectively, if there is a carry-out from bit 7 or bit 3, and cleared otherwise. When adding unsigned integers, the carry flag indicates an overflow occurred.

OV is set if there is a carry-out of bit 6 but not out of bit 7, or a carry-out of bit 7 but not out of bit 6; otherwise OV is cleared. When adding signed integers, OV indicates a negative number is produced as the sum of 2 positive operands, or a positive sum from 2 negative operands.

Four source-operand addressing modes are allowed: register, direct, register-indirect, or immediate.

Example: The accumulator holds 0C3H (11000011B) and register 0 holds 0AAH (10101010B) with the carry flag set. The instruction,

```
ADDC A,R0
```

leaves 6EH (01101110B) in the accumulator with AC cleared and both the carry flag and OV set to 1.

ADDC A,Rn

Bytes: 1

Cycles: 1

Encoding: 00110rrr

Operation: (A) ←(A) + (C) + (Rn)

ADDC A,direct

Bytes: 2

Cycles: 1

Encoding: 00110101 aaaaaaaa

Operation: (A) ←(A) + (C) + (direct)

ADDC A,@Ri

Bytes: 1

Cycles: 1

Encoding: 0011011i

Operation: (A) ←(A) + (C) + (Ri)

ADDC A,#data

Bytes: 2

Cycles: 1

Encoding: 00110100 dddddddd

Operation: (A) ←(A) + (C) + #data

AJMP addr11

Function: Absolute Jump

Description: AJMP transfers program execution to the indicated address, which is formed at run-time by concatenating the high-order 5 bits of the PC (*after* incrementing the PC twice), opcode bits 7–5, and the second byte of the instruction. The destination must therefore be within the same 2K block of program memory as the first byte of the instruction following AJMP.

Example: The label "JMPADR" is at program memory location 0123H. The instruction,

```
AJMP JMPADR
```

is at location 0345H and loads the PC with 0123H.

Bytes: 2

Cycles: 2

Encoding: aaa00001 aaaaaaaa

Note: aaa = A10–A8 and aaaaaaaa = A7–A0 of the destination address.

Operation: (PC) ←(PC) + 2
(PC10–PC0) ←page address

ANL <dest-byte>,<src-byte>

Function: Logical-AND for byte variables

Description: ANL performs the bitwise logical-AND operation between the variables indicated and stores the results in the destination variable. No flags are affected.

The 2 operands allow 6 addressing mode combinations. When the destination is the accumulator, the source can use register, direct, register-indirect, or immediate addressing; when the destination is a direct address, the source can be the accumulator or immediate data.

Note: When this instruction is used to modify an output port, the value used as the original port data is read from the output data latch, *not* the input pins.

Example: If the accumulator holds 0C3H (11000011B) and register 0 holds 55H (01010101B) then the instruction,

```
ANL  A,R0
```

leaves 41H (01000001H) in the accumulator.

When the destination is a directly addressed byte, this instruction clears combinations of bits in any RAM location or hardware register. The mask byte determining the pattern of bits to be cleared is either a constant contained in the instruction or a value computed in the accumulator at run-time. The instruction,

```
ANL  P1,#01110011B
```

clears bits 7, 3, and 2 of output port 1.

ANL A,Rn

Bytes: 1
Cycles: 1
Encoding: 01011rrr
Operation: (A) ←(A) AND (Rn)

ANL A,direct

Bytes: 2
Cycles: 1
Encoding: 01010101 aaaaaaaa
Operation: (A) ←(A) AND (direct)

ANL A,@Ri

Bytes: 1
Cycles: 1
Encoding: 0101011i
Operation: (A) ←(A) AND((Ri))

ANL A,#data

Bytes: 2
Cycles: 1
Encoding: 01010100 dddddddd
Operation: (A) ←(A) AND #data

ANL direct,A

Bytes: 2
Cycles: 1
Encoding: 01010010 aaaaaaaa
Operation: (direct) ←(direct) AND (A)

ANL direct,#data

Bytes: 3
Cycles: 2
Encoding: 01010011 aaaaaaaa dddddddd
Operation: (direct) ←(direct) AND #data

ANL C,<src-bit>

Function: Logical-AND for bit variables

Description: If the Boolean value of the source bit is a logical 0, then clear the carry flag; otherwise leave the carry flag in its current state. A slash (/) preceding the operand in the assembly language program indicates that the logical complement of the addressed bit is used as the source value, *but the source bit itself is not affected.* No other flags are affected.

Only direct addressing is allowed for the source operand.

Example: Set the carry flag if, and only if, P1.0 = 1, ACC.7 = 1, and OV = 0:

```
MOV  C,P1.0   ;LOAD C WITH INPUT PIN STATE
ANL  C,ACC.7  ;AND CARRY WITH ACC BIT 7
ANL  C,/OV    ;AND WITH IN VERSE OF OV  FLAG
```

ANL C,bit

Bytes: 2

Cycles: 2

Encoding: 10000010 bbbbbbbb

Operation: (C) ←(C) AND (bit)

ANL C,/bit

Bytes: 2

Cycles: 2

Encoding: 10110000 bbbbbbbb

Operation: (C) ←(C) AND NOT(bit)

CALL (See ACALL, or LCALL)

CJNE <dest-byte>,<src-byte>,rel

Function: Compare and Jump if Not Equal

Description: CJNE compares the magnitudes of the first 2 operands, and branches if their values are not equal. The branch destination is computed by adding the signed relative-displacement in the last instruction byte to the PC, after incrementing the PC to the start of the next instruction. The carry flag is set if the unsigned integer value of <dest-byte>is less than the unsigned integer value of <src-byte>; otherwise, the carry flag is cleared. Neither operand is affected.

The first 2 operands allow 4 addressing mode combinations: the accumulator may be compared with any directly addressed byte or immediate data, and any indirect RAM location or working register can be compared with an immediat constant.

Example: The accumulator contains 34H. Register 7 contains 56H. The first instruction in the sequence

```
                CJNE R7,#60H,NOT_EQ
;               ... ...                     ;R7 = 60H
NOT_EQ:         JC   REG_LOW                ;IF R7 < 60H
;               ... ...                     ;R7 > 60H
REG_LOW:        ... ...                     ;R7 < 60H
```

sets the carry flag and branches to the instruction at label NOT_EQ. By testing the carry flag, this instruction determines whether R7 is greater than or less than 60H.

If the data being presented to Port 1 is also 34H, then the instruction,

```
WAIT:           CJNE A,P1,WAIT
```

clears the carry flag and continues with the next instruction, since the accumulator does equal the data read from Port 1. (If some other value is inputted on P1, the program loops at this point until the P1 data changes to 34H.)

CJNE A,direct,rel

Bytes: 3

Cycles: 2

Encoding: 10110101 aaaaaaaa eeeeeeee

Operation: (PC) ←(PC) + 3
IF (A) <>(direct)
THEN
(PC) ←(PC) + relative address
IF (A) <(direct)
THEN
(C) ←1
ELSE
(C) ←0

CJNE A,#data,rel

Bytes: 3

Cycles: 2

Encoding: 10110100 dddddddd eeeeeeee

Operation:
(PC) ←(PC) + 3
IF (A) <>data
THEN
    (PC) ←(PC) + relative address
IF (A) <data
THEN
    (C) ←1
ELSE
    (C) ←0

CJNE Rn,#data,rel

Bytes: 3

Cycles: 2

Encoding: 10111rrr dddddddd eeeeeeee

Operation:
(PC) ←(PC) + 3
IF (Rn) <>data
THEN
    (PC) ←(PC) + relative address
IF (Rn) <data
THEN
    (C) ←1
ELSE
    (C) ←0

CJNE @Ri,#data,rel

Bytes: 3

Cycles: 2

Encoding: 1011011i eeeeeeee

Operation:
(PC) ←(PC) + 3
IF ((Ri)) <>data
THEN
    (PC) ←(PC) + relative address
IF ((Ri)) <data
THEN
    (C) ←1
ELSE
    (C) ←0

CLR A

Function: Clear Accumulator

Description: The accumulator is cleared (all bits set to 0). No flags are affected.

Example: The accumulator contains 5CH (01011100B). The instruction,

```
CLR   A
```

leaves the accumulator set to 00H (00000000B).

Bytes: 1

Cycles: 1

Encoding: 11100100

Operation: (A) ←0

CLR bit

Function: Clear bit

Description: The indicated bit is cleared (reset to 0). No other flags are affected. CLR can operate on the carry flag or any directly addressable bit.

Example: Port 1 has previously been written with 5DH (01011101B). The instruction,

```
CLR   P1.2
```

leaves the port set to 59H (01011001B).

CLR C

Bytes: 1

Cycles: 1

Encoding: 11000011

Operation: (C) ←0

CLR bit

Bytes: 2

Cycles: 1

Encoding: 11000010 bbbbbbbb

Operation: (bit) ←0

CPL A

Function: Complement Accumulator

Description: Each bit of the accumulator is logically complemented (1's complement). Bits that previously contained a 1 are changed to a 0 and vice versa. No flags are affected.

Example: The accumulator contains 5CH (01011100B). The instruction,

```
CPL   A
```

leaves the accumulator set to 0A3H (10100011B).

Bytes: 1

Cycles: 1

Encoding: 11110100

Operation: (A) ←NOT(A)

CPL bit

Function: Complement bit

Description: The bit variable specified is complemented. A bit that was a 1 is changed to 0 and vice versa. No other flags are affected. CLR can operate on the carry or any directly addressable bit.

Note: When this instruction is used to modify an output pin, the value used as the original data is from the output data latch, *not* the input pin.

Example: Port 1 has previously been written with 5BH (01011011B). The instructions,

```
CPL  P1.1
CPL  P1.2
```

leave the port set to 5BH (01011011B).

CPL C

Bytes: 1

Cycles: 1

Encoding: 10110011

Operation: (C) ←NOT(C)

CPL bit

Bytes: 2

Cycles: 1

Encoding: 10110010 bbbbbbbb

Operation: (bit) ←NOT(bit)

DA A

Function: Decimal-adjust Accumulator for Addition

Description: DA A adjusts the 8-bit value in the accumulator resulting from the earlier addition of 2 variables (each in packed-BCD format), producing 2 4-bit digits. Any ADD or ADDC instruction may be used to perform the addition.

If accumulator bits 3–0 are greater than 9 (xxxx1010- xxxx1111), or if the AC flag is 1, 6 is added to the accumulator, producing the proper BCD digit in the low- order nibble. This internal addition sets the carry flag if a carry-out of the low-order 4-bit field propagated through all high-order bits, but it does not clear the carry flag otherwise.

If the carry flag is now set, or if the 4 high-order bits now exceed 9 (1010xxxx–1111xxxx), these high-order bits are incremented by 6, producing the proper BCD digit in the high-order bits, but not clearing the carry. The carry flag thus indicates if the sum of the original 2 BCD variables is greater than 99, allowing precision decimal addition. OV is not affected.

All of the above occurs during 1 instruction cycle. Essentially, this instruction performs the decimal conversion by adding 00H, 06H, 60H, or 66H to the accumulator, depending on initial accumulator and PSW conditions.

Note: DA A *cannot* simply convert a hexadecimal number in the accumulator to BCD notation, nor does DA A apply to decimal subtraction.

Example: The accumulator holds the value 56H (01010110B), representing the packed BCD digits of the decimal number 56. Register 3 contains the value 67H (01100111B), representing the packed BCD digits of the decimal 67. The carry flag is set. The instructions,

```
ADDC A,R3
DA   A
```

first perform a standard 2s-complement binary addition, resulting in the value 0BEH (10111110B) in the accumulator. The carry and auxiliary-carry flag are cleared.

The decimal adjust instruction then alters the accumulator to the value 24H (00100100B), indicating the packed BCD digits of the decimal number 24, the low-order 2 digits of the decimal sum of 56, 67, and the carry-in. The carry flag is set by the decimal adjust instruction, indicating that a decimal overflow occurred. The true sum of 56, 67 and 1 is 124.

BCD variables can be incremented or decremented by adding 01H or 99H. If the accumulator initially holds 30H (representing the digits 30 decimal), then the instructions,

```
ADD  A,#99H
DA   A
```

leave the carry set and 29H in the accumulator, since 30 + 99 = 129. The low-order byte of the sum can be interpreted to mean 30 − 1 = 29.

Bytes: 1

Cycles: 1

Encoding: 11010100

Operation: (Assume the contents of the accumulator are BCD.)

IF [[(A3–A0) >9]AND [(AC) = 1]]
THEN (A3–A0) ←(A3–A0) + 6
AND
IF [[(A7–A4) >9]AND [(C) = 1]]
THEN (A7–A4) ←(A7–A4) + 6)

DEC byte

Function: Decrement

Description: The variable indicated is decremented by 1. An original value of 00H underflows to 0FFH. No flags are affected. Four operand addressing modes are allowed: accumulator, register, direct, or register-indirect.

Note: When this instruction is used to modify an output port, the value used as the original port data is from the output data latch, *not* the input pins.

Example: Register 0 contains 7FH (01111111B). Internal RAM locations 7EH and 7FH contain 00H and 40H, respectively. The instructions,

```
DEC   @R0
DEC   R0
DEC   @R0
```

leave register 0 set to 7EH and internal RAM locations 7EH and 7FH set to 0FFH and 3FH.

DEC A

Bytes: 1

Cycles: 1

Encoding: 00010100

Operation: (A) ←(A) − 1

DEC Rn

Bytes: 1

Cycles: 1

Encoding: 00011rrr

Operation: (Rn) ←(Rn) − 1

DEC direct

Bytes: 2
Cycles: 1
Encoding: 00010101 aaaaaaaa
Operation: (direct) ←(direct) − 1

DEC @Ri

Bytes: 1
Cycles: 1
Encoding: 0001011i
Operation: ((Ri)) ←((Ri)) − 1

DIV AB

Function: Divide

Description: DIV AB divides the unsigned 8-bit integer in the accumulator by the unsigned 8-bit integer in register B. The accumulator receives the integer part of the quotient; register B receives the integer remainder. The carry and OV flags are cleared.

Exception: If B originally contained 00H, the values returned in the accumulator and B-register are undefined and the overflow flag is set. The carry flag is cleared in any case.

Example: The accumulator contains 251 (0FBH or 11111011B) and B contains 18 (12H or 00010010B). The instruction,

```
DIV   AB
```

leaves 13 in the accumulator (0DH or 00001101B) and the value 17 (11H or 00010001B) in B, since 251 = 13 × 18 + 17. Carry and OV are both cleared.

Bytes: 1
Cycles: 4
Encoding: 10000100
Operation: (A) ←QUOTIENT OF (A)/(B)
(B) ←REMAINDER OF (A)/(B)

DJNZ <byte>,<rel-addr>

Function: Decrement and Jump if Not Zero

Description: DJNZ decrements the location indicated by the first operand, and branches to the address indicated by the second operand if the resulting value is not 0. An original value of 00H underflows to

0FFH. No flags are affected. The branch destination is computed by adding the signed relative-displacement value in the last instruction byte to the PC, after incrementing the PC to the first byte of the following instruction.

The location decremented may be a register or directly addressed byte.

Note: When this instruction is used to modify an output port, the value used as the original port data is read from the output data latch, *not* the input pins.

Example: Internal RAM locations 40H, 50H, and 60H contain the values 01H, 70H, and 15H, respectively. The instructions,

```
DJNZ 40H,LABEL1
DJNZ 50H,LABEL2
DJNZ 60H,LABEL3
```

cause a jump to the instruction at LABEL2 with the values 00H, 6FH, and 15H in the 3 RAM locations. The first jump is *not* taken because the result was 0.

This instruction provides a simple way to execute a program loop a given number of times, or to add a moderate time delay (from 2 to 512 machine cycles) with a single instruction. The instructions,

```
         MOV  R2,#8
TOGGLE:  CPL  P1.7
         DJNZ R2,TOGGLE
```

toggle P1.7 eight times, causing 4 output pulses to appear at bit 7 of output Port 1. Each pulse lasts 3 machine cycles; 2 for DJNZ and 1 to alter the pin.

DJNZ Rn,rel

Bytes: 2

Cycles: 2

Encoding: 11011rrr eeeeeeee

Operation: (PC) ←(PC) + 2
(Rn) ←(Rn) − 1
IF (Rn) <>0
THEN
(PC) ←(PC) + byte_2

DJNZ direct,rel

Bytes: 3

Cycles: 2

Encoding: 11010101 aaaaaaaa eeeeeeee

Operation: (PC) ←(PC) + 2
(direct) ←(direct) − 1
IF (direct) <>0
THEN
(PC) ←(PC) + byte_2

INC <byte>

Function: Increment

Description: INC increments the indicated variable by 1. An original value of 0FFH overflows to 00H. No flags are affected. Three addressing modes are allowed: register, direct, or register- indirect.

Note: When this instruction is used to modify an output port, the value used as the original port data is from the output data latch, *not* the input pins.

Example: Register 0 contains 7EH (01111110B). Internal RAM locations 7EH and 7FH contain 0FFH and 40H, respectively. The instructions,

```
INC   @R0
INC   R0
INC   @R0
```

leave register 0 set to 7FH and internal RAM locations 7EH and 7FH holding (respectively) 00H and 41H.

INC A

Bytes: 1

Cycles: 1

Encoding: 00000100

Operation: (A) ←(A) + 1

INC Rn

Bytes: 1

Cycles: 1

Encoding: 00001rrr

Operation: (Rn) ←(Rn) + 1

INC direct

Bytes: 2

Cycles: 1

Encoding: 00000101 aaaaaaaa

Operation: (direct) ←(direct) + 1

INC @Ri

Bytes: 1

Cycles: 1

Encoding: 0000011i

Operation: ((Ri)) ←((Ri)) + 1

INC DPTR

Function: Increment Data Pointer

Description: Increment the 16-bit data pointer by 1. A 16-bit increment (modulo 2^{16}) is performed; an overflow of the low-order byte of the data pointer (DPL) from 0FFH to 00H increments the high-order byte (DPH). No flags are affected.

This is the only 16-bit register that can be incremented.

Example: Registers DPH and DPL contain 12H and 0FEH, respectively. The instructions,

```
INC   DPTR
INC   DPTR
INC   DPTR
```

change DPH and DPL to 13H and 01H.

Bytes: 1

Cycles: 2

Encoding: 10100011

Operation: (DPTR) ←(DPTR) + 1

JB bit, rel

Function: Jump if Bit set

Description: If the indicated bit is a 1, jump to the address indicated; otherwise proceed with the next instruction. The branch destination is computed by adding the signed relative-displacement in the third instruction byte to the PC, after incrementing the PC to the first byte of the next instruction. *The bit tested is not modified.* No flags are affected.

Example: The data present at input port 1 are 11001010B. The accumulator holds 56H (01010110B). The instructions,

```
JB    P1.2,LABEL1
JB    ACC.2,LABEL2
```

cause program execution to branch to the instruction at LABEL2.

Bytes: 3

Cycles: 2

Encoding: 00100000 bbbbbbbb eeeeeeee

Operation: (PC) ←(PC) + 3
IF (bit) = 1
THEN
(PC) ←(PC) + byte_2

JBC bit,rel

Function: Jump if Bit set and Clear bit

Description: If the indicated bit is 1, clear it and branch to the address indicated; otherwise proceed with the next instruction. *The bit is not cleared if it is already a 0.* The branch destination is computed by adding the signed relative-displacement in the third instruction byte to the PC, after incrementing the PC to the first byte of the next instruction. No flags are affected.

Note: When this instruction is used to modify an output port, the value used as the original port data is read from the output data latch, *not* the input pins.

Example: The accumulator holds 56H (01010110B). The instructions,

```
JBC   ACC.3,LABEL1
JBC   ACC.2,LABEL2
```

cause program execution to continue at the instruction identified by LABEL2, with the accumulator modified to 52H (01010010B).

Bytes: 3

Cycles: 2

Encoding: 00010000 bbbbbbbb eeeeeeee

Operation: (PC) ←(PC) + 3
IF (bit) = 1
THEN
(bit) ←0
(PC) ←(PC) + byte_2

JC rel

Function: Jump if Carry is set

Description: If the carry flag is set, branch to the address indicated; otherwise proceed with the next instruction. The branch destination is computed by adding the signed relative-displacement in the second instruction byte to the PC, after incrementing the PC twice. No flags are affected.

Example: The carry flag is cleared. The instructions,

```
JC    LABEL1
CPL   C
JC    LABEL2
```

set the carry and cause program execution to continue at the instruction identified by LABEL2.

Bytes: 2

Cycles: 2

Encoding: 01000000 eeeeeeee

Operation: (PC) ←(PC) + 2
IF (C) = 1
THEN
(PC) + (PC) + byte_2

JMP <dest>(See SJMP, AJMP, or LJMP)
JMP @A+DPTR

Function: Jump indirect

Description: Add the 8-bit unsigned contents of the accumulator with the 16-bit pointer, and load the resulting sum to the program counter. This is the address for subsequent instruction fetches. Sixteen-bit addition is performed (modulo 2^{16}): a carry-out from the low-order 8 bits propagates through the higher-order bits. Neither the accumulator nor the data pointer is altered. No flags are affected.

Example: An even number from 0 to 6 is in the accumulator. The following instructions branch to 1 of 4 AJMP instructions in a jump table starting at JMP_TBL:

```
            MOV  DPTR,#JMP_TBL
            JMP  @A+DPTR
JMP_TBL:    AJMP LABEL0
            AJMP LABEL1
            AJMP LABEL2
            AJMP LABEL3
```

If the accumulator equals 04H when starting this sequence, execution jumps to LABEL2. Remember that AJMP is a 2-byte instruction, so the jump instruction starts at every other address.

Bytes: 1

Cycles: 2

Encoding: 01110011

Operation: (PC) ←(PC) + (A) + (DPTR)

JNB bit,rel

Function: Jump if Bit Not set

Description: If the indicated bit is a 0, branch to the indicated address; otherwise proceed with the next instruction. The branch destination is computed by adding the signed relative-displacement in the third instruction byte to the PC, after incrementing the PC to the first byte of the next instruction. *The bit tested is not modified.* No flags are affected.

Example: The data present at input port 1 are 11001010B. The accumulator holds 56H (01010110B). The instructions,

```
JNB  P1.3,LABEL1
JNB  ACC.3,LABEL2
```

cause program execution to continue at the instruction at LABEL2.

Bytes: 3

Cycles: 2

Encoding: 00110000 bbbbbbbb eeeeeeee

Operation: (PC) ←(PC) + 3
IF (bit) = 0
THEN
(PC) ←(PC) + byte_2

JNC rel

Function: Jump if Carry not set

Description: If the carry flag is a 0, branch to the address indicated; otherwise proceed with the next instruction. The branch destination is computed by adding the signed relative- displacement in the second instruction byte to the PC, after incrementing the PC twice to point to the next instruction. The carry flag is not modified.

Example: The carry flag is set. The instructions,

```
JNC  LABEL1
CPL  C
JNC  LABEL2
```

clear the carry flag and cause program execution to continue at the instruction identified by LABEL2.

Bytes: 2

Cycles: 2

Encoding: 01010000 eeeeeeee

Operation: (PC) ←(PC) + 2
IF (C) = 0
THEN
(PC) ←(PC) + byte_2

JNZ rel

Function: Jump if accumulator Not Zero

Description: If any bit of the accumulator is a 1, branch to the indicated address; otherwise proceed with the next instruction. The branch destination is computed by adding the signed relative-displacement in the second instruction byte to the PC, after incrementing the PC twice. The accumulator is not modified. No flags are affected.

Example: The accumulator originally holds 00H. The instructions,

```
JNZ  LABEL1
INC  A
JNZ  LABEL2
```

set the accumulator to 01H and continue at LABEL2.

Bytes: 2

Cycles: 2

Encoding: 01110000 eeeeeeee

Operation: (PC) ←(PC) + 2
IF (A) <>0
THEN
(PC) ←(PC) + byte_2

JZ rel

Function: Jump if accumulator Zero

Description: If all bits of the accumulator are 0, branch to the indicated address; otherwise proceed with the next instruction. The branch destination is computed by adding the signed relative-displacement in the second instruction byte to the PC, after incrementing the PC twice. The accumulator is not modified. No flags are affected.

Example: The accumulator originally holds 01H. The instructions,

```
JZ   LABEL1
DEC  A
JZ   LABEL2
```

change the accumulator to 00H and cause program execution to continue at the instruction identified by LABEL2.

Bytes: 2

Cycles: 2

Encoding: 01100000 eeeeeeee

Operation: (PC) ←(PC) + 2
IF (A) = 0
THEN
(PC) ←(PC) + byte_2

LCALL addr16

Function: Long Call to subroutine

Description: LCALL calls a subroutine located at the indicated address. The instruction adds 3 to the program counter to generate the address of the next instruction and then pushes the 16-bit result onto the stack (low-byte first), incrementing the stack pointer by 2. The high-order and low-order bytes of the PC are then loaded, respectively, with the second and third bytes of the LCALL instruction. Program execution continues with the instruction at this address. The subroutine may therefore begin anywhere in the full 64K-byte program memory address space. No flags are affected.

Example: Initially the stack pointer equals 07H. The label "SUBRTN" is assigned to program memory location 1234H. After executing the instruction,

```
LCALL    SUBRTN
```

at location 0123H, the stack pointer contains 09H, internal RAM locations 08H and 09H contain 26H and 01H, and the PC contains 1234H.

Bytes: 3

Cycles: 2

Encoding: 00010010 aaaaaaaa aaaaaaaa

Note: Byte 2 contains address bits 15–8, byte 3 contains address bits 7–0.

Operation: (PC) ←(PC) + 3
(SP) ←(SP) + 1
((SP)) ←(PC7–PC0)
(SP) ←(SP) + 1
((SP)) ←(PC15–PC8)
(PC) ←addr15–addr0

JMP addr16

Function: Long Jump

Description: LJMP causes an unconditional branch to the indicated address by loading the high-order and low-order bytes of the PC (respectively) with the second and third instruction bytes. The destination may therefore be anywhere in the full 64K program memory address space. No flags are affected.

Example: The label "JMPADR" is assigned to the instruction at program memory location 1234H. The instruction,

```
LJMP JMPADR
```

at location 0123H loads the program counter with 1234H.

Bytes: 3

Cycles: 2

Encoding: 00000010 aaaaaaaa aaaaaaaa

Note: Byte 2 contains address bits 15–8, byte 3 contains address bits 7–0.

Operation: (PC) ←addr15–addr0

MOV <dest-byte>,<src-byte>

Function: Move byte variable

Description: The byte variable indicated by the second operand is copied into the location specified by the first operand. The source byte is not affected. No other register or flag is affected.

This is by far the most flexible operation. Fifteen combinations of source and destination addressing modes are allowed.

Example: Internal RAM location 30H holds 40H. The value of RAM location 40H is 10H. The data present at input port 1 are 11001010B (0CAH). The instructions,

```
MOV   R0,#30H          ;RO ←30H
MOV   A,@R0            ;A ←40H
MOV   R1,A             ;R1 ←40H
MOV   B,@R1            ;B ←10H
MOV   @R1,P1           ;RAM(40H)
                       ; ←0CAH
MOV   P2,P1            ;P2 ←0CAH
```

leave the value 30H in register 0, 40H in both the accumulator and register 1, 10H in register B, and 0CAH (11001010B) both in RAM location 40H and output on port 2.

MOV A,Rn

Bytes: 1

Cycles: 1

Encoding: 11101rrr

Operation: (A) ←(Rn)

MOV A,direct

Bytes: 2
Cycles: 1
Encoding: 11100101 aaaaaaaa
Operation: (A) ←(direct)
Note: MOV A,ACC is not a valid instruction.

MOV A,@Ri

Bytes: 1
Cycles: 1
Encoding: 1110011i
Operation: (A) ←((Ri))

MOV A,#data

Bytes: 2
Cycles: 1
Encoding: 01110100 dddddddd
Operation: (A) ←#data

MOV Rn,A

Bytes: 1
Cycles: 1
Encoding: 11111rrr
Operation: (Rn) ←(A)

MOV Rn,direct

Bytes: 2
Cycles: 2
Encoding: 10101rrr
Operation: (Rn) ←(direct)

MOV Rn,#data

Bytes: 2
Cycles: 1
Encoding: 01111rrr dddddddd
Operation: (Rn) ←#data

MOV direct,A

Bytes: 2
Cycles: 1
Encoding: 11110101 aaaaaaaa
Operation: (direct) ←A

MOV direct,Rn

Bytes: 2
Cycles: 2
Encoding: 10001rrr aaaaaaaa
Operation: (direct) ←(Rn)

MOV direct,direct

Bytes: 3
Cycles: 2
Encoding: 10000101 aaaaaaaa aaaaaaaa
Note: Byte 2 contains the source address; byte 3 contains the destination address.
Operation: (direct) ←(direct)

MOV direct,@Ri

Bytes: 2
Cycles: 2
Encoding: 1000011i aaaaaaaa
Operation: (direct) ←((Ri))

MOV direct,#data

Bytes: 3
Cycles: 2
Encoding: 01110101 aaaaaaaa dddddddd
Operation: (direct) ←#data

MOV @Ri,A

Bytes: 1
Cycles: 1
Encoding: 1111011i
Operation: ((Ri)) ←A

MOV @Ri,direct

Bytes: 2
Cycles: 2
Encoding: 1010011i aaaaaaaa
Operation: ((Ri)) ←(direct)

MOV @Ri,#data

Bytes: 2
Cycles: 1
Encoding: 0111011i dddddddd
Operation: ((Ri)) ←#data

MOV <dest-bit>,<src-bit>

Function: Move bit variable

Description: The Boolean variable indicated by the second operand is copied into the location specified by the first operand. One of the operands must be the carry flag; the other may be any directly addressable bit. No other register or flag is affected.

Example: The carry flag is originally set. The data present at input Port 2 is 11000101B. The data previously written to output Port 1 are 35H (00110101B). The instructions,

```
MOV  P1.3,C
MOV  C,P3.3
MOV  P1.2,C
```

leave the carry cleared and change Port 1 to 39H (00111001B).

MOV C,bit

Bytes: 2
Cycles: 1
Encoding: 10100010 bbbbbbbb
Operation: (C) ←(bit)

MOV bit,C

Bytes: 2
Cycles: 2
Encoding: 10010010 bit address
Operation: (bit) ←(C)

MOV DPTR,#data16

Function: Load Data Pointer with a 16-bit constant

Description: The data pointer is loaded with the 16-bit constant indicated. The 16-bit constant is located in the second and third bytes of the instruction. The second byte (DPH) is the high-order byte, while the third byte (DPL) holds the low-order byte. No flags are affected.

Example: The instruction,

```
MOV  DPTR,#1234H
```

loads the value 1234H into the data pointer: DPH holds 12H and DPL holds 34H.

Bytes: 3

Cycles: 2

Encoding: 10010000 dddddddd dddddddd

Note: Byte 2 contains immediate data bits 15–8, byte 3 contains bits 7–0.

Operation: (DPTR) ←#data16

MOVC A,@A+<base-reg>

Function: Move code byte or constant byte

Description: The MOVC instructions load the accumulator with a code byte or constant byte from program memory. The address of the byte fetched is the sum of the original unsigned 8-bit accumulator contents and the contents of a 16-bit base register, which may be either the data pointer or the PC. In the latter case, the PC is incremented to the address of the following instruction before being added to the accumulator; otherwise the base register is not altered. Sixteen-bit addition is performed so a carry-out from the low-order 8 bits may propagate through higher-order bits. No flags are affected.

Example: A value between 0 and 3 is in the accumulator. The following subroutine translates the value in the accumulator to 1 of 4 values defined by the DB (define byte) directive.

```
REL_PC:     INC  A
            MOVC A,@A + PC
            RET
            DB   66H
            DB   77H
            DB   88H
            DB   99H
```

If the subroutine is called with the accumulator equal to 01H, it returns with 77H in the accumulator. The INC A before the MOVC instruction is needed to "get around" the RET instruction above the table. If several bytes of code separate the MOVC from the table, the corresponding number should be added to the accumulator instead.

MOVC A,@A+DPTR

Bytes: 1
Cycles: 2
Encoding: 10010011
Operation: (A) ←((A) + (DPTR))

MOVC A,@A+PC

Bytes: 1
Cycles: 2
Encoding: 10000011
Operation: (PC) ←(PC) + 1
(A) ←((A) + (PC))

MOVX <dest-byte>,<src-byte>

Function: Move External

Description: The MOVX instructions transfer data between the accumulator and a byte of external data memory; hence the "X" appended to MOV. There are two types of instructions, differing in whether they provide an 8-bit or 16-bit indirect address to the external data RAM.

In the first type, the contents of R0 or R1 in the current register bank provide an 8-bit address multiplexed with data on P0. Eight bits are sufficient for external I/O expansion decoding or for a relatively small RAM array. For somewhat larger arrays, any output port pins can be used to output higher- order address bits. These pins would be controlled by an output instruction preceding the MOVX.

In the second type of MOVX instruction, the data pointer generates a 16-bit address. P2 outputs the high-order 8 address bits (the contents of DPH), while P0 multiplexes the low-order 8 bits (DPL) with data. The P2 special function register retains its previous contents while the P2 output buffers are emitting the contents of DPH. This form is faster and more efficient when accessing very large data arrays (up to 64K bytes), since no additional instructions are needed to set up the output ports.

It is possible in some situations to mix the two MOVX types. A large RAM array with its high-order address lines driven by P2 can be addressed via the data pointer, or with code to output high-order address bits to P2 followed by a MOVX instruction using R0 or R1.

Example: An external 256-byte RAM using multiplexed address/data lines (e.g., an Intel 8155 RAM/I/O/TIMER) is connected to the 8051 Port 0. Port 3 provides control lines for the external RAM. Ports 1 and 2 are used for normal I/O. Registers 0 and 1 contain 12H and 34H. Location 34H of the external RAM holds the value 56H. The instruction sequence

```
MOVX A,@R1
MOVX @R0,A
```

copies the value 56H into both the accumulator and external RAM location 12H.

MOVX A,@Ri

Bytes: 1

Cycles: 2

Encoding: 1110001i

Operation: (A) ←((Ri))

MOVX A,@DPTR

Bytes: 1

Cycles: 2

Encoding: 11100000

Operation: (A) ←((DPTR))

MOVX @Ri,A

Bytes: 1

Cycles: 2

Encoding: 11110011

Operation: ((Ri)) ←(A)

MOVX @DPTR,A

Bytes: 1

Cycles: 2

Encoding: 11110000

Operation: (DPTR) ←(A)

MUL AB

Function: Multiply

Description: MUL AB multiplies the unsigned 8-bit integers in the accumulator and register B. The low-order byte of the 16-bit product is left in the accumulator, and the high-order byte in B. If the product is greater than 255 (0FFH), the overflow flag is set; otherwise it is cleared. The carry flag is always cleared.

Example: Originally the accumulator holds the value 80 (50H). Register B holds the value 160 (0A0H). The instruction,

```
MUL  AB
```

gives the product 12,800 (3200H), so B is changed to 32H (00110010B) and the accumulator is cleared. The overflow flag is set, and carry is cleared.

Bytes: 1

Cycles: 4

Encoding: 10100100

Operation: (A) ←HIGH BYTE OF (A) $\times$(B)
(B) ←LOW BYTE OF (A) $\times$(B)

NOP

Function: No Operation

Description: Execution continues at the following instruction. Other than the PC, no register or flags are affected.

Example: It is desired to produce a low-going output pulse on bit 7 of Port 2 lasting exactly 5 cycles. A simple SETB/CLR sequence generates a 1-cycle pulse, so 4 additional cycles must be inserted. This may be done (assuming no interrupts are enabled) with the instructions,

```
CLR  P2.7
NOP
NOP
NOP
NOP
SETB P2.7
```

Bytes: 1

Cycles: 1

Encoding: 00000000

Operation: (PC) ←(PC) + 1

ORL <dest-byte>,<src-byte>

Function: Logical-OR for byte variables

Description: ORL performs the bitwise logical-OR operation between the indicated variables, storing the results in the destination byte. No flags are affected.

The 2 operands allow 6 addressing mode combinations. When the destination is the accumulator, the source can use register, direct, register-indirect, or immediate addressing; when the destination is a direct address, the source can be the accumulator or immediate data.

Note: When this instruction is used to modify an output port, the value used as the original port data is read from the output data latch, *not* the input pins.

Example: If the accumulator holds 0C3H (11000011B) and R0 holds 55H (01010101B) then the instruction,

```
ORL  A,R0
```

leaves the accumulator holding the value 0D7H (11010111B).

When the destination is a directly addressed byte, the instruction can set combinations of bits in any RAM location or hardware register. The pattern of bits to be set is determined by a mask byte, which may be either a constant data value in the instruction or a variable computed in the accumulator at run-time. The instruction,

```
ORL  P1,00110010B
```

sets bits 5, 4, and 1 of output Port 1.

ORL A,Rn

Bytes: 1

Cycles: 1

Encoding: 01001rrr

Operation: (A) ←(A) OR (Rn)

ORL A,direct

Bytes: 2

Cycles: 1

Encoding: 01000101 aaaaaaaa

Operation: (A) ←(A) OR (direct)

ORL A,@Ri

Bytes: 1
Cycles: 1
Encoding: 0100011i
Operation: (A) ←(A) OR ((Ri))

ORL A,#data

Bytes: 2
Cycles: 1
Encoding: 01000100 dddddddd
Operation: (A) ←(A) OR #data

ORL direct,A

Bytes: 2
Cycles: 1
Encoding: 01000010 aaaaaaaa
Operation: (direct) ←(direct) OR (A)

ORL direct,#data

Bytes: 3
Cycles: 2
Encoding: 01000011 aaaaaaaa dddddddd
Operation: (direct) ←(direct) OR #data

ORL C,<src-bit>

Function: Logical-OR for bit variables

Description: Set the carry flag if the Boolean value is a logical 1; leave the carry in its current state otherwise. A slash (/) preceding the operand in the assembly language indicates that the logical complement of the addressed bit is used as the source value, but the source bit itself is not affected. No other flags are affected.

Example: Set the carry flag if, and only if, P1.0 = 1, ACC.7 = 1, or OV = 0.

```
MOV  C,P1.0  ;LOAD CY WITH INPUT PIN P1.0
ORL  C,ACC.7 ;OR CY WITH THE ACC, BIT7
ORL  C,/OV   ;OR CY WITH INVERSE OF OV
```

ORL C,bit

Bytes: 2

Cycles: 2

Encoding: 01110010 bbbbbbbb

Operation: (C) ←(C) OR (bit)

ORL C,/bit

Bytes: 2

Cycles: 2

Encoding: 10100000 bbbbbbbb

Operation: (C) ←(C) OR NOT(bit)

POP direct

Function: Pop from stack

Description: The contents of the internal RAM location addressed by the stack pointer are read, and the stack pointer is decremented by 1. The value read is then transferred to the directly addressed byte indicated. No flags are affected.

Example: The stack pointer originally contains the value 32H, and internal RAM locations 30H through 32H contain the values 20H, 23H, and 01H, respectively. The instructions,

```
POP   DPH
POP   DPL
```

leave the stack pointer equal to the value 30H and the data pointer set to 0123H. At this point the instruction,

```
POP   SP
```

leaves the stack pointer set to 20H. Note that in this special case the stack pointer is decremented to 2FH before being loaded with the value popped (20H).

Bytes: 2

Cycles: 2

Encoding: 11010000 aaaaaaaa

Operation: (direct) ←((SP))
(SP) ←(SP) − 1

PUSH direct

Function: Push onto stack

Description: The stack pointer is incremented by 1. The contents of the indicated variable are then copied into the internal RAM location addressed by the stack pointer. Otherwise, no flags are affected.

Example: On entering an interrupt routine, the stack pointer contains 09H. The data pointer holds the value 0123H. The instructions,

```
PUSH DPL
PUSH DPH
```

leave the stack pointer set to 0BH and store 23H and 01H in internal RAM locations 0AH and 0BH, respectively.

Bytes: 2

Cycles: 2

Encoding: 11000000 aaaaaaaa

Operation: (SP) ←(SP) + 1
((SP)) ←(direct)

RET

Function: Return from subroutine

Description: RET pops the high- and low-order bytes of the PC successively from the stack, decrementing the stack pointer by 2. Program execution continues at the resulting address, generally the instruction immediately following an ACALL or LCALL. No flags are affected.

Example: The stack pointer originally contains the value 0BH. Internal RAM locations 0AH and 0BH contain the values 23H and 01H, respectively. The instruction,

```
RET
```

leaves the stack pointer equal to the value 09H. Program execution continues at location 0123H.

Bytes: 1

Cycles: 2

Encoding: 00100010

Operation: (PC15–PC8) ←((SP))
(SP) ←(SP) − 1
(PC7–PC0) ←((SP))
(SP) ←(SP) − 1

RETI

Function: Return from interrupt

Description: RETI pops the high- and low-order bytes of the PC successively from the stack, and restores the interrupt logic to accept additional interrupts at the same priority level as the one just processed. The stack pointer is left decremented by 2. No other registers are affected; the PSW is *not* automatically restored to its pre-interrupt status. Program execution continues at the resulting address, which is generally an instruction immediately after the point at which the interrupt request is detected. If a lower- or same-level interrupt is pending when the RETI instruction is executed, then one instruction is executed before the pending interrupt is processed.

Example: The stack pointer originally contains the value 0BH. An interrupt is detected during the instruction ending at location 0123H. Internal RAM locations 0AH and 0BH contain the values 23H and 01H, respectively. The instruction,

```
RETI
```

leaves the stack pointer equal to 09H and returns program execution to location 0123H.

Bytes: 1

Cycles: 2

Encoding: 00110010

Operation: (PC15–PC8) ←((SP))
(SP) ←(SP) − 1
(PC7–PC0) ←((SP))
(SP) ←(SP) − 1

RL A

Function: Rotate Accumulator Left

Description: The 8 bits in the accumulator are rotated 1 bit to the left. Bit 7 is rotated into the bit 0 position. No flags are affected.

Example: The accumulator holds the value 0C5H (11000101B). The instruction,

```
RL    A
```

leaves the accumulator holding the value 8BH (10001011B) with the carry unaffected.

Bytes: 1

Cycles: 1

Encoding: 00100011

Operation: (An + 1) ←(An), n = 0 − 6
(A0) ←(A7)

RLC A

Function: Rotate Accumulator Left through the Carry flag

Description: The 8 bits in the accumulator and the carry flag are together rotated 1 bit to the left. Bit 7 moves into the carry flag; the original state of the carry flag moves into the bit 0 position. No other flags are affected.

Example: The accumulator holds the value 0C5H (11000101B), and the carry is 0. The instruction,

```
RLC   A
```

leaves the accumulator holding the value 8BH (10001011B) with the carry set.

Bytes: 1

Cycles: 1

Encoding: 00110011

Operation: (An + 1) ←(An), n = 0 − 6
(A0) ←(C)
(C) ←(A7)

RR A

Function: Rotate Accumulator Right

Description: The 8 bits in the accumulator are rotated 1 bit to the right. Bit 0 is rotated into the bit 7 position. No flags are affected.

Example: The accumulator holds the value 0C5H (11000101B) with the carry unaffected. The instruction,

```
RR    A
```

leaves the accumulator holding the value 0E2H (11100010B) with the carry unaffected.

Bytes: 1

Cycles: 1

Encoding: 00000011

Operation: (An) ←(An + 1), n = 0 − 6
(A7) ←(A0)

RRC A

Function: Rotate Accumulator Right through Carry flag

Description: The 8 bits in the accumulator and the carry flag are together rotated 1 bit to the right. Bit 0 moves into the carry flag; the original value of the carry flag moves into the bit 7 position. No other flags are affected.

Example: The accumulator holds the value 0C5H (11000101B), the carry is 0. The instruction,

```
RRC  A
```

leaves the accumulator holding the value 62H (01100010B) with the carry set.

Bytes: 1

Cycles: 1

Encoding: 00010011

Operation: (An) ←(An + 1), n = 0 − 6
(A7) ←(C)
(C) ←(A0)

SETB <bit>

Function: Set Bit

Description: SETB sets the indicated bit to 1. SETB can operate on the carry flag or any directly addressable bit. No other flags are affected.

Example: The carry flag is cleared. Output Port 1 has been written with the value 34H (00110100B). The instructions,

```
SETB C
SETB P1.0
```

leave the carry flag set to 1 and change the data output on Port 1 to 35H (00110101B).

SETB C

Bytes: 1

Cycles: 1

Encoding: 11010011

Operation: (C) ←1

SETB bit

Bytes: 2

Cycles: 1

Encoding: 11010010 bbbbbbbb

Operation: (bit) ←1

SJMP rel

Function: Short Jump

Description: Program control branches unconditionally to the address indicated. The branch destination is computed by adding the signed displacement in the second instruction byte to the PC, after incrementing the PC twice. Therefore, the range of destinations allowed is from 128 bytes preceding this instruction to 127 bytes following it.

Example: The label "RELADR" is assigned to an instruction at program memory location 0123H. The instruction,

```
SJMP RELADR
```

assembles into location 0100H. After the instruction is executed, the PC contains the value 0123H.

(*Note:* Under the above conditions, the instruction following SJMP is at 0102H. Therefore, the displacement byte of the instruction is the relative offset (0123H − 0102H = 21H. Put another way, an SJMP with a displacement of 0FEH is a 1-instruction infinite loop.)

Bytes: 2

Cycles: 2

Encoding: 10000000 eeeeeeee

Operation: (PC) ←(PC) + 2
(PC) ←(PC) + byte_2

SUBB A,<src-byte>

Function: Subtract with borrow

Description: SUBB subtracts the indicated variable and the carry flag together from the accumulator, leaving the result in the accumulator. SUBB sets the carry (borrow) flag if a borrow is needed for bit 7 and clears C otherwise. (If C is set *before* executing a SUBB instruction, this indicates that a borrow is needed for the previous step in a multiple-precision subtraction, so the carry is subtracted from the accumulator along with the source operand.) AC is set if a borrow is needed for bit 3, and cleared otherwise. OV is set if a borrow is needed into bit 6, but not into bit 7, or into bit 7, but not into bit 6.

When subtracting signed integers, OV indicates that a negative number is produced when a negative value is subtracted from a positive value, or a positive number is produced when a positive number is subtracted from a negative number.

The source operand allows four addressing modes: register, direct, register-indirect, or immediate.

Example: The accumulator holds 0C9H (11001001B), register 2 holds 54H (01010100B), and the carry flag is set. The instruction,

```
SUBB A,R2
```

leaves the value 74H (01110100B) in the accumulator, with the carry flag and AC cleared but OV set.

Notice that 0C9H minus 54H is 75H. The difference between this and the above result is due to the carry (borrow) flag being set before the operation. If the state of the carry is not known before starting a single or multiple-precision subtraction, it should be explicitly cleared by a CLR C instruction.

SUBB A,Rn

Bytes: 1

Cycles: 1

Encoding: 10011rrr

Operation: (A) ←(A) −(C) −(Rn)

SUBB A,direct

Bytes: 2

Cycles: 1

Encoding: 10010101 aaaaaaaa

Operation: (A) ←(A) −(C) −(direct)

SUBB A,@Ri

Bytes: 1

Cycles: 1

Encoding: 1001011i

Operation: (A) ←(A) −(C) −((Ri))

SUBB A,#data

Bytes: 2

Cycles: 1

Encoding: 10010100 dddddddd

Operation: (A) ←(A) −(C) −#data

SWAP A

Function: Swap nibbles within the Accumulator

Description: SWAP A interchanges the low- and high-order nibbles (4-bit fields) of the accumulator (bits 3–0 and bits 7–4). The operation can also be thought of as a 4-bit rotate instruction. No flags are affected.

Example: The accumulator holds the value 0C5H (11000101B). The instruction,

```
SWAP A
```

leaves the accumulator holding the value 5CH (01011100B).

Bytes: 1

Cycles: 1

Encoding: 11000100

Operation: (A3–A0) ↔ (A7–A4)

XCH A,<byte>

Function: Exchange Accumulator with byte variable

Description: XCH loads the accumulator with the contents of the indicated variable, at the same time writing the original accumulator contents to the indicated variable. The source/destination operand can use register, direct, or register-indirect addressing.

Example: R0 contains the address 20H. The accumulator holds the value 3FH (00111111B). Internal RAM location 20H holds the value 75H (01110101B). The instruction,

```
XCH  A,@R0
```

leaves RAM location 20H holding the values 3FH (00111111B) and 75H (01110101B) in the accumulator.

XCH A,Rn

Bytes: 1

Cycles: 1

Encoding: 11001rrr

Operation: (A) ↔ (Rn)

XCH A,direct

Bytes: 2

Cycles: 1

Encoding: 11000101 aaaaaaaa

Operation: (A) ↔ (direct)

XCH A,@Ri

Bytes: 1
Cycles: 1
Encoding: 1100011i
Operation: (A) ↔ ((Ri))

XCHD A,@Ri

Function: Exchange Digit

Description: XCHD exchanges the low-order nibble of the accumulator (bits 0–3), generally representing a hexadecimal or BCD digit, with that of the internal RAM location indirectly addressed by the specified register. The high-order nibbles (bits 7–4) of each register are not affected. No flags are affected.

Example: R0 contains the address 20H. The accumulator holds the value 36H (00110110B). Internal RAM location 20H holds the value 75H (01110101B). The instruction,

```
XCHD A,@R0
```

leaves RAM location 20H holding the value 76H (01110110B) and 35H (00110101B) in the accumulator.

Bytes: 1
Cycles: 1
Encoding: 1101011i
Operation: (A3–A0) ↔ ((Ri3–Ri0))

XRL <dest-byte>,<src-byte>

Function: Logical Exclusive-OR for byte variables

Description: XRL performs the bitwise logical exclusive-OR operation between the indicated variables, storing the results in the destination. No flags are affected.

The 2 operands allow 6 addressing mode combinations. When the destination is the accumulator, the source can use register, direct, register-indirect, or immediate addressing; when the destination is a direct address, the source can be the accumulator or immediate data.

Note: When this instruction is used to modify an output port, the value used as the original port data is read from the output data latch, *not* the input pins.

Example: If the accumulator holds 0C3H (11000011B) and register 0 holds 0AAH (10101010B), then the instruction,

```
XRL  A,R0
```

leaves the accumulator holding the value 69H (01101001B).

When the destination is a directly addressed byte, this instruction can complement combinations of bits in any RAM location or hardware register. The pattern of bits to be complemented is then determined in the accumulator at run-time. The instruction,

```
XRL  P1,#00110001B
```

complements bits 5, 4, and 0 of output Port 1.

XRL A,Rn

Bytes: 1
Cycles: 1
Encoding: 01101rrr
Operation: $(A) \leftarrow (A) \oplus (Rn)$

XRL A,direct

Bytes: 2
Cycles: 1
Encoding: 01100101 aaaaaaaa
Operation: $(A) \leftarrow (A) \oplus (direct)$

XRL A,@Ri

Bytes: 1
Cycles: 1
Encoding: 0110011i
Operation: $(A) \leftarrow (A) \oplus ((Ri))$

XRL A,#data

Bytes: 2
Cycles: 1
Encoding: 01100100 dddddddd
Operation: $(A) \leftarrow (A) \oplus \#data$

XRL direct,A

Bytes: 2
Cycles: 1

Encoding: 01100010 aaaaaaaa

Operation: (direct) ←(direct) ⊕(A)

XRL direct,#data

Bytes: 3

Cycles: 2

Encoding: 01100011 aaaaaaaa dddddddd

Operation: (direct) ←(direct) ⊕#data

D

APPENDIX D
SPECIAL FUNCTION REGISTERS[1]

The 8051's special function registers are shown in the SFR memory map in Figure D–1. Blank locations are reserved for future products and should not be written to. The SFRs identified with an asterisk contain bits that are defined as mode or control bits. These registers and their bit definitions are described on the following pages.

Some bits are identified as "not implemented." User software should not write 1s to these bits, since they may be used in future MCS-51™products to invoke new features. In that case, the reset or inactive value of the new bit will be 0, and its active value will be 1.

PCON (POWER CONTROL REGISTER)

Symbol: PCON

Function: Power Control and miscellaneous features

Bit Address: 87H

Bit Addressable: No

Summary:

7	6	5	4	3	2	1	0
SMOD	–	–	–	GF1	GF0	PD	IDL

[1]Adapted from *8-Bit Embedded Controllers* (270645). Santa Clara, CA: Intel Corporation, 1991, by permission of Intel Corporation.

FIGURE D–1
Special function register memory map

8 Bytes

F8									FF
F0	B								F7
E8									EF
E0	ACC								E7
D8									DF
D0	PSW*								D7
C8	T2CON*		RCAP2L	RCAP2H	TL2	TH2			CF
C0									C7
B8	IP*								BF
B0	P3								B7
A8	IE*								AF
A0	P2								A7
98	SCON*	SBUF							9F
90	P1								97
88	TCON*	TMOD*	TL0	TL1	TH0	TH1			8F
80	P0	SP	DPL	DPH				PCON*	87

↑ Bit addressable

*SFRs containing mode or control bits

Bit Definitions:

Bit Symbol	Bit Description
SMOD	Double baud rate. If timer 1 is used to generate baud rate and SMOD = 1, the baud rate is doubled when the serial port is used in modes 1, 2, or 3.
	—Not implemented; reserved for future use.
	—Not implemented; reserved for future use.
	—Not implemented; reserved for future use.
GF1	General purpose flag bit 1.
GF0	General purpose flag bit 0.

---

[2]If 1s are written to PD and IDL at the same time, PD takes precedence.

PD — Power down bit. Setting this bit activates power down operation in the CMOS version of the 8051.[2]

IDL — Idle mode bit. Setting this bit activates idle mode operation in the CMOS versions of the 8051.[2]

TCON (TIMER/COUNTER CONTROL REGISTER)

Symbol: TCON

Function: Timer/Counter Control

Bit Address: 88H

Bit Addressable: Yes

Summary:

7	6	5	4	3	2	1	0
TF1	TR1	TF0	TR0	IE1	IT1	IE0	IT0

Bit Definitions:

Bit Symbol	Position	Bit Address	Description
TF1	TCON.7	8FH	Timer 1 overflow flag. Set by hardware when the timer/counter 1 overflows; cleared by software or by hardware as processor vectors to the interrupt service routine.
TR1	TCON.6	8EH	Timer 1 run control bit. Set/cleared by software to turn timer/counter 1 ON/OFF.
TF0	TCON.5	8DH	Timer 0 overflow flag. (See TF1)
TR0	TCON.4	8CH	Timer 0 run control fit. (See TR1)
IE1	TCON.3	8BH	External interrupt 1 edge flag. Set by hardware when external interrupt edge is detected; cleared by hardware when interrupt is processed.
IT1	TCON.2	8AH	Interrupt 1 type control bit. Set/cleared by software to specify falling-edge/low-level triggered external interrupt.
IE0	TCON.1	89H	External interrupt 0 edge flag. (See IE1)
IT0	TCON.0	88H	Interrupt 0 type control bit. (See IT1)

TMOD (TIMER/COUNTER MODE CONTROL REGISTER)

Symbol: TMOD

Function: Timer/Counter Mode Control

Bit Address: 89H

Bit Addressable: No

Summary:

7	6	5	4	3	2	1	0
GATE	$C/\overline{T}$	M1	M0	GATE	$C/\overline{T}$	M1	M0
Timer 1				Timer 0			

Bit Definitions:

Bit Symbol	Description
GATE	Gate bit. When TRx (in TCON) is set and Gate = 1, the timer/counter will run only while INTx pin is high (hardware control). When GATE = 0, the timer/counter will run only while TRx = 1 (software control).
$C/\overline{T}$	Counter/timer selector. Cleared for timer operation (input from the internal system clock). Set for counter operation (input from Tx input pin).
M1	Mode selector 1. (See Table D–1)
M0	Mode selector 0. (See Table D–1)

TABLE D–1

M1	M0	MODE	DESCRIPTION
0	0	0	13-bit timer (MCS-48 compatibility)
0	1	1	16-bit timer/counter
1	0	2	8-bit auto-reload timer/counter
1	1	3	Timer 0: TL0 is an 8-bit timer/counter controlled by the standard timer 0 control bits. TH0 is an 8-bit timer and is controlled by timer 1 control bits. Timer 1: timer/counter 1 stopped.

SCON (SERIAL PORT CONTROL REGISTER)

Symbol: SCON
Function: Serial Port Control
Bit Address: 98H
Bit Addressable: Yes
Summary:

7	6	5	4	3	2	1	0
SM0	SM1	SM2	REN	TB8	RB8	TI	RI

Bit Definitions:

Bit Symbol	Position	Bit Address	Description
SM0	SCON.7	9FH	Serial port mode bit 0. (See Table D–2)
SM1	SCON.6	9EH	Serial port mode bit 1. (See Table D–2)
SM2	SCON.5	9DH	Serial port mode bit 2. Enable the multiprocessor communication feature in modes 2 and 3. In mode 2 or 3, if SM2 is set to 1, then RI will not be activated if the received 9th data bit (RB8) is 0. In Mode 1, if SM2 = 1 then RI will not be activated if a valid stop bit was not received. In mode 0, SM2 should be 0.
REN	SCON.4	9CH	Receiver enable. Set/cleared by software to enable/disable reception.
TB8	SCON.3	9BH	Transmit bit 8. The 9th bit that will be transmitted in modes 2 and 3. Set/cleared by software.

TABLE D–2

SM0	SM1	MODE	DESCRIPTION	BAUD RATE
0	0	0	Shift Register	$F_{osc} \div 12$*
0	1	1	8-bit UART	Variable
1	0	2	9-bit UART	$F_{OSC} \div 64$ or $F_{osc} \div 32$
1	1	3	9-bit UART	Variable

*F_{osc} is the oscillator frequency of the 8051 IC. Typically, this derived from a crystal source of 12 MHz.

RB8	SCON.2	9AH	Receive bit 8. In modes 2 and 3, RB8 is the 9th data bit that was received. In mode 1, if SM2 = 0, RB8 is the stop bit that was received. In mode 0, TB8 is not used.
TI	SCON.1	99H	Transmit interrupt. Set by hardware at the end of the 8th bit time in mode 0, or at the beginning of the stop bit in the other modes. Must be cleared by software.
RI	SCON.0	98H	Receive interrupt. Set by hardware at the end of the 8th bit time in mode 0, or halfway through the stop bit time in the other modes (except see SM2). Must be cleared by software.

IE (INTERRUPT ENABLE REGISTER)

Symbol: IE

Function: Interrupt Enable

Bit Address: A8H

Bit Addressable: Yes

Summary:

7	6	5	4	3	2	1	0
EA	–	ET2	ES	ET1	EX1	ET0	EX0

Bit Definitions:

Bit Symbol	Position	Bit Address	Description (1 = enable, 0 = disable)
EA	IE.7	AFH	Enable/disable all interrupts. If EA = 0, no interrupt will be acknowledged. If EA = 1, each interrupt source is individually enabled or disabled by setting or clearing its enable bit.
—	IE.6	AEH	Not implemented; reserved for future use.
ET2	IE.5	ADH	Enable/disable timer 2 overflow or capture interrupt (80x2 only).
ES	IE.4	ACH	Enable/disable serial port interrupt.
ET1	IE.3	ABH	Enable/disable timer 1 overflow interrupt.

EX1	IE.2	AAH	Enable/disable external interrupt 1.
ET0	IE.1	A9H	Enable/disable timer 0 overflow interrupt.
EX0	IE.0	A8H	Enable/disable external interrupt 0.

IP (INTERRUPT PRIORITY REGISTER)

Symbol: IP
Function: Interrupt Priority
Bit Address: B8H
Bit Addressable: Yes
Summary:

7	6	5	4	3	2	1	0
—	—	PT2	PS	PT1	PX1	PT0	PX0

Bit Definitions:

Bit Symbol	Position	Bit Address	Description (1 = high priority, 0 = low priority)
—	IP.7	BFH	Not implemented; reserved for future use.
—	IP.6	BEH	Not implemented; reserved for future use.
PT2	IP.5	BDH	Timer 2 interrupt priority level (80x2 only).
PS	IP.4	BCH	Serial port interrupt priority level.
PT1	IP.3	BBH	Timer 1 interrupt priority level.
PX1	IP.2	BAH	External interrupt 1 priority level.
PT0	IP.1	B9H	Timer 0 interrupt priority level.
PX0	IP.0	B8H	External interrupt 0 priority level.

T2CON (TIMER/COUNTER 2 CONTROL REGISTER)

Symbol: T2CON
Function: Timer/Counter 2 Control
Bit Address: C8H
Bit Addressable: Yes

Summary:

7	6	5	4	3	2	1	0
TF2	EXF2	RCLK	TCLK	EXEN2	TR2	C/$\overline{T2}$	CP/$\overline{RL2}$

Bit Definitions:

Bit Symbol	Position	Bit Address	Description
TF2	T2CON.7	CFH	Timer 2 overflow flag. Set by hardware and cleared by software. TF2 cannot be set when either RCLK = 1 or TCLK = 1.
EXF2	T2CON.6	CEH	Timer 2 external flag. Set when either a capture or reload is caused by a negative transition on T2EX and EXEN2 = 1. When timer 2 interrupt is enabled, EXF2 = 1 will cause the CPU to vector to the timer interrupt service routine. EXF2 must be cleared by software.
RCLK	T2CON.5	CDH	Receive clock. When set, causes the serial port to use timer 2 overflow pulses for its receive clock in modes 1 and 3. RCLK = 0 causes timer 1 overflow to be used for the receive clock.
TCLK	T2CON.4	CCH	Transmit clock. When set, causes the serial port to use timer overflow pulses for its transmit clock in modes 1 and 3. TCLK = 0 causes timer 1 overflows to be used for the transmit clock.
EXEN2	T2CON.3	CBH	Timer 2 external enable flag. When set, allows a capture of reload to occur as a result of negative transition on T2EX if timer 2 is not being used to clock to serial port. EXEN2 = 0 causes timer 2 to ignore events at T2EX.
TR2	T2CON.2	CAH	Timer 2 run bit. Software START/STOP control for timer 2; A logic 1 starts the timer.
C/$\overline{T2}$	T2CON.1	C9H	Counter/timer select for timer 2. 0 = internal timer, 1 = external event counter (falling edge triggered).
CP/$\overline{RL2}$	T2CON.0	C8H	Capture/reload flag. When set, captures will occur on negative transitions at T2EX if EXEN2 = 1. When cleared, auto-reloads will

occur either with timer 2 overflows or on negative transitions at T2EX when EXEN2 = 1. When either RCLK = 1 or TCLK = 1, this bit is ignored and the timer is forced to auto-reload on timer 2 overflows.

PSW (PROGRAM STATUS WORD)

Symbol: PSW

Function: Program Status

Bit Address: D0H

Bit Addressable: Yes

Summary:

7	6	5	4	3	2	1	0
CY	AC	F0	RS1	RS0	OV	–	P

Bit Definitions:

Bit Symbol	Position	Bit Address	Description
CY	PSW.7	D7H	Carry flag. Set if there is a carry-out of bit 7 during an add, or set if there is a borrow into bit 7 during a subtract.
AC	PSW.6	D6H	Auxiliary carry flag. Set during add instructions if there is a carry-out of bit 3 into bit 4 or if the result in the lower nibble is in the range 0AH to 0FH.
F0	PSW.5	D5H	Flag 0. Available to the user for general purposes.
RS1	PSW.4	D4H	Register bank select bit 1. (See Table D–3)

TABLE D–3

RS1	RS0	BANK	ACTIVE ADDRESSES
0	0	0	00H–07H
0	1	1	08H–0FH
1	0	2	10H–17H
1	1	3	18H–1FH

RS0	PSW.3	D3H	Register bank select bit 0. (See Table D–3)
OV	PSW.2	D2H	Overflow flag. Set after an addition or subtraction operation if there was an arithmetic overflow (i.e., the signed result is greater than 127 or less than −128).
—	PSW.1	D1H	—
P	PSW.0	D0H	Parity flag. Set/cleared by hardware each instruction cycle to indicate an odd/even number of "1" bits in the accumulator.

E

APPENDIX E
8051 DATA SHEET[1]

[1]Reprinted with permission of Intel Corporation from *8-Bit Embedded Controllers* (270645). Santa Clara, CA.: Intel Corporation, 1991.

MCS®-51
8-BIT CONTROL-ORIENTED MICROCOMPUTERS
8031/8051
8031AH/8051AH
8032AH/8052AH
8751H/8751H-8

- **High Performance HMOS Process**
- **Internal Timers/Event Counters**
- **2-Level Interrupt Priority Structure**
- **32 I/O Lines (Four 8-Bit Ports)**
- **64K Program Memory Space**
- **Security Feature Protects EPROM Parts Against Software Piracy**
- **Boolean Processor**
- **Bit-Addressable RAM**
- **Programmable Full Duplex Serial Channel**
- **111 Instructions (64 Single-Cycle)**
- **64K Data Memory Space**

The MCS®-51 products are optimized for control applications. Byte-processing and numerical operations on small data structures are facilitated by a variety of fast addressing modes for accessing the internal RAM. The instruction set provides a convenient menu of 8-bit arithmetic instructions, including multiply and divide instructions. Extensive on-chip support is provided for one-bit variables as a separate data type, allowing direct bit manipulation and testing in control and logic systems that require Boolean processing.

The 8051 is the original member of the MCS-51 family. The 8051AH is identical to the 8051, but it is fabricated with HMOS II technology.

The 8751H is an EPROM version of the 8051AH; that is, the on-chip Program Memory can be electrically programmed, and can be erased by exposure to ultraviolet light. It is fully compatible with its predecessor, the 8751-8, but incorporates two new features: a Program Memory Security bit that can be used to protect the EPROM against unauthorized read-out, and a programmable baud rate modification bit (SMOD). The 8751H-8 is identical to the 8751H but only operates up to 8 MHz.

The 8052AH is an enhanced version of the 8051AH. It is backwards compatible with the 8051AH and is fabricated with HMOS II technology. The 8052AH enhancements are listed in the table below. Also refer to this table for the ROM, ROMless, and EPROM versions of each product.

Device	Internal Memory		Timers/ Event Counters	Interrupts
	Program	Data		
8052AH	8K x 8 ROM	256 x 8 RAM	3 x 16-Bit	6
8051AH	4K x 8 ROM	128 x 8 RAM	2 x 16-Bit	5
8051	4K x 8 ROM	128 x 8 RAM	2 x 16-Bit	5
8032AH	none	256 x 8 RAM	3 x 16-Bit	6
8031AH	none	128 x 8 RAM	2 x 16-Bit	5
8031	none	128 x 8 RAM	2 x 16-Bit	5
8751H	4K x 8 EPROM	128 x 8 RAM	2 x 16-Bit	5
8751H-8	4K x 8 EPROM	128 x 8 RAM	2 x 16-Bit	5

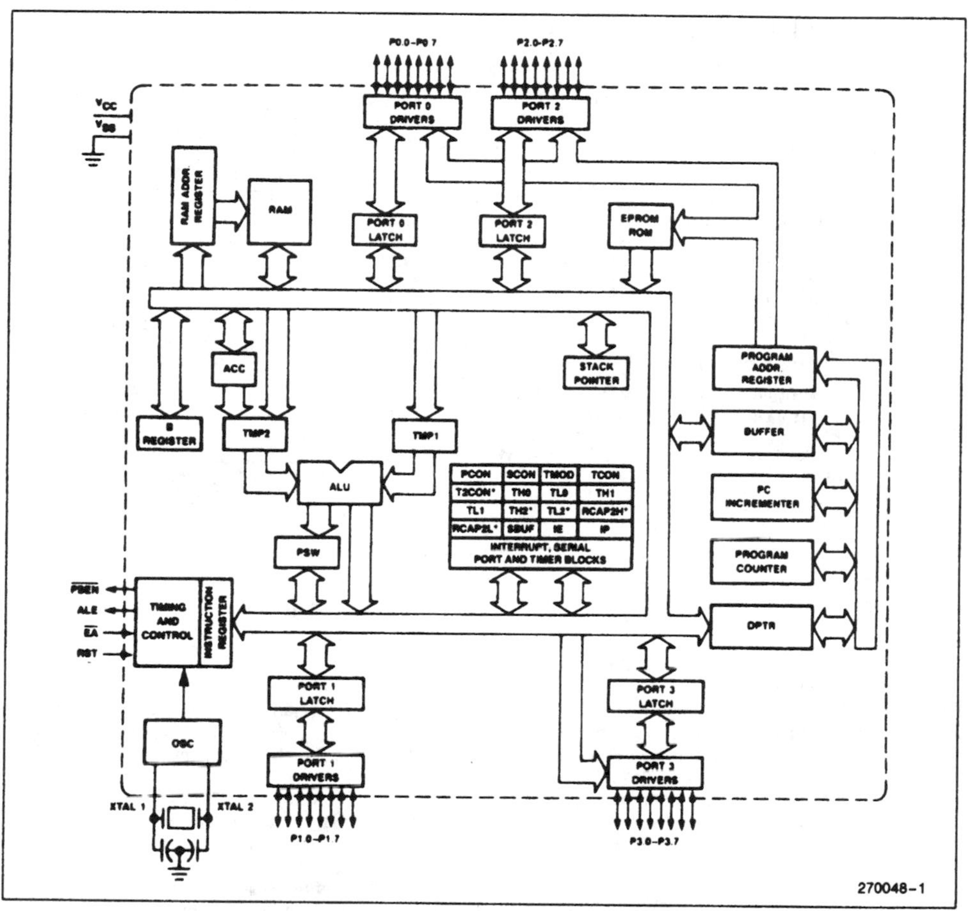

Figure 1. MCS®-51 Block Diagram

PACKAGES

Part	Prefix	Package Type
8051AH/ 8031AH	P D N	40-Pin Plastic DIP 40-Pin CERDIP 44-Pin PLCC
8052AH/ 8032AH	P D N	40-Pin Plastic DIP 40-Pin CERDIP 44-Pin PLCC
8751H/ 8751H-8	D	40-Pin CERDIP

PIN DESCRIPTIONS

V_{CC}: Supply voltage.

V_{SS}: Circuit ground.

Port 0: Port 0 is an 8-bit open drain bidirectional I/O port. As an output port each pin can sink 8 LS TTL inputs.

Port 0 pins that have 1s written to them float, and in that state can be used as high-impedance inputs.

Port 0 is also the multiplexed low-order address and data bus during accesses to external Program and Data Memory. In this application it uses strong internal pullups when emitting 1s and can source and sink 8 LS TTL inputs.

Port 0 also receives the code bytes during programming of the EPROM parts, and outputs the code bytes during program verification of the ROM and EPROM parts. External pullups are required during program verification.

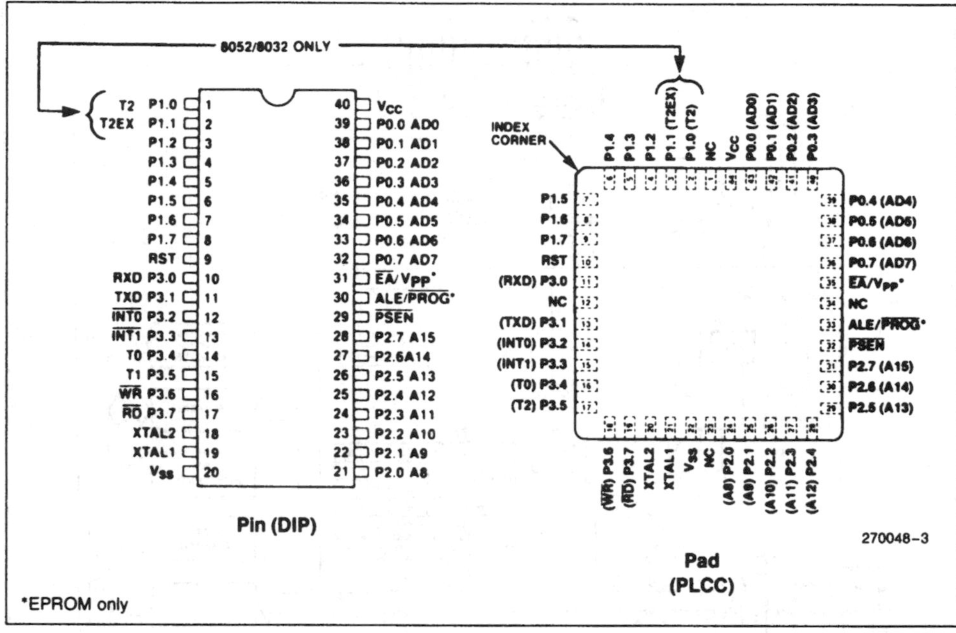

Figure 2. MCS®-51 Connections

Port 1: Port 1 is an 8-bit bidirectional I/O port with internal pullups. The Port 1 output buffers can sink/source 4 LS TTL inputs. Port 1 pins that have 1s written to them are pulled high by the internal pullups, and in that state can be used as inputs. As inputs, Port 1 pins that are externally being pulled low will source current (I_{IL} on the data sheet) because of the internal pullups.

Port 1 also receives the low-order address bytes during programming of the EPROM parts and during program verification of the ROM and EPROM parts.

In the 8032AH and 8052AH, Port 1 pins P1.0 and P1.1 also serve the T2 and T2EX functions, respectively.

Port 2: Port 2 is an 8-bit bidirectional I/O port with internal pullups. The Port 2 output buffers can sink/source 4 LS TTL inputs. Port 2 pins that have 1s written to them are pulled high by the internal pullups, and in that state can be used as inputs. As inputs, Port 2 pins that are externally being pulled low will source current (I_{IL} on the data sheet) because of the internal pullups.

Port 2 emits the high-order address byte during fetches from external Program Memory and during accesses to external Data Memory that use 16-bit addresses (MOVX @DPTR). In this application it uses strong internal pullups when emitting 1s. During accesses to external Data Memory that use 8-bit addresses (MOVX @Ri), Port 2 emits the contents of the P2 Special Function Register.

Port 2 also receives the high-order address bits during programming of the EPROM parts and during program verification of the ROM and EPROM parts.

Port 3: Port 3 is an 8-bit bidirectional I/O port with internal pullups. The Port 3 output buffers can sink/source 4 LS TTL inputs. Port 3 pins that have 1s written to them are pulled high by the internal pullups, and in that state can be used as inputs. As inputs, Port 3 pins that are externally being pulled low will source current (I_{IL} on the data sheet) because of the pullups.

Port 3 also serves the functions of various special features of the MCS-51 Family, as listed below:

Port Pin	Alternative Function
P3.0	RXD (serial input port)
P3.1	TXD (serial output port)
P3.2	$\overline{\text{INT0}}$ (external interrupt 0)
P3.3	$\overline{\text{INT1}}$ (external interrupt 1)
P3.4	T0 (Timer 0 external input)
P3.5	T1 (Timer 1 external input)
P3.6	$\overline{\text{WR}}$ (external data memory write strobe)
P3.7	$\overline{\text{RD}}$ (external data memory read strobe)

RST: Reset input. A high on this pin for two machine cycles while the oscillator is running resets the device.

ALE/$\overline{\text{PROG}}$: Address Latch Enable output pulse for latching the low byte of the address during accesses to external memory. This pin is also the program pulse input ($\overline{\text{PROG}}$) during programming of the EPROM parts.

In normal operation ALE is emitted at a constant rate of 1/6 the oscillator frequency, and may be used for external timing or clocking purposes. Note, however, that one ALE pulse is skipped during each access to external Data Memory.

$\overline{\text{PSEN}}$: Program Store Enable is the read strobe to external Program Memory.

When the device is executing code from external Program Memory, $\overline{\text{PSEN}}$ is activated twice each machine cycle, except that two $\overline{\text{PSEN}}$ activations are skipped during each access to external Data Memory.

$\overline{\text{EA}}$/V_{PP}: External Access enable $\overline{\text{EA}}$ must be strapped to V_{SS} in order to enable any MCS-51 device to fetch code from external Program memory locations starting at 0000H up to FFFFH. $\overline{\text{EA}}$ must be strapped to V_{CC} for internal program execution.

Note, however, that if the Security Bit in the EPROM devices is programmed, the device will not fetch code from any location in external Program Memory.

This pin also receives the 21V programming supply voltage (VPP) during programming of the EPROM parts.

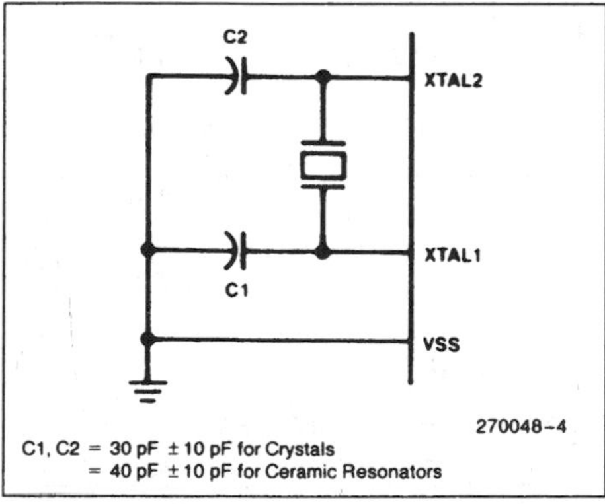

Figure 3. Oscillator Connections

XTAL1: Input to the inverting oscillator amplifier.

XTAL2: Output from the inverting oscillator amplifier.

OSCILLATOR CHARACTERISTICS

XTAL1 and XTAL2 are the input and output, respectively, of an inverting amplifier which can be configured for use as an on-chip oscillator, as shown in Figure 3. Either a quartz crystal or ceramic resonator may be used. More detailed information concerning the use of the on-chip oscillator is available in Application Note AP-155, "Oscillators for Microcontrollers."

To drive the device from an external clock source, XTAL1 should be grounded, while XTAL2 is driven, as shown in Figure 4. There are no requirements on the duty cycle of the external clock signal, since the input to the internal clocking circuitry is through a divide-by-two flip-flop, but minimum and maximum high and low times specified on the Data Sheet must be observed.

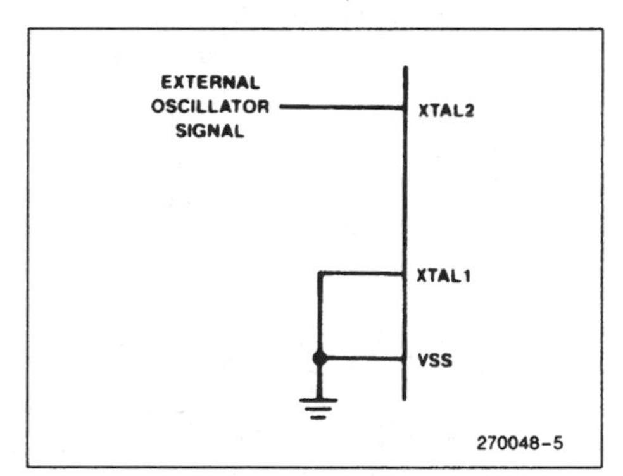

Figure 4. External Drive Configuration

DESIGN CONSIDERATIONS

If an 8751BH or 8752BH may replace an 8751H in a future design, the user should carefully compare both data sheets for DC or AC Characteristic differences. Note that the V_{IH} and I_{IH} specifications for the $\overline{\text{EA}}$ pin differ significantly between the devices.

Exposure to light when the EPROM device is in operation may cause logic errors. For this reason, it is suggested that an opaque label be placed over the window when the die is exposed to ambient light.

ABSOLUTE MAXIMUM RATINGS*

Ambient Temperature Under Bias ......0°C to 70°C
Storage Temperature .... −65°C to +150°C
Voltage on $\overline{EA}/V_{PP}$ Pin to V_{SS} ... −0.5V to +21.5V
Voltage on Any Other Pin to V_{SS} .... −0.5V to +7V
Power Dissipation.........................1.5W

NOTICE: This is a production data sheet. The specifications are subject to change without notice.

**WARNING: Stressing the device beyond the "Absolute Maximum Ratings" may cause permanent damage. These are stress ratings only. Operation beyond the "Operating Conditions" is not recommended and extended exposure beyond the "Operating Conditions" may affect device reliability.*

Operating Conditions: T_A (Under Bias) = 0°C to +70°C; V_{CC} = 5V ±10%; V_{SS} = 0V

D.C. CHARACTERISTICS (Under Operating Conditions)

Symbol	Parameter	Min	Max	Units	Test Conditions
V_{IL}	Input Low Voltage (Except $\overline{EA}$ Pin of 8751H & 8751H-8)	−0.5	0.8	V	
V_{IL1}	Input Low Voltage to $\overline{EA}$ Pin of 8751H & 8751H-8	0	0.7	V	
V_{IH}	Input High Voltage (Except XTAL2, RST)	2.0	V_{CC} + 0.5	V	
V_{IH1}	Input High Voltage to XTAL2, RST	2.5	V_{CC} + 0.5	V	XTAL1 = V_{SS}
V_{OL}	Output Low Voltage (Ports 1, 2, 3)*		0.45	V	I_{OL} = 1.6 mA
V_{OL1}	Output Low Voltage (Port 0, ALE, $\overline{PSEN}$)*				
	8751H, 8751H-8		0.60 0.45	V V	I_{OL} = 3.2 mA I_{OL} = 2.4 mA
	All Others		0.45	V	I_{OL} = 3.2 mA
V_{OH}	Output High Voltage (Ports 1, 2, 3, ALE, $\overline{PSEN}$)	2.4		V	I_{OH} = −80 μA
V_{OH1}	Output High Voltage (Port 0 in External Bus Mode)	2.4		V	I_{OH} = −400 μA
I_{IL}	Logical 0 Input Current (Ports 1, 2, 3, RST) 8032AH, 8052AH All Others		−800 −500	μA μA	V_{IN} = 0.45V V_{IN} = 0.45V
I_{IL1}	Logical 0 Input Current to $\overline{EA}$ Pin of 8751H & 8751H-8 Only		−15	mA	V_{IN} = 0.45V
I_{IL2}	Logical 0 Input Current (XTAL2)		−3.2	mA	V_{IN} = 0.45V
I_{LI}	Input Leakage Current (Port 0) 8751H & 8751H-8 All Others		±100 ±10	μA μA	0.45 ≤ V_{IN} ≤ V_{CC} 0.45 ≤ V_{IN} ≤ V_{CC}
I_{IH}	Logical 1 Input Current to $\overline{EA}$ Pin of 8751H & 8751H-8		500	μA	V_{IN} = 2.4V
I_{IH1}	Input Current to RST to Activate Reset		500	μA	V_{IN} < (V_{CC} − 1.5V)
I_{CC}	Power Supply Current: 8031/8051 8031AH/8051AH 8032AH/8052AH 8751H/8751H-8		160 125 175 250	mA mA mA mA	All Outputs Disconnected; $\overline{EA}$ = V_{CC}
C_{IO}	Pin Capacitance		10	pF	Test freq = 1 MHz

***NOTE:**
Capacitive loading on Ports 0 and 2 may cause spurious noise pulses to be superimposed on the V_{OL}s of ALE and Ports 1 and 3. The noise is due to external bus capacitance discharging into the Port 0 and Port 2 pins when these pins make 1-to-0 transitions during bus operations. In the worst cases (capacitive loading > 100 pF), the noise pulse on the ALE line may exceed 0.8V. In such cases it may be desirable to qualify ALE with a Schmitt Trigger, or use an address latch with a Schmitt Trigger STROBE input.

A.C. CHARACTERISTICS Under Operating Conditions;

Load Capacitance for Port 0, ALE, and PSEN = 100 pF;
Load Capacitance for All Other Outputs = 80 pF

Symbol	Parameter	12 MHz Oscillator		Variable Oscillator		Units
		Min	Max	Min	Max	
1/TCLCL	Oscillator Frequency			3.5	12.0	MHz
TLHLL	ALE Pulse Width	127		2TCLCL−40		ns
TAVLL	Address Valid to ALE Low	43		TCLCL−40		ns
TLLAX	Address Hold after ALE Low	48		TCLCL−35		ns
TLLIV	ALE Low to Valid Instr In 8751H All Others		 183 233		 4TCLCL−150 4TCLCL−100	 ns ns
TLLPL	ALE Low to $\overline{PSEN}$ Low	58		TCLCL−25		ns
TPLPH	$\overline{PSEN}$ Pulse Width 8751H All Others	 190 215		 3TCLCL−60 3TCLCL−35		 ns ns
TPLIV	$\overline{PSEN}$ Low to Valid Instr In 8751H All Others		 100 125		 3TCLCL−150 3TCLCL−125	 ns ns
TPXIX	Input Instr Hold after $\overline{PSEN}$	0		0		ns
TPXIZ	Input Instr Float after $\overline{PSEN}$		63		TCLCL−20	ns
TPXAV	$\overline{PSEN}$ to Address Valid	75		TCLCL−8		ns
TAVIV	Address to Valid Instr In 8751H All Others		 267 302		 5TCLCL−150 5TCLCL−115	 ns ns
TPLAZ	$\overline{PSEN}$ Low to Address Float		20		20	ns
TRLRH	$\overline{RD}$ Pulse Width	400		6TCLCL−100		ns
TWLWH	$\overline{WR}$ Pulse Width	400		6TCLCL−100		ns
TRLDV	$\overline{RD}$ Low to Valid Data In		252		5TCLCL−165	ns
TRHDX	Data Hold after $\overline{RD}$	0		0		ns
TRHDZ	Data Float after $\overline{RD}$		97		2TCLCL−70	ns
TLLDV	ALE Low to Valid Data In		517		8TCLCL−150	ns
TAVDV	Address to Valid Data In		585		9TCLCL−165	ns
TLLWL	ALE Low to $\overline{RD}$ or $\overline{WR}$ Low	200	300	3TCLCL−50	3TCLCL+50	ns
TAVWL	Address to $\overline{RD}$ or $\overline{WR}$ Low	203		4TCLCL−130		ns
TQVWX	Data Valid to $\overline{WR}$ Transition 8751H All Others	 13 23		 TCLCL−70 TCLCL−60		 ns ns
TQVWH	Data Valid to $\overline{WR}$ High	433		7TCLCL−150		ns
TWHQX	Data Hold after $\overline{WR}$	33		TCLCL−50		ns
TRLAZ	$\overline{RD}$ Low to Address Float		20		20	ns
TWHLH	$\overline{RD}$ or $\overline{WR}$ High to ALE High 8751H All Others	 33 43	 133 123	 TCLCL−50 TCLCL−40	 TCLCL+50 TCLCL+40	 ns ns

NOTE:
*This table does not include the 8751-8 A.C. characteristics (see next page).

This Table is only for the 8751H-8

A.C. CHARACTERISTICS

Under Operating Conditions;
Load Capacitance for Port 0, ALE, and PSEN = 100 pF;
Load Capacitance for All Other Outputs = 80 pF

Symbol	Parameter	8 MHz Oscillator		Variable Oscillator		Units
		Min	Max	Min	Max	
1/TCLCL	Oscillator Frequency			3.5	8.0	MHz
TLHLL	ALE Pulse Width	210		2TCLCL−40		ns
TAVLL	Address Valid to ALE Low	85		TCLCL−40		ns
TLLAX	Address Hold after ALE Low	90		TCLCL−35		ns
TLLIV	ALE Low to Valid Instr In		350		4TCLCL−150	ns
TLLPL	ALE Low to $\overline{\text{PSEN}}$ Low	100		TCLCL−25		ns
TPLPH	$\overline{\text{PSEN}}$ Pulse Width	315		3TCLCL−60		ns
TPLIV	$\overline{\text{PSEN}}$ Low to Valid Instr In		225		3TCLCL−150	ns
TPXIX	Input Instr Hold after $\overline{\text{PSEN}}$	0		0		ns
TPXIZ	Input Instr Float after $\overline{\text{PSEN}}$		105		TCLCL−20	ns
TPXAV	$\overline{\text{PSEN}}$ to Address Valid	117		TCLCL−8		ns
TAVIV	Address to Valid Instr In		475		5TCLCL−150	ns
TPLAZ	$\overline{\text{PSEN}}$ Low to Address Float		20		20	ns
TRLRH	$\overline{\text{RD}}$ Pulse Width	650		6TCLCL−100		ns
TWLWH	$\overline{\text{WR}}$ Pulse Width	650		6TCLCL−100		ns
TRLDV	$\overline{\text{RD}}$ Low to Valid Data In		460		5TCLCL−165	ns
TRHDX	Data Hold after $\overline{\text{RD}}$	0		0		ns
TRHDZ	Data Float after $\overline{\text{RD}}$		180		2TCLCL−70	ns
TLLDV	ALE Low to Valid Data In		850		8TCLCL−150	ns
TAVDV	Address to Valid Data In		960		9TCLCL−165	ns
TLLWL	ALE Low to $\overline{\text{RD}}$ or $\overline{\text{WR}}$ Low	325	425	3TCLCL−50	3TCLCL+50	ns
TAVWL	Address to $\overline{\text{RD}}$ or $\overline{\text{WR}}$ Low	370		4TCLCL−130		ns
TQVWX	Data Valid to $\overline{\text{WR}}$ Transition	55		TCLCL−70		ns
TQVWH	Data Valid to $\overline{\text{WR}}$ High	725		7TCLCL−150		ns
TWHQX	Data Hold after $\overline{\text{WR}}$	75		TCLCL−50		ns
TRLAZ	$\overline{\text{RD}}$ Low to Address Float		20		20	ns
TWHLH	$\overline{\text{RD}}$ or $\overline{\text{WR}}$ High to ALE High	75	175	TCLCL−50	TCLCL+50	ns

EXTERNAL PROGRAM MEMORY READ CYCLE

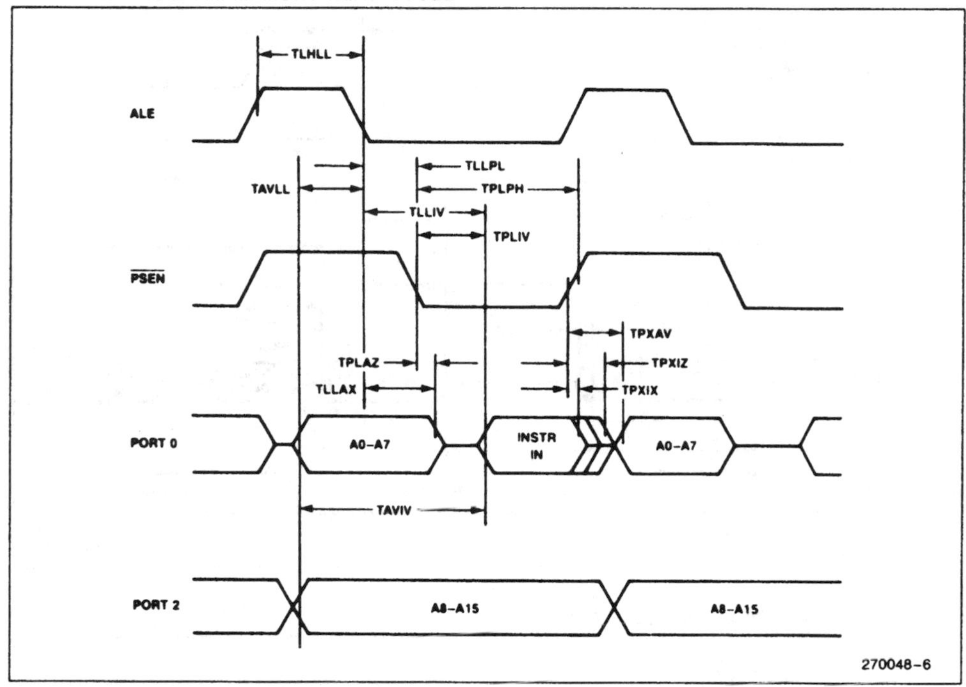

EXTERNAL DATA MEMORY READ CYCLE

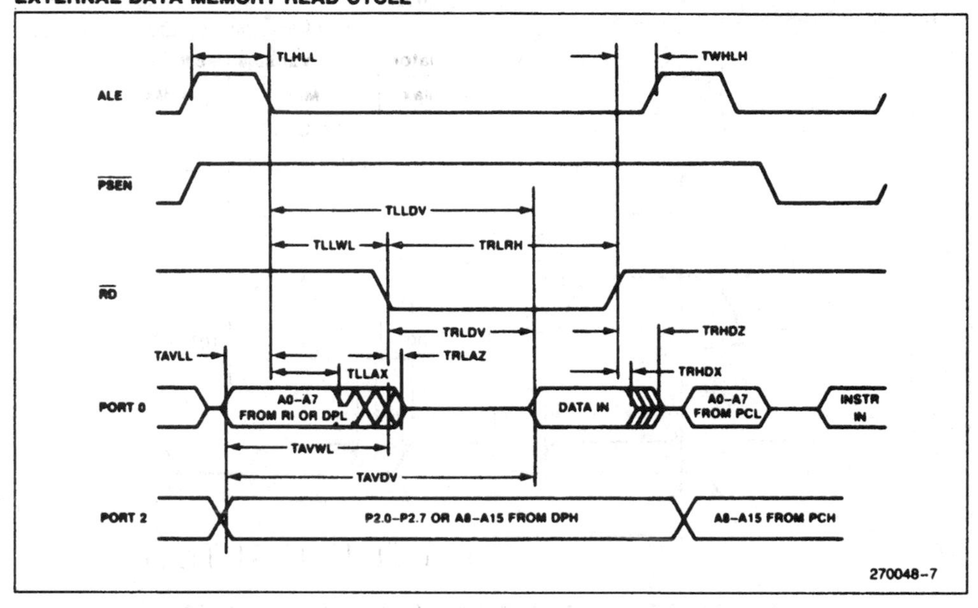

EXTERNAL DATA MEMORY WRITE CYCLE

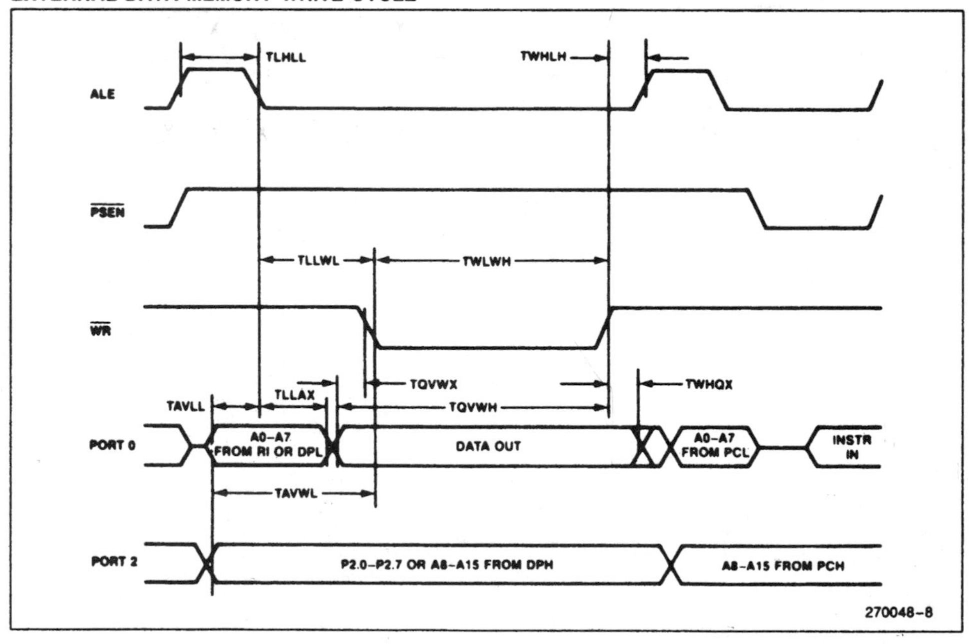

SERIAL PORT TIMING—SHIFT REGISTER MODE

Test Conditions: T_A = 0°C to 70°C; VCC = 5V ±10%; VSS = 0V; Load Capacitance = 80 pF

Symbol	Parameter	12 MHz Oscillator		Variable Oscillator		Units
		Min	Max	Min	Max	
TXLXL	Serial Port Clock Cycle Time	1.0		12TCLCL		μs
TQVXH	Output Data Setup to Clock Rising Edge	700		10TCLCL − 133		ns
TXHQX	Output Data Hold after Clock Rising Edge	50		2TCLCL − 117		ns
TXHDX	Input Data Hold after Clock Rising Edge	0		0		ns
TXHDV	Clock Rising Edge to Input Data Valid		700		10TCLCL − 133	ns

SHIFT REGISTER TIMING WAVEFORMS

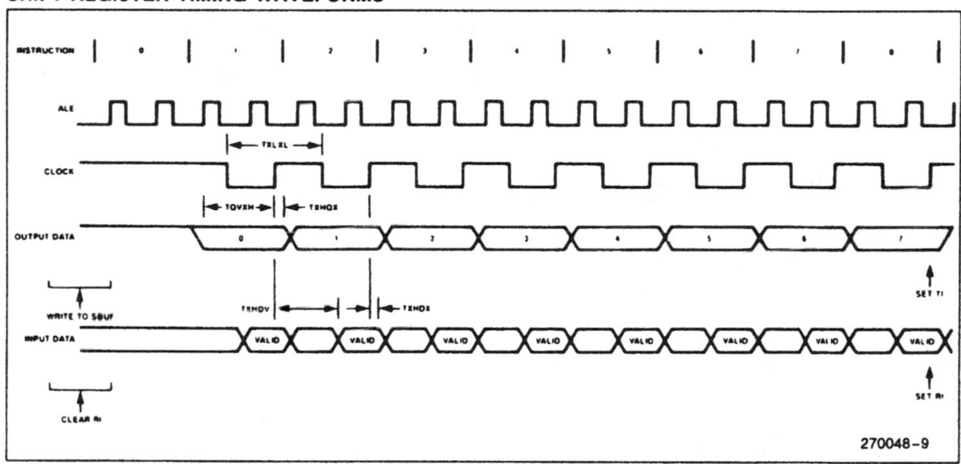

270048-9

EXTERNAL CLOCK DRIVE

Symbol	Parameter	Min	Max	Units
1/TCLCL	Oscillator Frequency (except 8751H-8) 8751H-8	3.5 3.5	12 8	MHz MHz
TCHCX	High Time	20		ns
TCLCX	Low Time	20		ns
TCLCH	Rise Time		20	ns
TCHCL	Fall Time		20	ns

EXTERNAL CLOCK DRIVE WAVEFORM

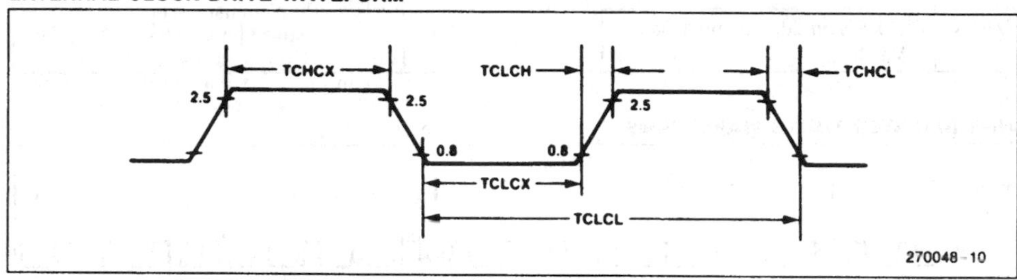

A.C. TESTING INPUT, OUTPUT WAVEFORM

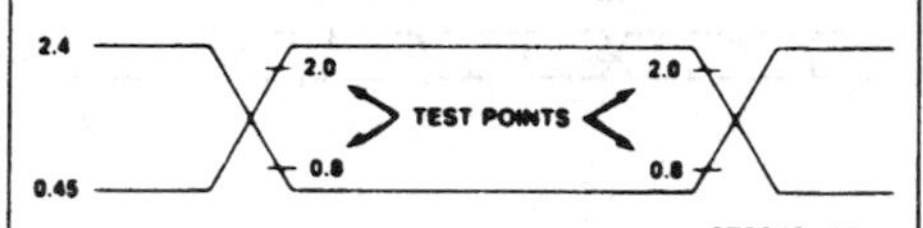

270048-11

A.C. Testing: Inputs are driven at 2.4V for a Logic "1" and 0.45V for a Logic "0". Timing measurements are made at 2.0V for a Logic "1" and 0.8V for a Logic "0".

EPROM CHARACTERISTICS

Table 3. EPROM Programming Modes

Mode	RST	PSEN	ALE	EA	P2.7	P2.6	P2.5	P2.4
Program	1	0	0*	VPP	1	0	X	X
Inhibit	1	0	1	X	1	0	X	X
Verify	1	0	1	1	0	0	X	X
Security Set	1	0	0*	VPP	1	1	X	X

NOTE:
"1" = logic high for that pin
"0" = logic low for that pin
"X" = "don't care"
"VPP" = +21V ±0.5V
*ALE is pulsed low for 50 ms.

Programming the EPROM

To be programmed, the part must be running with a 4 to 6 MHz oscillator. (The reason the oscillator needs to be running is that the internal bus is being used to transfer address and program data to appropriate internal registers.) The address of an EPROM location to be programmed is applied to Port 1 and pins P2.0–P2.3 of Port 2, while the code byte to be programmed into that location is applied to Port 0. The other Port 2 pins, and RST, PSEN, and EA should be held at the "Program" levels indicated in Table 3. ALE is **pulsed** low for 50 ms to program the code byte into the addressed EPROM location. The setup is shown in Figure 5.

Normally EA is held at a logic high until just before ALE is to be pulsed. Then EA is raised to +21V, ALE is pulsed, and then EA is returned to a logic high. Waveforms and detailed timing specifications are shown in later sections of this data sheet.

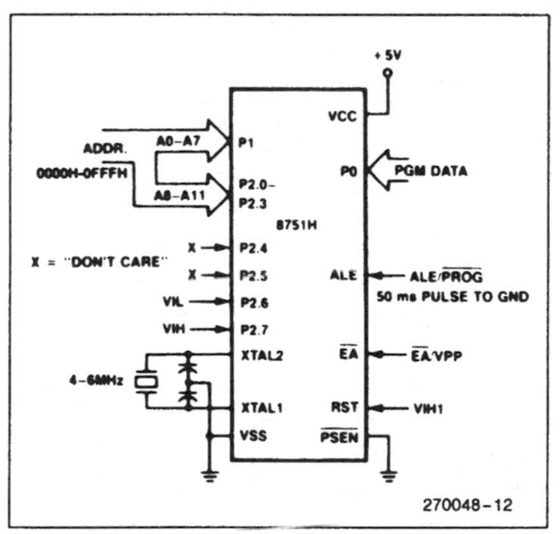

Figure 5. Programming Configuration

Note that the EA/VPP pin must not be allowed to go above the maximum specified VPP level of 21.5V for any amount of time. Even a narrow glitch above that voltage level can cause permanent damage to the device. The VPP source should be well regulated and free of glitches.

Program Verification

If the Security Bit has not been programmed, the on-chip Program Memory can be read out for verification purposes, if desired, either during or after the programming operation. The address of the Program Memory location to be read is applied to Port 1 and pins P2.0–P2.3. The other pins should be held at the "Verify" levels indicated in Table 3. The contents of the addressed location will come out on Port 0. External pullups are required on Port 0 for this operation.

The setup, which is shown in Figure 6, is the same as for programming the EPROM except that pin P2.7 is held at a logic low, or may be used as an active-low read strobe.

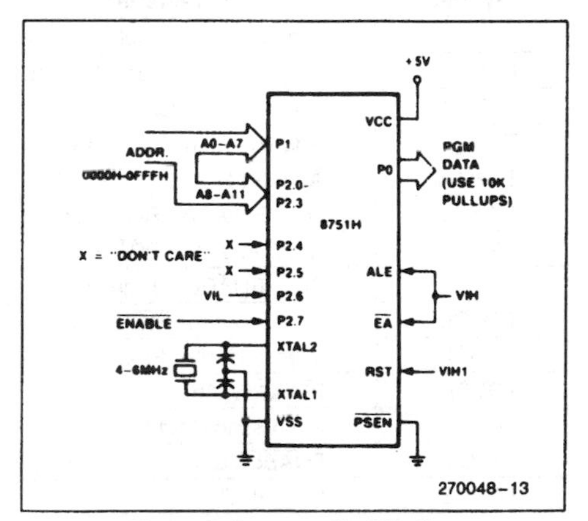

Figure 6. Program Verification

EPROM Security

The security feature consists of a "locking" bit which when programmed denies electrical access by any external means to the on-chip Program Memory. The bit is programmed as shown in Figure 7. The setup and procedure are the same as for normal EPROM programming, except that P2.6 is held at a logic high. Port 0, Port 1, and pins P2.0–P2.3 may be in any state. The other pins should be held at the "Security" levels indicated in Table 3.

Once the Security Bit has been programmed, it can be cleared only by full erasure of the Program Memory. While it is programmed, the internal Program Memory can not be read out, the device can not be further programmed, and it **can not execute out of external program memory**. Erasing the EPROM, thus clearing the Security Bit, restores the device's full functionality. It can then be reprogrammed.

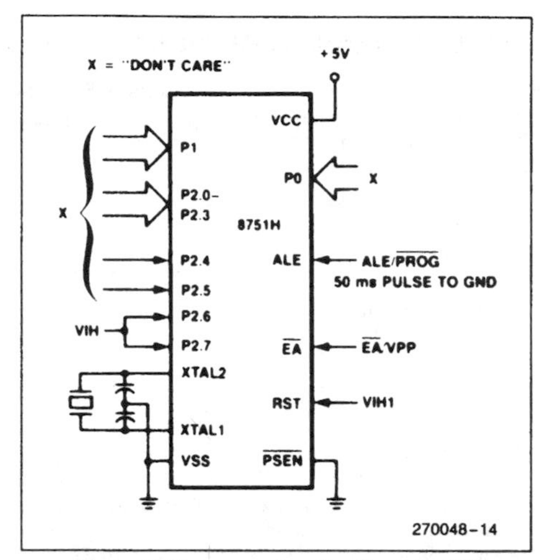

Figure 7. Programming the Security Bit

Erasure Characteristics

Erasure of the EPROM begins to occur when the chip is exposed to light with wavelengths shorter than approximately 4,000 Angstroms. Since sunlight and fluorescent lighting have wavelengths in this range, exposure to these light sources over an extended time (about 1 week in sunlight, or 3 years in room-level fluorescent lighting) could cause inadvertent erasure. If an application subjects the device to this type of exposure, it is suggested that an opaque label be placed over the window.

The recommended erasure procedure is exposure to ultraviolet light (at 2537 Angstroms) to an integrated dose of at least 15 W-sec/cm^2. Exposing the EPROM to an ultraviolet lamp of 12,000 μW/cm^2 rating for 20 to 30 minutes, at a distance of about 1 inch, should be sufficient.

Erasure leaves the array in an all 1s state.

EPROM PROGRAMMING AND VERIFICATION CHARACTERISTICS

T_A = 21°C to 27°C; VCC = 5V ±10%; VSS = 0V

Symbol	Parameter	Min	Max	Units
VPP	Programming Supply Voltage	20.5	21.5	V
IPP	Programming Supply Current		30	mA
1/TCLCL	Oscillator Frequency	4	6	MHz
TAVGL	Address Setup to $\overline{\text{PROG}}$ Low	48TCLCL		
TGHAX	Address Hold after $\overline{\text{PROG}}$	48TCLCL		
TDVGL	Data Setup to $\overline{\text{PROG}}$ Low	48TCLCL		
TGHDX	Data Hold after $\overline{\text{PROG}}$	48TCLCL		
TEHSH	P2.7 ($\overline{\text{ENABLE}}$) High to VPP	48TCLCL		
TSHGL	VPP Setup to $\overline{\text{PROG}}$ Low	10		µs
TGHSL	VPP Hold after $\overline{\text{PROG}}$	10		µs
TGLGH	$\overline{\text{PROG}}$ Width	45	55	ms
TAVQV	Address to Data Valid		48TCLCL	
TELQV	$\overline{\text{ENABLE}}$ Low to Data Valid		48TCLCL	
TEHQZ	Data Float after $\overline{\text{ENABLE}}$	0	48TCLCL	

EPROM PROGRAMMING AND VERIFICATION WAVEFORMS

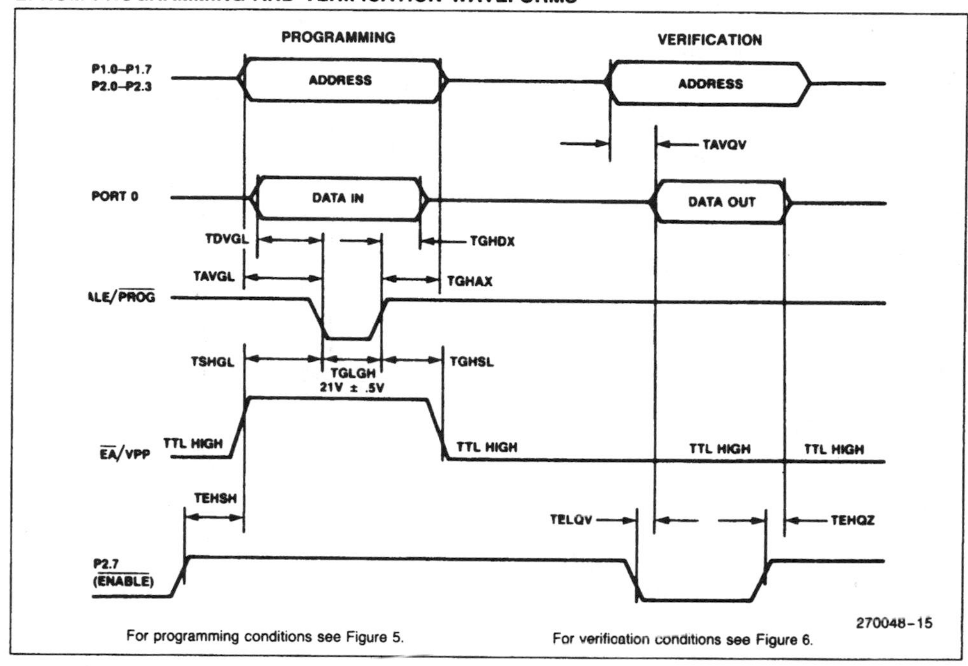

For programming conditions see Figure 5. For verification conditions see Figure 6.

DATA SHEET REVISION HISTORY

The following are the key differences between this and the -004 version of this data sheet.

1. Data sheet status changed from "Preliminary" to "Production".
2. LCC package offering deleted.
3. Maximum Ratings Warning and Data Sheet Revision History revised.

The following are the key differences between this and the -003 version of this data sheet:

1. Introduction was expanded to include product descriptions.
2. Package table was added.
3. Design Considerations added.
4. Test Conditions for I_{IL1} and I_{IH} specifications added to the DC Characteristics.
5. Data Sheet Revision History added.

F

APPENDIX F
ASCII CODE CHART

Bits				7	0	0	0	0	1	1	1	1
				6	0	0	1	1	0	0	1	1
4	3	2	1	5	0	1	0	1	0	1	0	1
0	0	0	0		NUL	DLE	SP	0	@	P	\	p
0	0	0	1		SOH	DC1	!	1	A	Q	a	q
0	0	1	0		STX	DC2	"	2	B	R	b	r
0	0	1	1		ETX	DC3	#	3	C	S	c	s
0	1	0	0		EOT	DC4	$	4	D	T	d	t
0	1	0	1		ENQ	NAK	%	5	E	U	e	u
0	1	1	0		ACK	SYN	&	6	F	V	f	v
0	1	1	1		BEL	ETB	'	7	G	W	g	w
1	0	0	0		BS	CAN	(	8	H	X	h	x
1	0	0	1		HT	EM	)	9	I	Y	i	y
1	0	1	0		LF	SUB	*	:	J	Z	j	z
1	0	1	1		VT	ESC	+	;	K	[	k	{
1	1	0	0		FF	FS	'	<	L	\	l	:
1	1	0	1		CR	GS	–	=	M	]	m	}
1	1	1	0		SO	RS	.	>	N	∧	n	~
1	1	1	1		SI	US	/	?	O	–	o	DEL

FIGURE F–1
ASCII code chart

G

APPENDIX G
MON51—AN 8051 MONITOR PROGRAM

This appendix contains the listing for an 8051 monitor program (MON51) along with a general description of its design and operation. Many of the concepts in assembly language programming developed earlier in short examples (see Chapter 7) can be reinforced by reviewing this appendix. The source and listing files for MON51 are contained on the diskette accompanying this text.

MON51 is a monitor program written for the 8051 microcontroller and, more specifically, for the SBC-51 single-board computer (see Chapter 10). We begin with a description of the purpose of a monitor program, and then give a summary of MON51 commands. The overall operation of MON51 and some design details are also described. The final pages of this appendix contain listings of each assembled source file, the listing created by RL51, and a dump of MON51 in Intel hexadecimal format.

A monitor program is not an operating system. It is a small program with commands that provide a primitive level of system operation and user interaction. The major difference is that operating systems are found on larger computers with disk drives, while monitor programs are found on small systems, such as the SBC-51, with a keyboard (or keypad) for input and a CRT (or LEDs) for output. Some single-board computers also provide mass storage using an audio cassette interface.

MON51 is an 8051 assembly language program approximately two kilobytes in length. It was written on an Intel iPDS100 development system using an editor (CREDIT), a cross assembler (ASM51), and a linker/locator (RL51). Testing and debugging were performed using a hardware emulator (EMV51) connected to an Intel SDK-51™single-board computer. It is worth mentioning that hardware emulators are such powerful tools that the first version of MON51 burned into EPROM was essentially "bug-free."

MON51 was developed using modular programming techniques. Ten source files were used, including nine program files and one macro definition file. To keep the listing files relatively short for this appendix, the "create symbol table" option was switched off in each source file using the $NOSYMBOLS assembler control. (See MAIN.LST, line 4 on p. 314.) However, since the $DEBUG assembler control was also used, the listing file created by the linker/locator, RL51, contains a complete, absolute symbol table for the entire program. The linker/locator absolute output was placed in V12 (for "version 12") and the listing file was placed in V12.M51. The OH (object-to-hex) utility was used to convert the absolute output to an Intel hex file, V12.HEX, suitable for printing, downloading to a target system, or burning into EPROM. (Spaces have been inserted into the hex file to improve readability.)

The files appearing in this appendix are identified below.

COMMANDS AND SPECIFICATIONS

MON51 uses the 8051's on-chip serial port for input/output. It requires an RS232C VDT operating at 2400 baud with seven data bits, one stop bit, and odd parity. Command line editing is possible using the BACKSPACE or DELETE keys to back up and correct mistakes. VT100 escape sequences are used (see below).

The following four commands are supported:

DUMP	dump the contents of a range of memory locations to the console
SET	examine single memory locations and set their contents to a new value
GO	go to user program at a specified address
LOAD	load an Intel hex-formatted file

COMMAND FORMAT

Commands are entered using the following format:

<c1><c2><p1>,<p2>,<p3>,<p4><CR>

where <c1>and <c2>form a 2-character command
<p1>to <p4>are command parameters
<CR>is the RETURN key

The following general comments describe command entry on MON51:

- When the system is powered up or reset, the "V12>" prompt appears on the VDT.
- Commands are entered on the keyboard and followed by pressing the RETURN key.
- During command entry, mistakes may be corrected using the BACKSPACE or DELETE key.
- Maximum line length is 20 characters.
- CONTROL-C terminates a command at any time and returns control to the prompt.
- An empty command line (RETURN key only) defaults to the SET command.
- Each command uses from 0 to 4 parameters entered in hexadecimal. (*Note:* The "H" is not needed.)
- Parameters are separated by commas.
- Any parameter may be omitted by entering two commas.
- Parameters omitted will default to their previous value.
- CONTROL-P toggles an enable bit for printer output. (*Note:* If printer output is enabled, the printer must be connected and on-line; otherwise MON51 waits for the printer and no output is sent to the console.)

COMMAND SYNTAX

DUMP

Format: D<m><start>,<end><CR>

Where: D is the command identifier

<m>is the memory space selector as follows:

B—bit-addressable memory

C—code memory

I—internal data memory

R—special function registers

X—external data memory

<start>is the start address

<end>is the end address

<CR>is the RETURN key

(*Note:* Output to the console may be suspended by entering CONTROL-S (XOFF). CONTROL-Q (XON) resumes output.)

Example: DX8000,8100<CR>dumps the contents of 257 external data locations from addresses 8000H to 8100H

SET

Format: S<m><address><CR>followed by one of
<value>
<SP>
—
<CR>
Q

Where: S is the command identifier
<m>is the memory space selector as follows:
B—bit-addressable memory
C—code memory
I—internal data memory
R—special function registers
X—external data memory
<address>is the address of a memory location to examine and/or set to a value
<value>is a data value to write into memory
<SP>is the space bar used to write same value into the next location
−examine previous location
<CR>is the return key used to examine the next location
Q quit and return to the MON51 prompt

Example: SI90<CR>followed by A5<CR>sets Port 1 to A5H

GO

Format: GO<address><CR>

Where: GO is the command identifier
<address>is the address in code memory to begin execution
<CR>is the RETURN key

Example: GO8000<CR>loads 8000H into the program counter

LO

Format: LO<CR>

Where: LO is the command identifier
<CR>is the RETURN key

Example: LO<CR> The message "<host download to SBC-51>" appears. MON51 receives an Intel hex-formatted file at the serial port and writes the bytes received into external data memory. When the transfer is complete, the message "EOF—file download OK" appears. *Note:* Use the DUMP command to verify that the transfer was successful.

GENERAL OPERATION OF MON51

The general operation of MON51 is described by the following steps:

1. Initialize registers and memory locations.
2. Send prompt to VDT display.
3. Get a command line from VDT keyboard.
4. Decode command.
5. Execute command.
6. Go to step 2.

Although numerous subroutines and program modules participate in the overall operation of MON51, the basic framework to implement the above steps is found in MAIN.LST. The instructions to initialize registers and memory locations (step 1) are found in lines 171 to 200 (see pp. 315–16). The prompt is sent to the VDT display in lines 201 and 202, and then a command is inputted from the VDT keyboard in lines 203 and 204 (steps 3 and 4). The command is decoded in lines 205 to 224. MON51 is designed to support 26 commands, one for each letter in the alphabet. Only four are implemented, however, while the others are for future expansion or custom applications. Decoding the command entails (a) using the first character in the input line buffer to look up the command's address from the table in lines 227 to 252, (b) placing the command's address on the stack (1 byte at a time), and (c) popping the address into the program counter using the RET instruction (line 225).

The code for the DUMP, SET, and LOAD commands is contained in separate files, while the code for the GO command is found in MAIN.LST in lines 268 to 274. Numerous subroutines are used throughout MON51 and can be found in the various listings provided.

During command decoding, the subroutine GETPAR is called (line 224). This subroutine gets parameters from the input line (which are stored in the line buffer as ASCII characters) and converts them to binary, placing the results in internal RAM at four 16-bit locations starting at the label PARMTR (see GETPAR.LST, line 58 on p. 319). These parameters are needed by the commands DUMP, SET, and GO.

After each command is executed, control is passed back to MON51 at the label GETCMD (line 200 in MAIN.LST) and the above steps are repeated. User programs may terminate using the address of GETCMD (00BCH) as the destination of an LJMP instruction.

MON51 contains numerous comments, and each subroutine and file contains a comment block explaining the general operation of the section of code that follows. Readers interested in understanding the intricacies of MON51 operation are directed to the listing files and comment lines for further details. In the following section, the overall design of MON51 is discussed.

THE DESIGN OF MON51

Developing an understanding of the "design" (as opposed to the operation) is also important, since MON51 incorporates—on a relatively large scale—many of the concepts developed earlier in brief examples. The following paragraphs elaborate on key design features that can be adopted in developing software for 8051-based designs. In a sense, then, the present appendix is like a case study—describing the design details of an existing 8051 software product.

Assembler controls, as mentioned in Chapter 7, are used primarily to control the format and content of the listing files created by ASM51 and RL51. At the top of each of the nine program listings, several assembler controls appear. These were chosen primarily to produce listing files suitable for reproduction in this appendix. Note in MAIN.LST, for example, that the control $NOLIST is used in line 6. This was used because line 7 (not listed) contained the control $INCLUDE(MACROS.SRC), which directed ASM51 to the file containing the macro definitions (pp. 341–42). Turning the "list" option off and on just before and after the $INCLUDE statement prevented ASM51 from needlessly putting the macro definitions into MAIN.LST.

Many of the assembler directives supported by ASM51 appear throughout the MON51 listing files. Recall (Section 7.5 Assembler Directives) that ASM51 supports five categories of directives:

- Assembler state control
- Symbol definition
- Storage initialization/reservation
- Program linkage
- Segment selection

Examples from each of these categories appear in the listings.

Assembler State Control. The only assembler state control directives used in MON51 are END and USING. The ORG directive was never used, since the code and data segments were designed to be relocatable with addresses established only at link-time.[1] An instance of USING can be found in GETPAR.LST, line 108 on p. 320.

Symbol Definitions. Many symbols are defined throughout MON51, as, for example, in MAIN.LST, lines 115 to 127 (pp. 314–15). Note in particular the definitions for EPROM, ONCHIP, and BITRAM. EPROM is defined as the name for a code segment. Similarly, ONCHIP is defined as the data segment and BITRAM is defined as the bit segment. All three are, by definition, relocatable. Recall that the segment type "DATA" corresponds to the 8051 on-chip data space accessible by direct addressing (00H to 7FH). All instructions in MON51 go in the EPROM code segment, all on-chip data locations are defined in the ONCHIP data segment, and all bit locations are defined in the BITRAM segment.

Storage Initialization/Reservation. The define byte directive (DB) is used throughout MON51 to initialize code memory with byte constants. Usually, the definition is for ASCII character strings sent to the console as part of MON51's normal operation. For example, the prompt characters for MON51 are defined in MAIN.LST on lines 288 and 289.

---

[1]The only exception to this is the absolute definition of the stack using the DSEG directive (see p. 317, line 288).

Other definitions include the escape sequence sent to the console to back up the cursor. MON51 is designed to work with VT100 compatible terminals. If the BACKSPACE or DELETE key is pressed while entering a line, the escape sequence <ESC>[D (or hexadecimal bytes 1BH, 5BH, and 44H)] must be sent to the terminal. These bytes are defined as a null-terminated ASCII string (see IO.LST, line 144 on p. 323) using the DB directive. Once the code for the BACKSPACE or DELETE key is detected (see lines 17 and 18 on p. 321 and lines 133 and 135 on p. 323), the pointer to the input line buffer is decremented twice, and the "back-up cursor" escape sequence is sent to the console using the output string (OUTSTR) subroutine. (See lines 136 to 139 on p. 323.)

Storage locations are reserved using either the define storage (DS) or define bit (DBIT) directives. The first instance of a DS directive defines a 24-byte stack (see line 288 in MAIN.LST) in an absolute data segment starting at address 08H in internal RAM. MON51 does not use register banks 1 to 3, so internal RAM addresses 08H to 1FH have been reserved for the stack. The default value of 07H for the stack pointer ensures that the first write to the stack is at address 08H. A 20-byte buffer for the command line is reserved at line 296 in MAIN.LST. Several bit locations are reserved in the BITRAM bit segment at the end of MAIN.LST.

Program Linkage. The PUBLIC and EXTRN directives are used near the beginning of most source files. PUBLIC declares which symbols defined in a particular file are to be made available for use in other source files. EXTRN declares which symbols are used in the current source file, but are defined in another source file. The NAME directive is not used in MON51; therefore, each file is a module with the file name serving as the module name.

Several of the files for MON51 hold the source code for subroutines used elsewhere. For example, IO.LST (pp. 321–25) contains the source code for the input/output subroutines INCHAR, OUTCHR, INLINE, OUTSTR, OUTHEX, and OUT2HX. These subroutines are all declared as "public" in line 26 (p. 321). Other files that use these subroutines must contain a statement declaring these symbols as external code address symbols (e.g., MAIN.LST, line 108 p. 314). Where these subroutines are called, the address for the LCALL or ACALL instruction is unknown by ASM51, so "F" appears in the listing file to indicate that linking/locating is required to "fix" the instruction. For example, in MAIN.LST the prompt is sent to the console in line 202 using ACALL OUTSTR. This instruction appears in the listing as 1100H with an "F" beside it. The linker/locator, RL51, will determine the correct absolute address of the OUTSTR subroutines and fix the instruction. In this example, since the addressing mode is absolute within the current 2K page, the fix substitutes 11 bits into the instruction—3 bits in the upper byte and all 8 bits in the lower byte.

Segment Selection. The segment selection directives are RSEG to select a relocatable segment, and the five directives to select an absolute segment of a specified memory space (DSEG, BSEG, XSEG, CSEG, and ISEG). The RSEG is used at least once in each file to begin the EPROM code segment (e.g., MAIN.LST, line 129, p. 315). Other instances of RSEG appear as necessary to select the ONCHIP data segment (e.g., MAIN.LST, line 295) or the BITRAM bit segment (e.g., MAIN.LST, line 306, p. 318). The only instance of an absolute segment is the definition of the stack at absolute internal RAM address 08H (MAIN.LST, line 287, p. 317).

OPTION JUMPERS

The SBC-51 has three jumpers on Port 1 that are read by software to evoke special features. MON51 reads these jumpers after a system reset and sets or clears corresponding bit locations (see MAIN.LST lines 171 to 176). Each jumper has a special purpose as outlined below.

Jumper	Installed	Purpose
X3	NO	normal execution
	YES	jump to 2000H upon reset
X4	NO	interrupt jump table at 80xxH
	YES	interrupt jump table at 20xxH
X13	NO	normal execution
	YES	software UART (see below)

Caution should be exercised when interfacing I/O circuitry to Port 1. If the attached interface presents a logic 0 to the corresponding pin during a system reset, this is interpreted by MON51 as if the jumper were installed, and the option listed above takes effect.

Reset Entry Point. Jumper X3 can be installed to force a jump to 2000H immediately after a system reset (see MAIN.LST, line 199). This is useful in order to have both the MON51 EPROM and a user EPROM installed simultaneously and to select which executes upon reset. If, for example, the user program does not use a VDT, then it is inconvenient to power up in "MON51 mode" just to enter GO2000H to evoke the user program. The installation of jumper X3 avoids the need for this.

Interrupt Jump Tables. Although MON51 does not use interrupts, user applications that coexist with MON51 can adopt an interrupt-driven design. Since MON51 resides in EPROM starting at address 0000H *and* since all interrupts vector to locations near the bottom of memory, a jump table was devised to allow user applications executing at other addresses to employ interrupts. Based on the default hardware configuration of the SBC-51, user applications generally execute either in RAM starting at 8000H or in EPROM starting at 2000H. If a user application enables interrupts and subsequently an interrupt occurs and is accepted, the interrupt vector address (i.e., entry point) is one of 0003H, 000BH, and 0013H, etc., depending on the source of the interrupt (see Table 6–4). At each interrupt vector address in MON51, there is a short instruction sequence that jumps to a location either in the user EPROM at 20xxH or in RAM at 80xxH, depending on whether or not jumper X4 is installed on the SBC-51 (see MAIN.LST, lines 150 to 183). The correct entry point for each interrupt is listed below.

	Interrupt Entry Point	
Interrupt Source	**X4 Installed**	**X4 Not Installed**
(default entry point)	2000H	8000H
External 0	2003H	8003H
Timer 0	2006H	8006H

External 1	2009H	8009H
Timer 1	200CH	800CH
Serial Port	200FH	800FH
Timer 2	2012H	8012H

The entry points above are spaced three locations apart, leaving just enough room for an LJMP instruction to the interrupt service routine. If operation is from a 12 MHz crystal, 6μs is added to the interrupt latency because of the overhead of this scheme (due to the JNB and LJMP instructions in MON51 and the LJMP instruction in the user program). Of course, if user applications do not use interrupts, programs may begin at 2000H or 8000H and proceed through the above entry points.

For an example of a user application using interrupts that coexists with MON51, see the MC14499 interface project in Chapter 10.

Software UART. When jumper X13 is installed, MON51 implements a software UART for serial input/output instead of using the 8051's on-chip serial port. This feature is useful for developing interfaces to serial devices other than the default terminal. Note that the baud rate is 1200 if the software UART is used. The operation of the software UART is described in IO.LST (p. 325, ll. 219–278).

THE LINK MAP AND SYMBOL TABLE

One of the most important listings is that produced by the linker/locator, RL51. This listing, V12.M51, (pp. 342–44), contains a link map and symbol table. The link map shows the starting address and length of the three relocatable segments used by MON51 (p. 343). The addresses are absolute at this stage, since the segments have been located by RL51 based on the specifications provided in the invocation line (p. 342). The symbol table gives the absolute values assigned to all relocatable symbols (as assigned by RL51 at link-time). It is partitioned alphabetically by module (i.e., file) in the order linked.

As an example, the absolute address of the OUTSTR subroutine is determined by looking up OUTSTR in the IO module (p. 343–44). The address of OUTSTR in MON51 version 12 is 0282H.

INTEL HEX FILE

The final listing is that produced by the OH (object-to-hex) conversion utility. The listing V12.HEX (pp. 344–46) contains the machine-language bytes of the MON51 program in Intel hexadecimal format. For example, the OUTSTR subroutine starts at address 0282H. On p. 345, the first three bytes of this subroutine appear as E4H, 93H, and 60H. Referring directly to the OUTSTR subroutine in the IO.LST confirms that these values are correct (see p. 324, ll. 156–158).

MAIN MODULE FOR 8051 MONITOR PROGRAM

```
DOS 3.31 (038-N) MCS-51 MACRO ASSEMBLER, V2.2
OBJECT MODULE PLACED IN MAIN.OBJ
ASSEMBLER INVOKED BY:  C:\ASM51\ASM51.EXE MAIN.SRC EP

LOC  OBJ            LINE     SOURCE
                       1     $debug
                       2     $title(*** MAIN MODULE FOR 8051 MONITOR PROGRAM ***)
                       3     $pagewidth(98)
                       4     $nosymbols               ;symbol table in .M51 file created by RL51
                       5     $nopaging
                       6     $nolist                  ;next line contains $include(macros.src)
                      76     ;previous line contains $list
                      77     ;*******************************************************************
                      78     ;                                                                  *
                      79     ;                            M O N 5 1                             *
                      80     ;                                                                  *
                      81     ;                   (An 8051 Monitor Program)                      *
                      82     ;                                                                  *
                      83     ; Copyright (c) I. Scott MacKenzie, 1988, 1991                     *
                      84     ; Dept. of Computing & Information Science                         *
                      85     ; University of Guelph                                             *
                      86     ; Guelph, Ontariao                                                 *
                      87     ; Canada NIG 2WI                                                   *
                      88     ;                                                                  *
                      89     ; VERSION 12 - April 1991                                          *
                      90     ;                                                                  *
                      91     ; COMMANDS: D - dump memory to console                             *
                      92     ;           G - go to user program                                 *
                      93     ;           L - load hex file                                      *
                      94     ;           S - set memory to value                                *
                      95     ;                                                                  *
                      96     ; RESET OPTIONS:                                                   *
                      97     ;                                                                  *
                      98     ; JUMPER  INSTALLED  RESULT                                        *
                      99     ; ------  ---------  ----------------------------------------       *
                     100     ;   X3       NO      Execute MON51                                 *
                     101     ;   X3       YES     Jump to 2000H upon reset                      *
                     102     ;   X4       NO      Interrupt jump table in user RAM (8000H)      *
                     103     ;   X4       YES     Interrupt jump table in user EPROM (2000H)    *
                     104     ;   X13      NO      Serial I/O using P3.1 (TXD) and P3.0 (RXD)    *
                     105     ;   X13      YES     Serial I/O using P1.7 (TXD) and P1.6 (RXD) and *
                     106     ;                    interrupts (except if X4 installed)           *
                     107     ;*******************************************************************
                     108             extrn   code(outchr, inline, outstr, htoa, atoh, getpar)
                     109             extrn   code(load, dump, setcmd, serial_io_using_interrupts)
                     110             extrn   data(parmtr, tb_count)
                     111             extrn   bit(t_flag, r_flag, r_idle)
                     112             public  getcmd, linbuf, buflen, endbuf, notice
                     113             public  error, p_bit, riot, ram, rom, x13_bit
                     114
 0014                115     buflen  equu    20        ;maximum line length
 0007                116     bel     equ     07H       ;ASCII bell code
 000D                117     cr      equ     0DH       ;ASCII carriage return code
 000A                118     lf      equ     0AH       ;ASCII line feed code
 0100                119     riot    xdata   0100h     ;8155 RAM/IO/TIMER SBC51
 8000                120     ram     equ     8000h     ;RAM address for user software
 2000                121     rom     code    2000h     ;EPROM address for user firmware
                     122     eprom   segment code
                     123     onchip  segment data
                     124     bitram  segment bit
 0095                125     x13     bit     p1.5      ;reset option
```

```
  0094              126    x3      bit     p1.4
  0093              127    x4      bit     p1.3
                    128
 ----               129            rseg    eprom
 0000 020000    F   130    main:   ljmp    skip            ;jump above interrupt vectors
 0003 300034    F   131            jnb     x4_bit,rom3     ;external 0 interrupt entry point
 0006 028003        132            ljmp    ram+3
 0009 0000          133            dw      0
 000B 30002F    F   134            jnb     x4_bit,rom6     ;timer 0 interrupt entry point
 000E 028006        135            ljmp    ram+6
 0011 0000          136            dw      0
 0013 30002A    F   137            jnb     x4_bit,rom9     ;external 1 interrupt entry point
 0016 028009        138            ljmp    ram+9
 0019 0000          139            dw      0
 001B 300025    F   140            jnb     x4_bit,rom12    ;timer 1 interrupt entry point
 001E 020000    F   141            ljmp    check           ;check jumper X13 first
 0021 0000          142            dw      0
 0023 300020    F   143            jnb     x4_bit,rom15    ;serial port interrupt entry point
 0026 02800F        144            ljmp    ram+15
 0029 0000          145            dw      0
 002B 30001B    F   146            jnb     x4_bit,rom18    ;timer 2 interrupt entry point
 002E 028012        147            ljmp    ram+18
                    148
                    149    ;*******************************************************************
                    150    ; Timer 1 interrupts are used for serial I/O if jumper X13 is      *
                    151    ; installed.  P1.7 is used for TXD and P1.6 is used for RXD.       *
                    152    ; (See IO module)                                                   *
                    153    ;*******************************************************************
 0031 200003    F   154    check:  jb      x13_bit,ram12   ;if X13 installed, serial I/O using
 0034 020000    F   155            ljmp    serial_io_using_interrupts      ;interrupts
 0037 02800C        156    ram12:  ljmp    ram+12          ;if X13 not installed, use RAM table
 003A 022003        157    rom3:   ljmp    rom+3           ;if X4 is installed, interrupts
 003D 022006        158    rom6:   ljmp    rom+6           ; are directed to a jump table in
 0040 022009        159    rom9:   ljmp    rom+9           ; user EPROM at 2003H (otherwise
 0043 02200C        160    rom12:  ljmp    rom+12          ; directed to jump table in user
 0046 02200F        161    rom15:  ljmp    rom+15          ; RAM at 8003H; see above)
 0049 022012        162    rom18:  ljmp    rom+18
                    163
 004C 436F7079      164    notice: db      'Copyright (c) I. Scott MacKenzie, 1988, 1991'
 0050 72696768
 0054 74202863
 0058 2920492E
 005C 2053636F
 0060 7474204D
 0064 61634B65
 0068 6E7A6965
 006C 2C203139
 0070 38382C20
 0074 31393931
                    165
                    166    ;*******************************************************************
                    167    ; Copy state of pins used for reset jumpers to flag bits.  Reset   *
                    168    ; jumpers are only read upon reset.  Subsequent tests are on the   *
                    169    ; flag bits.                                                        *
                    170    ;*******************************************************************
 0078 A294          171    skip:   mov     c,x3
 007A 9200      F   172            mov     x3_bit,c
 007C A293          173            mov     c,x4
 007E 9200      F   174            mov     x4_bit,c
 0080 A295          175            mov     c,x13
 0082 9200      F   176            mov     x13_bit,c
 0084 200003    F   177            jb      x3_bit,skip4    ;If x3 jumper installed,
 0087 022000        178            ljmp    rom             ; jump to user EPROM at 2000H
 008A 200017    F   179    skip4:  jb      x13_bit,skip5   ;If x13 jumper installed,
```

```
008D C200    F   180              clr    r_flag          ; r_flag = 0 (no character waiting)
008F D200    F   181              setb   r_idle          ; r_idle = 1 (receiver idle)
0091 D200    F   182              setb   t_flag          ; t_flag = 1 (ready to transmit)
0093 75000B  F   183              mov    tb_count,#11    ; 11 bits sent (start + 8 + stop)
0096 758920      184              mov    tmod,#20H       ; 8-bit auto-reload mode
0099 758D98      185              mov    th1,#-104       ; 1 / (8 x 0.0012) us for 1200 baud
009C D28E        186              setb   tr1             ; start timer 1 (interrupts coming)
009E D2AF        187              setb   ea              ; turn on timer 1 interrupts - serial
00A0 D2AB        188              setb   et1             ; I/O uses P1.7 (TXD) and P1.6 (RXD)
00A2 800B        189              sjmp   skip6           ; and skip over serial port init
00A4 759852      190     skip5:   mov    scon,#01010010B ;If x13 not installed, initialize
00A7 758DF3      191              mov    th1,#0F3H       ; for 2400 baud reload value
00AA 758920      192              mov    tmod,#20H       ;auto reload mode
00AD D28E        193              setb   tr1             ;start timer
00AF C200    F   194     skip6:   clr    p_bit           ;default = no printer output
00B1 900100      195              mov    dptr,#riot      ;address of 8155
00B4 7401        196              mov    a,#1            ;Port A = output
00B6 F0          197              movx   @dptr,a         ;initialize printer port
00B7 900000  F   198              mov    dptr,#hello     ;send hello message to console
00BA 1100    F   199              acall  outstr
00BC 758107      200     getcmd:  mov    sp,#stack - 1   ;initialize stack pointer
00BF 900000  F   201              mov    dptr,#prompt    ;send prompt to console
00C2 1100    F   202              acall  outstr
00C4 7800    F   203              mov    r0,#linbuf      ;R0 points to line buffer
00C6 1100    F   204              acall  inline          ;input command line from console
00C8 E500    F   205              mov    a,linbuf        ;get first character
00CA B40D03      206              cjne   a,#cr,skip7     ;empty line?
00CD 020000  F   207              jmp    setcmd          ;yes: default to "set" command
                 208     skip7:                          ;check for alphabetic character
00D0 B44100      209 +2           cjne   a,#'A',$+3      ;JOR
00D3 404C        210 +2           jc     error
00D5 B45B00      211 +2           cjne   a,#'Z'+1,$+3
00D8 5047        212 +2           jnc    error
00DA 541F        213              anl    a,#1fh          ;reduce to 5 bits
00DC 14          214              dec    a               ;reduce ASCII 'A' to 0
00DD 23          215              rl     a               ;adjust to word boundary
00DE F8          216              mov    r0,a            ;save
00DF 04          217              inc    a
00E0 900000  F   218              mov    dptr,#table     ;command address table
00E3 93          219              movc   a,@a+dptr       ;get address low byte of command
00E4 C0E0        220              push   acc             ;save on stack
00E6 E8          221              mov    a,r0            ;restore accumulator
00E7 93          222              movc   a,@a+dptr       ;get address high byte of command
00E8 C0E0        223              push   acc             ;sneaky: pushing address on stack
00EA 1100    F   224              acall  getpar          ;get parameters from linbuf
00EC 22          225              ret                    ;pop command address into PC and
                 226                                     ; off we go!
00ED 0000    F   227     table:   dw     error           ;A (Note: most commands undefined)
00EF 0000    F   228              dw     error           ;B
00F1 0000    F   229              dw     error           ;C
00F3 0000    F   230              dw     dump            ;Dump memory locations
00F5 0000    F   231              dw     error           ;E
00F7 0000    F   232              dw     error           ;F
00F9 0000    F   233              dw     go              ;Go to user program
00FB 0000    F   234              dw     error           ;H
00FD 0000    F   235              dw     error           ;I
00FF 0000    F   236              dw     error           ;J
0101 0000    F   237              dw     error           ;K
0103 0000    F   238              dw     load            ;Load hex file
0105 0000    F   239              dw     error           ;M
0107 0000    F   240              dw     error           ;N
0109 0000    F   241              dw     error           ;O
010B 0000    F   242              dw     error           ;P
010D 0000    F   243              dw     error           ;Q
```

```
010F 0000      F    244            dw     error             ;R
0111 0000      F    245            dw     setcmd            ;Set memory to value
0113 0000      F    246            dw     error             ;T
0115 0000      F    247            dw     error             ;U
0117 0000      F    248            dw     error             ;V
0119 0000      F    249            dw     error             ;W
011B 0000      F    250            dw     error             ;X
011D 0000      F    251            dw     error             ;Y
011F 0000      F    252            dw     error             ;Z
0121 900000    F    253    error:  mov    dptr,#emess       ;unimplemented command
0124 1100      F    254            acall  outstr
0126 E500      F    255            mov    a,linbuf          ;send out 1st two characters in
0128 1100      F    256            acall  outchr            ; line buffer too
012A E500      F    257            mov    a,linbuf + 1
012C 1100      F    258            acall  outchr
012E 808C           259            jmp    getcmd
                    260
                    261    ;*****************************************************************
                    262    ; GO to user program command.  This command is small enough to    *
                    263    ; include with the main module.                                   *
                    264    ;                                                                 *
                    265    ;  FORMAT:  GO<address>                                           *
                    266    ;                                                                 *
                    267    ;*****************************************************************
0130 7400      F    268    go:     mov    a,#low(getcmd)   ;if "ret" from user program,
0132 C0E0           269            push   acc               ; save "getcmd" address on stack
0134 7400      F    270            mov    a,#high(getcmd)
0136 C0E0           271            push   acc
0138 C000      F    272            push   parmtr + 1        ;push address of user program on
013A C000      F    273            push   parmtr            ; stack
013C 22             274            ret                      ;pop address into PC (Off we go!)
                    275
                    276    ;*****************************************************************
                    277    ; ASCII null-terminated strings                                   *
                    278    ;*****************************************************************
013D 0D             279    hello:  db     cr,'MON51',0
013E 4D4F4E35
0142 31
0143 00
0144 0D             280    prompt: db     cr,'V12>',0
0145 5631323E
0149 00
014A 07             281    emess:  db     bel,cr,'Error: Invalid Command - ',0
014B 0D
014C 4572726F
0150 723A2049
0154 6E76616C
0158 69642043
015C 6F6D6D61
0160 6E64202D
0164 20
0165 00
                    282
                    283    ;*****************************************************************
                    284    ; Reserve 24 bytes for the stack in an absolute segment starting at *
                    285    ; 08H.  (Note: MON51 does not use register banks 1-3.)            *
                    286    ;*****************************************************************
----                287                   dseg at 8
0008                288    stack:         ds      24
                    289
                    290    ;*****************************************************************
                    291    ; Create a line buffer for monitor commands in a relocatable data *
                    292    ; segment.  See linker/locator listing (.M51) for absolute address *
                    293    ; assigned.                                                       *
                    294    ;*****************************************************************
```

```
0000                296     linbuf:         ds      buflen          ;input command line goes here
  0014              297     endbuf          equ     $
                    298
                    299     ;*****************************************************************
                    300     ; Create 1-bit flags.  If p_bit flag set, OUTCHR transmits  to the *
                    301     ; console and the printer; if clear, output only sent to the       *
                    302     ; console (default).  "x" bits are set to the state of the reset   *
                    303     ; jumpers as read immediately after reset.  Consult the link map   *
                    304     ; to determine the exact (i.e., absolute) location of these bits.  *
                    305     ;*****************************************************************
----                306                     rseg    bitram
0000                307     p_bit:          dbit    1
0001                308     x3_bit:         dbit    1
0002                309     x4_bit:         dbit    1
0003                310     x13_bit:        dbit    1
                    311                     end

REGISTER BANK(S) USED: 0

ASSEMBLY COMPLETE, NO ERRORS FOUND
```

GET PARAMETERS FROM INPUT LINE

```
DOS 3.31 (038-N) MCS-51 MACRO ASSEMBLER, V2.2
OBJECT MODULE PLACED IN GETPAR.OBJ
ASSEMBLER INVOKED BY:  C:\ASM51\ASM51.EXE GETPAR.SRC EP


LOC  OBJ            LINE     SOURCE

                       1     $debug
                       2     $title(*** GET PARAMETERS FROM INPUT LINE ***)
                       3     $pagewidth(98)
                       4     $nosymbols
                       5     $nopaging
                       6     ;*****************************************************************
                       7     ;                                                                 *
                       8     ; GETPAR.SRC - GET PARAMETERS                                     *
                       9     ;                                                                 *
                      10     ; This routine scans the characters in the input buffer "linbuf"  *
                      11     ; and extracts the parameters used by the monitor commands.  The  *
                      12     ; input line must be of the form:                                 *
                      13     ;                                                                 *
                      14     ;  <C1><C2><P1>,<P2>,<P3>,<P4><CR>0                               *
                      15     ;                                                                 *
                      16     ; where <C1> and <C2> are the 2-character command and <P1> to <P4> *
                      17     ; are up to 4 parameters required by the commands.  Scanning begins *
                      18     ; after the 2-character command (i.e., at linbuf+2).              *
                      19     ;                                                                 *
                      20     ; Parameters are entered as hex ASCII codes separated by          *
                      21     ; commas.  The hex ASCII codes are converted to 16-bit binary     *
                      22     ; values and placed in memory beginning at "parmtr" in the order  *
                      23     ; high byte/low byte.                                             *
                      24     ;                                                                 *
                      25     ; Any parameters omitted retain their value from the previous     *
                      26     ; command.  Any illegal hex ASCII code terminates the scanning.   *
                      27     ;                                                                 *
                      28     ;*****************************************************************
                      29             extrn code(ishex, atoh)
                      30             extrn data(linbuf)
```

```
                        31              public getpar, parmtr, reg_sp, bit_sp, int_sp, ext_sp, cde_sp
                        32
  0004                  33      numpar  equ     4               ;4 input parameters
                        34      eprom   segment code
                        35      onchip  segment data
                        36      bitram  segment bit
                        37
----                    38              rseg    eprom
0000 7800        F      39      getpar: mov     r0,#linbuf+2    ;R0 points to input line data
0002 7900        F      40              mov     r1,#parmtr      ;R1 points to parameters destination
0004 7F04               41              mov     r7,#numpar      ;use R7 as counter
0006 B62C02             42      getp2:  cjne    @r0,#',',getp3  ;parameter missing?
0009 8008               43              sjmp    getp4           ;yes: try to convert next parameter
000B 1100        F      44      getp3:  acall   gethx           ;no:  convert it
000D 300008      F      45              jnb     found,getp5     ;if illegal byte, exit
0010 B62C05             46              cjne    @r0,#',',getp5  ;next character should be ','
0013 08                 47      getp4:  inc     r0
0014 09                 48              inc     r1
0015 09                 49              inc     r1              ;point to next parameter destination
0016 DFEE               50              djnz    r7,getp2        ; and let's try for another
0018 1100        F      51      getp5:  acall   space           ;determine memory space selected
001A 22                 52              ret
                        53
                        54      ;*******************************************************************
                        55      ; Create space for the parameters in the "onchip" data segment.    *
                        56      ;*******************************************************************
----                    57              rseg    onchip
0000                    58      parmtr: ds      numpar*2        ;2 bytes required for each parameter
                        59
                        60      ;*******************************************************************
                        61      ; GETHX - GET HeX word from "linbuf"                               *
                        62      ;                                                                  *
                        63      ;       enter:  R0 points to string                                *
                        64      ;               R1 points to 2-byte destination                    *
                        65      ;       exit:   hex word in memory                                 *
                        66      ;               R0 points to next non-hex character in string      *
                        67      ;               'found' = 1 if legal value found or 0 otherwise    *
                        68      ;       uses:   atoh, insrt, ishex                                 *
                        69      ;                                                                  *
                        70      ;*******************************************************************
----                    71              rseg    eprom
001B C200        F      72      gethx:  clr     found           ;default = no value found
001D E6                 73              mov     a,@r0
001E 1100        F      74              acall   ishex           ;any value?
0020 5013               75              jnc     gethx9          ;no:  return
0022 D200        F      76      gethx2: setb    found           ;set value found flag
0024 E4                 77              clr     a
0025 F7                 78              mov     @r1,a           ;clear destination value
0026 09                 79              inc     r1
0027 F7                 80              mov     @r1,a
0028 19                 81              dec     r1
0029 E6                 82      gethx3: mov     a,@r0           ;get value
002A 1100        F      83              acall   ishex           ;value found?
002C 5007               84              jnc     gethx9          ;no:  return
002E 1100        F      85              acall   atoh            ;yes: attempt to convert it
0030 1100        F      86              acall   insrt           ; and insert into parameter
0032 08                 87              inc     r0              ;increment buffer pointer
0033 80F4               88              sjmp    gethx3          ;repeat until illegal value
0035 22                 89      gethx9: ret
                        90
                        91      ;*******************************************************************
                        92      ; Create space for a "found" bit which is set as the command line  *
                        93      ; is scanned.                                                      *
                        94      ;*******************************************************************
```

```
----                    95              rseg    bitram
0000                    96      found:  dbit    1
                        97
                        98      ;*********************************************************************
                        99      ; INSRT - INSeRT nibble from A into 16-bit value pointed at by R1    *
                       100      ;                                                                    *
                       101      ;       enter:  ACC.0 to ACC.3 contains hex nibble                   *
                       102      ;               R1 points to 2-byte RAM location                     *
                       103      ;       exit:   16-bit value shifted left 4 and nibble in ACC        *
                       104      ;                inserted into least significant digit               *
                       105      ;                                                                    *
                       106      ;*********************************************************************
----                   107              rseg    eprom
                       108              using   0
0036 C007              109      insrt:  push    AR7             ;use R7 as counter
0038 C000              110              push    AR0             ;use R0 to point to lsb
003A A801              111              mov     r0,1            ;R0 <-- R1
003C 08                112              inc     r0
003D 7F04              113              mov     r7,#4           ;4 rotates
003F C4                114              swap    a
0040 33                115      insrt2: rlc     a
0041 C0E0              116              push    acc
0043 E6                117              mov     a,@r0
0044 33                118              rlc     a
0045 F6                119              mov     @r0,a
0046 E7                120              mov     a,@r1
0047 33                121              rlc     a
0048 F7                122              mov     @r1,a
0049 D0E0              123              pop     acc
004B DFF3              124              djnz    r7,insrt2
004D D000              125              pop     AR0             ;R0
004F D007              126              pop     AR7             ;R7
0051 22                127              ret
                       128
                       129      ;*********************************************************************
                       130      ; SPACE - determine memory SPACE selector as per second character    *
                       131      ;         in command line as follows:                                *
                       132      ;                                                                    *
                       133      ;               B = bit space                                        *
                       134      ;               C = code space                                       *
                       135      ;               I = internal data                                    *
                       136      ;               R = Special Functions Registers                      *
                       137      ;               X = external data                                    *
                       138      ;                                                                    *
                       139      ;*********************************************************************
0052 C200       F      140      space:  clr     bit_sp          ;default: no space selected
0054 C200       F      141              clr     cde_sp          ; clear all bits
0056 C200       F      142              clr     int_sp
0058 C200       F      143              clr     reg_sp
005A C200       F      144              clr     ext_sp
005C E500       F      145              mov     a,linbuf+1      ;memory space selector in command
005E B44202            146              cjne    a,#'B',spac2    ;determine space and set selector bit
0061 D200       F      147              setb    bit_sp
0063 B44302            148      spac2:  cjne    a,#'C',spac3
0066 D200       F      149              setb    cde_sp
0068 B44902            150      spac3:  cjne    a,#'I',spac4
006B D200       F      151              setb    int_sp
006D B45202            152      spac4:  cjne    a,#'R',spac5
0070 D200       F      153              setb    reg_sp
0072 B45802            154      spac5:  cjne    a,#'X',spac6
0075 D200       F      155              setb    ext_sp
0077 22                156      spac6:  ret
                       157
```

```
                        158     ;*********************************************************************
                        159     ; Create space for the memory space selector bits in the 8031       *
                        160     ; bit-addressable space.  See link map for exact address of each of *
                        161     ; these bits.                                                       *
                        162     ;*********************************************************************
----                    163             rseg    bitram
0001                    164     bit_sp: dbit    1
0002                    165     cde_sp: dbit    1
0003                    166     int_sp: dbit    1
0004                    167     reg_sp: dbit    1
0005                    168     ext_sp: dbit    1
                        169             end

REGISTER BANK(S) USED: 0

ASSEMBLY COMPLETE, NO ERRORS FOUND
```

INPUT/OUTPUT ROUTINES

```
DOS 3.31 (038-N) MCS-51 MACRO ASSEMBLER, V2.2
OBJECT MODULE PLACED IN IO.OBJ
ASSEMBLER INVOKED BY:  C:\ASM51\ASM51.EXE IO.SRC EP


LOC  OBJ                LINE    SOURCE

                          1     $debug
                          2     $title(*** INPUT/OUTPUT ROUTINES ***)
                          3     $pagewidth(98)
                          4     $nopaging
                          5     $nosymbols
                          6     ;*********************************************************************
                          7     ;                                                                   *
                          8     ; IO.SRC - INPUT/OUTPUT SUBROUTINES                                 *
                          9     ;                                                         SM 03/91 *
                         10     ;*********************************************************************
  0007                   11     bel     equ     07H     ;ASCII bell code
  000D                   12     cr      equ     0DH     ;ASCII carriage return code
  001B                   13     esc     equ     1BH     ;ASCII escape code
  000A                   14     lf      equ     0AH     ;ASCII line feed code
  0003                   15     etx     equ     3       ;control-C (terminate input, get next cmd)
  0010                   16     dle     equ     10h     ;control-P (output to printer as well)
  007F                   17     del     equ     7FH     ;delete
  0004                   18     bs      equ     4       ;back space (VT100)
  0097                   19     TX_pin  bit     p1.7
  0096                   20     RX_pin  bit     p1.6
                         21
                         22             extrn   code(getcmd, htoa, touppr)
                         23             extrn   data(linbuf)
                         24             extrn   bit(p_bit, x13_bit)
                         25             extrn   number(endbuf, riot)
                         26             public  inchar, outchr, inline, outstr, outhex, out2hx
                         27             public  serial_io_using_interrupts, t_flag, r_flag
                         28             public  tb_count, r_idle
                         29
                         30     eprom   segment code
                         31     onchip  segment data
                         32     bitram  segment bit
                         33
```

```
                        34  ;*****************************************************************
                        35  ; The following buffers & flags are needed for character I/O using *
                        36  ; interrupts.  Interrupts are used only if jumper X13 is installed. *
                        37  ; Jumpers X9 and X10 should be removed and jumpers X11 and X12     *
                        38  ; should be installed.  This connects the RS232 TXD and RXD lines   *
                        39  ; to P1.7 and P1.6 (rather than to P3.1 and P3.0).                   *
                        40  ;*****************************************************************
----                    41                  rseg    onchip
0000                    42  t_buff:         ds      1       ;transmit_buffer (character to send)
0001                    43  r_buff:         ds      1       ;receive_buffer (character received)
0002                    44  ti_count:       ds      1       ;transmit_interrupt_count
0003                    45  ri_count:       ds      1       ;receive_interrupt_count
0004                    46  tb_count:       ds      1       ;transmit_bit_count
0005                    47  rb_count:       ds      1       ;receive_bit_count
----                    48                  rseg    bitram
0000                    49  t_flag:         dbit    1       ;1 = end of character transmission
0001                    50  r_flag:         dbit    1       ;1 = end of character reception
0002                    51  r_idle:         dbit    1       ;1 = idle state
                        52
                        53  ;*****************************************************************
                        54  ; OUTCHR -  OUTput CHaRacter to serial port                         *
                        55  ;           with odd parity added in ACC.7                          *
                        56  ;                                                                   *
                        57  ;  enter:   ASCII code in accumulator                               *
                        58  ;  exit:    character written to SBUF; <cr> sent as <cr><lf>; all   *
                        59  ;           registers intact                                        *
                        60  ;                                                                   *
                        61  ;*****************************************************************
----                    62          rseg    eprom
0000 C0E0               63  outchr: push    acc
0002 A2D0               64  nl:     mov     c,p             ;add odd parity
0004 B3                 65          cpl     c
0005 92E7               66          mov     acc.7,c
0007 20000D      F      67          jb      x13_bit,out1    ;if X13 installed, use interrupts
000A 3000FD      F      68          jnb     t_flag,$        ;wait for transmitter ready
000D C2AB               69          clr     et1             ;begin "critical section"
000F C200        F      70          clr     t_flag          ;>>>  clear flag and ...
0011 F500        F      71          mov     t_buff,a        ;>>>  load data to transmit
0013 D2AB               72          setb    et1             ;end "critical section"
0015 8007               73          sjmp    out2
                        74                                  ;if X13 not installed, test TI
0017 3099FD             75  out1:   jnb     ti,$            ;wait for transmitter ready
001A C299               76          clr     ti              ;clear TI flag
001C F599               77          mov     sbuf,a          ;done!
001E C2E7               78  out2:   clr     acc.7           ;remove parity bit
0020 B40D04             79          cjne    a,#cr,out3
0023 740A               80          mov     a,#lf
0025 80DB               81          jmp     nl              ;if <cr>, send <lf> as well
0027 D0E0               82  out3:   pop     acc
0029 300003      F      83          jnb     p_bit,out4      ;if printer control = off, exit
002C 120000      F      84          call    pchar           ;if on, print character
002F 22                 85  out4:   ret
                        86
                        87  ;*****************************************************************
                        88  ; INCHR -   INput CHaRacter from serial port                        *
                        89  ;                                                                   *
                        90  ;  enter:   no conditions                                           *
                        91  ;  exit:    ASCII code in ACC.0 to ACC.6; ACC.7 cleared; control-C  *
                        92  ;           aborts to prompt                                        *
                        93  ;                                                                   *
                        94  ;*****************************************************************
0030 20000D      F      95  inchar: jb      x13_bit,in1     ;if X13 installed, use interrupts
0033 3000FD      F      96          jnb     r_flag,$        ;wait for receive_flag to be set
0036 C2AB               97          clr     et1             ;begin "critical section"
```

```
0038 C200       F     98          clr    r_flag          ;>>> clear receive_flag and ...
003A E500       F     99          mov    a,r_buff        ;>>> read receive_buffer
003C D2AB            100          setb   et1             ;end "critical section"
003E 8007            101          sjmp   in2
                     102                                 ;if X13 not installed, test RI
0040 3098FD          103  in1:    jnb    ri,$            ;wait for receiver interrupt
0043 C298            104          clr    ri              ;clear RI flag
0045 E599            105          mov    a,sbuf          ;done!
0047 C2E7            106  in2:    clr    acc.7           ;clear parity bit (no error checking)
0049 B40303          107          cjne   a,#etx,in3      ;if control-C,
004C 020000     F    108          jmp    getcmd          ; warm start
004F 22              109  in3:    ret
                     110
                     111  ;*********************************************************************
                     112  ; INLINE -  INput LINE of characters                                 *
                     113  ;           line must end with <cr>; maximum size set by buflen;     *
                     114  ;           <bs> or <del> deletes previous character                 *
                     115  ;                                                                    *
                     116  ;  enter:   R0 points to input buffer in internal data RAM           *
                     117  ;  exit:    ASCII codes in internal RAM; 0 stored at end of line     *
                     118  ;  uses:    inchar, outchr                                           *
                     119  ;                                                                    *
                     120  ;*********************************************************************
0050 120000     F    121  inline: call   inchar          ;get character from console
0053 120000     F    122          call   touppr          ;convert to uppercase, if necessary
0056 B41004          123          cjne   a,#dle,inlin7   ;is it control-P?
0059 B200       F    124          cpl    p_bit           ;yes: toggle printing flag bit
005B 80F3            125          jmp    inline
005D 120000     F    126  inlin7: call   outchr          ;echo character to port
0060 F6              127          mov    @r0,a           ;put in buffer
0061 08              128          inc    r0
0062 B80008     F    129          cjne   r0,#endbuf,inlin2 ;overflow?
0065 900000     F    130          mov    dptr,#inlin8
0068 1100       F    131          acall  outstr
006A 020000     F    132          jmp    getcmd
006D B40402          133  inlin2: cjne   a,#bs,inlin3    ;back space?
0070 8003            134          sjmp   inlin5
0072 B47F0A          135  inlin3: cjne   a,#del,inlin6   ;delete?
0075 18              136  inlin5: dec    r0              ;yes: back up pointer twice
0076 18              137          dec    r0
0077 900000     F    138          mov    dptr,#backup    ;back up cursor
007A 120000     F    139          call   outstr
007D 80D1            140          jmp    inline
007F B40DCE          141  inlin6: cjne   a,#cr,inline    ;character = <cr> ?
0082 7600            142          mov    @r0,#0          ;yes: store 0
0084 22              143          ret                    ; and return
0085 1B              144  backup: db     esc,'[D',0      ;VT100 sequence to backup cursor
0086 5B44
0088 00
0089 07              145  inlin8: db     bel,cr,'Error: Command too long',cr,0
008A 0D
008B 4572726F
008F 723A2043
0093 6F6D6D61
0097 6E642074
009B 6F6F206C
009F 6F6E67
00A2 0D
00A3 00
                     146
                     147  ;*********************************************************************
                     148  ; OUTSTR -  OUTput a STRing of characters                            *
                     149  ;                                                                    *
                     150  ;  enter:   DPTR points to character string in external code         *
                     151  ;           memory; string must end with 00H                         *
```

```
                         152         ;  exit:    characters written to SBUF                              *
                         153         ;  uses:    outchr                                                  *
                         154         ;                                                                   *
                         155         ;*******************************************************************
00A4 E4                  156         outstr: clr     a
00A5 93                  157                 movc    a,@a+dptr       ;get ASCII code
00A6 6006                158                 jz      outst2          ;if last code, done
00A8 120000       F      159                 call    outchr          ;if not last code, send it
00AB A3                  160                 inc     dptr            ;point to next code
00AC 80F6                161                 sjmp    outstr         ; and get next character
00AE 22                  162         outst2: ret
                         163
                         164         ;*******************************************************************
                         165         ; OUT2HX - OUTput 2 HeX characters to serial port                   *
                         166         ; OUTHEX - OUTput 1 HEX character to serial port                    *
                         167         ;                                                                   *
                         168         ;  enter:   accumulator contains byte of data                        *
                         169         ;  exit:    nibbles converted to ASCII & sent out serial port;       *
                         170         ;           OUT2HX sends both nibbles; OUTHEX only sends lower       *
                         171         ;           nibble; all registers intact                             *
                         172         ;  uses:    outchr, htoa                                             *
                         173         ;                                                                   *
                         174         ;*******************************************************************
00AF C0E0                175         out2hx: push    acc             ;save data
00B1 C4                  176                 swap    a               ;send high nibble first
00B2 540F                177                 anl     a,#0FH          ;make sure upper nibble clear
00B4 120000       F      178                 call    htoa            ;convert hex nibble to ASCII
00B7 120000       F      179                 call    outchr          ;send to serial port
00BA D0E0                180                 pop     acc
00BC C0E0                181         outhex: push    acc             ;send low nibble
00BE 540F                182                 anl     a,#0FH
00C0 120000       F      183                 call    htoa
00C3 120000       F      184                 call    outchr
00C6 D0E0                185                 pop     acc
00C8 22                  186                 ret
                         187
                         188         ;*******************************************************************
                         189         ; PCHAR -   Print CHARacter                                         *
                         190         ;           send character to Centronics interface on 8155;         *
                         191         ;           implements handshaking for -ACK, BUSY, and -STROBE      *
                         192         ;                                                                   *
                         193         ;  enter:   ASCII code in accumulator                                *
                         194         ;  exit:    character written to 8155 Port A                         *
                         195         ;                                                                   *
                         196         ;*******************************************************************
  0092                   197         strobe  equ     92h             ;8051 port 1 interface line for
  0091                   198         busy    equ     91h             ; printer interface
  0090                   199         ack     equ     90h
                         200
00C9 C0E0                201         pchar:  push    acc
00CB C083                202                 push    dph
00CD C082                203                 push    dpl
00CF 2091FD              204         wait:   jb      busy,wait       ;wait for printer ready, i.e.,
00D2 3090FA              205                 jnb     ack,wait        ; BUSY = 0 AND -ACK = 1
00D5 900000       F      206                 mov     dptr,#riot + 1  ;8155 Port A address
00D8 C2E7                207                 clr     acc.7           ;clear 8th bit, if set
00DA B40D02              208                 cjne    a,#cr,pchar2    ;substitute <lf> for <cr>
00DD 740A                209                 mov     a,#lf
00DF F0                  210         pchar2: movx    @dptr,a         ;send data to printer
00E0 D292                211                 setb    strobe          ;toggle strobe bit
00E2 C292                212                 clr     strobe
00E4 D292                213                 setb    strobe
00E6 D082                214                 pop     dpl
00E8 D083                215                 pop     dph
```

```
00EA D0E0          216              pop     acc
00EC 22            217              ret
                   218
                   219      ;****************************************************************
                   220      ; The following code, which executes upon a Timer 1 interrupt,    *
                   221      ; implements a full duplex software UART.  Timer 1 interrupts occur *
                   222      ; at a rate of 8 times the baud rate, or every 1 / (8 x 0.0012) =  *
                   223      ; 104 microseconds.  For transmit, a bit is output on P1.7 every 8  *
                   224      ; interrupts.  For receive, a bit is read in P1.6 every 8          *
                   225      ; interrupts starting 12 interrupts after the start bit is         *
                   226      ; detected.  At 12 MHz, worse case execution time for this routine  *
                   227      ; is 53 us.  (This occurs for the last interrupt for simultaneous   *
                   228      ; transmit and receive operations.)                                *
                   229      ;****************************************************************
                   230      serial_io_using_interrupts:
00ED C0E0          231              push    acc
00EF C0D0          232              push    psw
00F1 A200     F    233              mov     c,r_idle        ;if r_idle = 1 & p1.6 = 1, no RX
00F3 8296          234              anl     c,RX_pin        ; activity, therefore ...
00F5 4027          235              jc      tx1             ; check for TX activity
00F7 30000A   F    236              jnb     r_idle,rx1      ;else, if r_idle = 0 & p1.6 = x,
                   237                                      ; reception in progress, check it out
                   238                                      ;else, r_idle = 1 & p1.6 = 0, there-
                   239                                      ; fore, start bit detected
00FA C200     F    240              clr     r_idle          ;clear r_idle (begin reception)
00FC 75000C   F    241              mov     ri_count,#12    ;11 interrupts to first data bit
00FF 750009   F    242              mov     rb_count,#9     ;9 bits received (includes stop bit)
0102 801A          243              sjmp    tx1
0104 D50017   F    244      rx1:    djnz    ri_count,tx1    ;decrement receive_interrupt_count
0107 750008   F    245              mov     ri_count,#8     ;if 8th interrupt, reload and
010A E500     F    246              mov     a,r_buff        ; read next serial bit on RXD (P1.6)
010C A296          247              mov     c,RX_pin
010E 13            248              rrc     a
010F F500     F    249              mov     r_buff,a
0111 D5000A   F    250              djnz    rb_count,tx1    ;decrement receive_bit_count
0114 750009   F    251              mov     rb_count,#9     ;if 9th bit received, done
0117 33            252              rlc     a               ;re-align bits (discard stop bit) and
0118 F500     F    253              mov     r_buff,a        ; save ASCII code in receive_buffer
011A D200     F    254              setb    r_idle
011C D200     F    255              setb    r_flag          ;done! (now check for TX activity)
011E 20002C   F    256      tx1:    jb      t_flag,tx4      ;if t_flag = 1, nothing to tramsmit
0121 E500     F    257              mov     a,tb_count      ;check transmit_bit_count
0123 B40B09        258              cjne    a,#11,tx2       ;if 11, first time here, therefore
0126 C297          259              clr     TX_pin          ; begin with start bit (p1.7 = 0)
0128 1500     F    260              dec     tb_count        ; decrement transmit_bit_count
012A 750008   F    261              mov     ti_count,#8     ;8 interrupts for each data bit
012D 801E          262              sjmp    tx4
012F D5001B   F    263      tx2:    djnz    ti_count,tx4    ;if transmit_interrupt_count not 0
0132 750008   F    264              mov     ti_count,#8     ; exit, otherwise reset count to 8 &
0135 E500     F    265              mov     a,t_buff        ; get next data bit to transmit
0137 D3            266              setb    c               ;(ensures 9th bit = stop bit)
0138 13            267              rrc     a               ;right rotate (sends LSB first)
0139 9297          268              mov     TX_pin,c        ;put bit on p1.7 (TXD)
013B F500     F    269              mov     t_buff,a        ;save in t_buff for next rotate
013D E500     F    270              mov     a,tb_count      ;check for stop bit (stretch count)
013F B40203        271              cjne    a,#2,tx3        ;if count = 1, stop bit just sent
0142 750010   F    272              mov     ti_count,#16    ;stretch count to 16 (2 stop bits)
0145 D50005   F    273      tx3:    djnz    tb_count,tx4    ;if last bit,
0148 75000B   F    274              mov     tb_count,#11    ; reset count and
014B D200     F    275              setb    t_flag          ; hoist flag (done!)
014D D0D0          276      tx4:    pop     psw
014F D0E0          277              pop     acc
0151 32            278              reti
                   279              end
```

```
REGISTER BANK(S) USED: 0

ASSEMBLY COMPLETE, NO ERRORS FOUND
```

CONVERSION SUBROUTINES

```
DOS 3.31 (038-N) MCS-51 MACRO ASSEMBLER, V2.2
OBJECT MODULE PLACED IN CONVRT.OBJ
ASSEMBLER INVOKED BY:  C:\ASM51\ASM51.EXE CONVRT.SRC EP


LOC  OBJ            LINE     SOURCE

                       1     $debug
                       2     $title(*** CONVERSION SUBROUTINES ***)
                       3     $pagewidth(98)
                       4     $nopaging
                       5     $nosymbols
                       6     $nolist                  ;next line contains $include(macros.src)
                      76     ;previous line contains $list
                      77     ;*******************************************************************
                      78     ;                                                                  *
                      79     ; CONVRT.SRC - CONVERSION SUBROUTINES                              *
                      80     ;                                                                  *
                      81     ;*******************************************************************
                      82             public  atoh, htoa, touppr, tolowr
                      83             extrn code(isalph)
                      84
                      85     eprom   segment code
----                  86             rseg    eprom
                      87
                      88     ;*******************************************************************
                      89     ; ATOH - Ascii TO Hex conversion                                   *
                      90     ;                                                                  *
                      91     ;       enter:  ASCII code in accumulator (assume hex character)   *
                      92     ;       exit:   hex nibble in ACC.0 - ACC.3; ACC.4 - ACC.7 cleared *
                      93     ;                                                                  *
                      94     ;*******************************************************************
0000 C2E7             95     atoh:   clr     acc.7           ;ensure parity bit is off
                      96                                     ;'0' to '9'?
0002 B43A00           97 +2          cjne  a,#'9'+1,$+3           ;JLE
0005 4002             98 +2          jc      atoh2
0007 2409             99             add     a,#9            ;no:  adjust for range A-F
0009 540F            100     atoh2:  anl     a,#0FH          ;yes: convert directly
000B 22              101             ret
                     102
                     103     ;*******************************************************************
                     104     ; HTOA - Hex TO Ascii conversion                                   *
                     105     ;                                                                  *
                     106     ;       enter:  hex nibble in ACC.0 to ACC.3                       *
                     107     ;       exit:   ASCII code in accumulator                          *
                     108     ;                                                                  *
                     109     ;*******************************************************************
000C 540F            110     htoa:   anl     a,#0FH          ;ensure upper nibble clear
                     111                                     ;'A' to 'F'?
000E B40A00          112 +2          cjne  a,#0AH,$+3             ;JLT
0011 4002            113 +2          jc      htoa2
0013 2407            114             add     a,#7            ;yes: add extra
0015 2430            115     htoa2:  add     a,#'0'          ;no:  convert directly
0017 22              116             ret
                     117
```

```
                                118   ;**********************************************************************
                                119   ; TOUPPR - convert character TO UPPeRcase                             *
                                120   ;                                                                      *
                                121   ;       enter:  ASCII code in accumulator                             *
                                122   ;       exit:   converted to uppercase if alphabetic (left as is      *
                                123   ;                otherwise)                                            *
                                124   ;       uses:   isalph                                                 *
                                125   ;                                                                      *
                                126   ;**********************************************************************
0018 120000    F                127   touppr: call    isalph          ;is character alphabetic?
001B 5002                       128           jnc     skip            ;no:  leave as is
001D C2E5                       129           clr     acc.5           ;yes: convert to uppercase by
001F 22                         130   skip:   ret                     ;      clearing bit 5
                                131
                                132   ;**********************************************************************
                                133   ; TOLOWR - convert character TO LOWeRcase                             *
                                134   ;                                                                      *
                                135   ;       enter:  ASCII code in accumulator                             *
                                136   ;       exit:   converter to lowercase if alphabetic (left as is      *
                                137   ;                otherwise)                                            *
                                138   ;       uses:   isalph                                                 *
                                139   ;                                                                      *
                                140   ;**********************************************************************
0020 120000    F                141   tolowr: call    isalph          ;is character alphabetic?
0023 5002                       142           jnc     skip2           ;no:  leave as is
0025 D2E5                       143           setb    acc.5           ;yes: convert to lowercase by setting
0027 22                         144   skip2:  ret                     ;      bit 5
                                145           end

REGISTER BANK(S) USED: 0

ASSEMBLY COMPLETE, NO ERRORS FOUND
```

LOAD INTEL HEX FILE

```
DOS 3.31 (038-N) MCS-51 MACRO ASSEMBLER, V2.2
OBJECT MODULE PLACED IN LOAD.OBJ
ASSEMBLER INVOKED BY:  C:\ASM51\ASM51.EXE LOAD.SRC EP

LOC  OBJ            LINE     SOURCE
                       1     $debug
                       2     $title(*** LOAD INTEL HEX FILE ***)
                       3     $pagewidth(98)
                       4     $nopaging
                       5     $nosymbols
                       6     ;**********************************************************************
                       7     ;                                                                      *
                       8     ; LOAD.SRC - LOAD INTEL HEX FILE COMMAND                               *
                       9     ;                                                                      *
                      10     ;  FORMAT: LO - load and return to MON51                               *
                      11     ;          LG - load and go to user program in RAM                     *
                      12     ;                                                                      *
                      13     ;**********************************************************************
                      14             extrn   code (inchar, outstr, atoh, getcmd,touppr)
                      15             extrn   data (linbuf,parmtr)
                      16             extrn   number (ram)
                      17             public  load
                      18
 0007                 19     bel     equ     07H             ;ASCII bell code
 000D                 20     cr      equ     0DH             ;ASCII carraige code
```

```
 000A                      21    lf       equ      0AH          ;ASCII line feed code
 001A                      22    eof      equ      1AH          ;ASCII control-z (end-of-file)
 0003                      23    cancel   equ      03H          ;ASCII control-c (cancel)
                           24    eprom    segment  code
----                       25             rseg     eprom
                           26
0000 900000      F         27    load:    mov      dptr,#mess1
0003 1100        F         28             acall    outstr
0005 1100        F         29    load1:   acall    inchar
0007 B41A02                30             cjne     a,#eof,skip1
000A 8031                  31             sjmp     done
000C B40303                32    skip1:   cjne     a,#cancel,skip2
000F 020000      F         33             jmp      getcmd
0012 B43AF0                34    skip2:   cjne     a,#':',load1   ;each record begins with a ':'
0015 7900                  35             mov      r1,#0          ;initialize checksum to 0
0017 1100        F         36             acall    gethex         ;get byte count from serial port
0019 F5F0                  37             mov      b,a            ;use B as byte counter
001B 6020                  38             jz       done           ;if B = 0, done
001D 05F0                  39    load2:   inc      b              ;no:
001F 1100        F         40             acall    gethex         ;get address high byte
0021 F583                  41             mov      dph,a
0023 1100        F         42             acall    gethex         ;get address low byte
0025 F582                  43             mov      dpl,a
0027 1100        F         44             acall    gethex         ;get record type (ignore)
0029 1100        F         45    load4:   acall    gethex         ;get data byte
002B F0                    46             movx     @dptr,a        ;store in 8031 external memory
002C A3                    47             inc      dptr
002D 15F0                  48             dec      b              ;repeat until count = 0
002F E5F0                  49             mov      a,b
0031 70F6                  50             jnz      load4
0033 E9                    51             mov      a,r1           ;checksum should be zero
0034 60CF                  52             jz       load1          ;if so, get next record
0036 900000      F         53    error:   mov      dptr,#mess4    ;if not, error
0039 1100        F         54             acall    outstr
003B 800F                  55             sjmp     wait
003D E500        F         56    done:    mov      a,linbuf+1     ;Load and Go command?
003F 1100        F         57             acall    touppr
0041 B44703                58             cjne     a,#'G',skip    ;no:  normal end to command
0044 020000      F         59             ljmp     ram            ;yes: go to user program
0047 900000      F         60    skip:    mov      dptr,#mess3
004A 1100        F         61             acall    outstr
004C 1100        F         62    wait:    acall    inchar         ;wait for eof before returning
004E B41AFB                63             cjne     a,#eof,wait
0051 020000      F         64             jmp      getcmd
                           65
                           66    ;*****************************************************************
                           67    ; Get two characters from serial port and form a hex byte.  Also  *
                           68    ; add byte to checksum in R1.                                      *
                           69    ;*****************************************************************
0054 1100        F         70    gethex:  acall    inchar         ;get first character
0056 1100        F         71             acall    atoh           ;convert to hex
0058 C4                    72             swap     a              ;put in upper nibble
0059 F8                    73             mov      r0,a           ;save it
005A 1100        F         74             acall    inchar         ;get second character
005C 1100        F         75             acall    atoh           ;convert to hex
005E 48                    76             orl      a,r0           ;OR with first nibble
005F FA                    77             mov      r2,a           ;save byte
0060 29                    78             add      a,r1           ;add byte to checksum
0061 F9                    79             mov      r1,a           ;restore checksum in R1
0062 EA                    80             mov      a,r2           ;retrieve byte
0063 22                    81             ret
0064 486F7374              82    mess1:   db       'Host download to SBC-51',cr
0068 20646F77
006C 6E6C6F61
```

```
0070 6420746F
0074 20534243
0078 2D3531
007B 0D
007C 5E5A203D         83          db      '^Z = end-of-file',cr,'^C = cancel',cr,0
0080 20656E64
0084 2D6F662D
0088 66696C65
008C 0D
008D 5E43203D
0091 2063616E
0095 63656C
0098 0D
0099 00
009A 656F6620         84   mess3: db      'eof - file downloaded OK',0
009E 2D206669
00A2 6C652064
00A6 6F776E6C
00AA 6F616465
00AE 64204F4B
00B2 00
00B3 07               85   mess4: db      bel,cr,'Error: ^Z Terminates',0
00B4 0D
00B5 4572726F
00B9 723A205E
00BD 5A205465
00C1 726D696E
00C5 61746573
00C9 00
                      86          end

REGISTER BANK(S) USED: 0

ASSEMBLY COMPLETE, NO ERRORS FOUND
```

DUMP MEMORY TO CONSOLE

```
DOS 3.31 (038-N) MCS-51 MACRO ASSEMBLER, V2.2
OBJECT MODULE PLACED IN DUMP.OBJ
ASSEMBLER INVOKED BY:  C:\ASM51\ASM51.EXE DUMP.SRC EP


LOC  OBJ            LINE     SOURCE

                       1     $debug
                       2     $title(*** DUMP MEMORY TO CONSOLE ***)
                       3     $pagewidth(98)
                       4     $nopaging
                       5     $nosymbols
                       6     ;*******************************************************************
                       7     ;                                                                  *
                       8     ; DUMP MEMORY COMMAND                                              *
                       9     ;                                                                  *
                      10     ;       FORMAT: D<char><start>,<end>                               *
                      11     ;                                                                  *
                      12     ;               D      - dump memory command                       *
                      13     ;               <char> - memory space selector as follows:         *
                      14     ;                          I = internal data RAM                   *
                      15     ;                          X = external data RAM                   *
                      16     ;                          C = code memory                         *
                      17     ;                          B = bit address space                   *
```

```
                                18      ;                               R = SFRs                                    *
                                19      ;                 <start> - starting address to dump                        *
                                20      ;                 <end>   - ending address to dump                          *
                                21      ;                                                                           *
                                22      ;****************************************************************************
                                23              extrn code (getcmd, outchr, htoa, out2hx, rsfr, isgrph)
                                24              extrn code (getval, inchar, outhex)
                                25              extrn data (parmtr, linbuf)
                                26              extrn bit (bit_sp, int_sp, reg_sp, x13_bit, r_flag)
                                27              public dump, outdat, outadd
                                28
                                29      onchip  segment data
                                30      eprom   segment code
  000D                          31      cr      equ     0DH
  0013                          32      xoff    equ     13H
  0020                          33      space   equ     ' '
                                34
----                            35              rseg    eprom
0000 850083       F             36      dump:   mov     dph,parmtr      ;starting address to dump
0003 850082       F             37              mov     dpl,parmtr+1
0006 750000       F             38      dump1:  mov     ascbuf,#0       ;initialize ASCII string to NULL
0009 7900         F             39              mov     r1,#ascbuf      ;R1 = pointer into ASCII string
000B 200005       F             40              jb      x13_bit,dump1a  ;check keyboard status, which is
000E 30000C       F             41              jnb     r_flag,dump2    ; r_flag if x13 jumper installed, or
0011 8003                       42              sjmp    dump1b
0013 309807                     43      dump1a: jnb     ri,dump2        ; ri if x13 jumper not installed
0016 1100         F             44      dump1b: acall   inchar
0018 B41302                     45              cjne    a,#xoff,dump2   ;if Control-S
001B 1100         F             46              acall   inchar          ; wait for next key
001D 1100         F             47      dump2:  acall   getval          ;read byte of data
001F C0E0                       48              push    acc
0021 1100         F             49              acall   outadd
0023 D0E0                       50      dump3:  pop     acc
0025 1100         F             51              acall   insert          ;add to ASCII buffer
0027 1100         F             52              acall   outdat          ;send to console
0029 40DB                       53              jc      dump1           ;if C, new line
002B 1100         F             54              acall   getval
002D C0E0                       55              push    acc
002F 80F2                       56              sjmp     dump3
                                57
                                58      ;****************************************************************************
                                59      ; OUTADD - OUTput ADDress to console                                        *
                                60      ;                                                                           *
                                61      ;       enter:  DPTR contains address; one memory space selector            *
                                62      ;                bit set                                                    *
                                63      ;       exit:   appropriate size (8-bit or 16-bit) address sent to          *
                                64      ;                console; followed by " = " if "set" command active         *
                                65      ;       uses:   out2hx, outchr                                              *
                                66      ;                                                                           *
                                67      ;****************************************************************************
0031 200020       F             68      outadd: jb      bit_sp,outad5   ;if bit, register, or internal space
0034 20001D       F             69              jb      reg_sp,outad5   ; only send 8-bit address
0037 20001A       F             70              jb      int_sp,outad5
003A E583                       71              mov     a,dph
003C 1100         F             72              acall   out2hx
003E E582                       73      outad2: mov     a,dpl
0040 1100         F             74              acall   out2hx
0042 E500         F             75              mov     a,linbuf        ;get command character
0044 B45308                     76              cjne    a,#'S',outad3
0047 7420                       77      outad4: mov     a,#space
0049 1100         F             78              acall   outchr
004B 743D                       79              mov     a,#'='          ;send '=' if "set" command
004D 1100         F             80              acall   outchr
004F 7420                       81      outad3: mov     a,#space
```

```
0051 1100      F      82             acall   outchr
0053 22               83             ret
0054 E583             84     outad5: mov     a,dph           ;address must be in range
0056 60E6             85             jz      outad2          ; 00H to FFH for bit, register,
0058 020000    F      86             jmp     getcmd          ; or internal memory spaces
                      87
                      88     ;*****************************************************************
                      89     ; OUTDAT - OUTput DATa to console                                *
                      90     ;                                                                *
                      91     ;       enter:  data in acc; one memory space selector bit set   *
                      92     ;       exit:   appropriate size value (1 bit or 8 bit) sent to  *
                      93     ;                console; ASCII buffer flushed if dump command and *
                      94     ;                end of boundary reached                         *
                      95     ;       uses:   outhex, outasc                                   *
                      96     ;                                                                *
                      97     ;*****************************************************************
005B C4               98     outdat: swap    a               ;this gets tricky!
005C 200002    F      99             jb      bit_sp,outdt2   ;if bit space,
005F 1100      F     100             acall   outhex          ;only send nibble
0061 C4              101     outdt2: swap    a
0062 1100      F     102             acall   outhex
0064 E500      F     103             mov     a,linbuf        ;if "set" command,
0066 6453            104             xrl     a,#'S'          ; don't check <end>
0068 60DD            105             jz      outad4          ; send ' = '
006A E583            106             mov     a,dph           ;DPTR = <end>?
006C 6500      F     107             xrl     a,parmtr+2
006E 7011            108             jnz     outdt3
0070 E582            109             mov     a,dpl
0072 6500      F     110             xrl     a,parmtr+3
0074 700B            111             jnz     outdt3
0076 200005    F     112             jb      reg_sp,outdt8   ;if bit or register space,
0079 200002    F     113             jb      bit_sp,outdt8   ; don't send ASCII
007C 1100      F     114             acall   outasc          ;yes: flush buffer
007E 020000    F     115     outdt8: jmp     getcmd          ;  & get next command
0081 A3              116     outdt3: inc     dptr            ;no:  onwards
0082 200012    F     117             jb      reg_sp,outdt7   ;if register space, one byte/line
0085 300006    F     118             jnb     bit_sp,outdt6   ;if bit space, check for 8-bit
0088 E582            119             mov     a,dpl           ; boundary
008A 5407            120             anl     a,#07H
008C 6009            121             jz      outdt7
008E E582            122     outdt6: mov     a,dpl           ;if internal, external, or code
0090 540F            123             anl     a,#0FH          ; space, check for 16-byte boundary
0092 C3              124             clr     c
0093 7007            125             jnz     outdt4
0095 1100      F     126             acall   outasc          ;if there, flush ASCII buffer
0097 740D            127     outdt7: mov     a,#cr
0099 1100      F     128             acall   outchr
009B D3              129             setb    c
009C 22              130     outdt4: ret
                     131
                     132     ;*****************************************************************
                     133     ; INSERT - INSERT ASCII code into buffer                         *
                     134     ;                                                                *
                     135     ;       enter:  byte of data in accumulator                      *
                     136     ;       exit:   ASCII code in buffer ('.' substituted for control *
                     137     ;                codes)                                          *
                     138     ;       uses:   isgrph                                           *
                     139     ;                                                                *
                     140     ;*****************************************************************
009D C0E0            141     insert: push    acc
009F 1100      F     142             acall   isgrph          ;is it a displayable character?
00A1 4002            143             jc      insrt2          ;yes: leave as is
00A3 742E            144             mov     a,#'.'          ;no:  substitute period
00A5 F7              145     insrt2: mov     @r1,a
```

```
00A6 09                  146              inc     r1
00A7 7700                147              mov     @r1,#0          ;null character at end
00A9 D0E0                148              pop     acc
00AB 22                  149              ret
                         150
                         151       ;*****************************************************************
                         152       ; OUTASC - OUTput ASCii codes to console                         *
                         153       ;                                                                 *
                         154       ;        enter:  -                                                *
                         155       ;        exit:   buffer of ASCII graphic codes sent to console    *
                         156       ;                                                                 *
                         157       ;*****************************************************************
00AC 7420                158       outasc: mov     a,#space
00AE 1100       F        159              acall   outchr
00B0 7900       F        160              mov     r1,#ascbuf
00B2 E7                  161       out3:  mov     a,@r1
00B3 7001                162              jnz     out2
00B5 22                  163              ret
00B6 1100       F        164       out2:  acall   outchr
00B8 09                  165              inc     r1
00B9 80F7                166              sjmp    out3
                         167
                         168       ;*****************************************************************
                         169       ; Create a buffer in internal RAM to hold ASCII codes to be dumped *
                         170       ; to console for DUMP command.                                    *
                         171       ;*****************************************************************
----                     172              rseg    onchip
0000                     173       ascbuf: ds      17              ;ascii buffer
                         174              end

REGISTER BANK(S) USED: 0
```

READ AND WRITE SFRs

```
DOS 3.31 (038-N) MCS-51 MACRO ASSEMBLER, V2.2
OBJECT MODULE PLACED IN SFR.OBJ
ASSEMBLER INVOKED BY:  C:\ASM51\ASM51.EXE SFR.SRC EP


LOC  OBJ                 LINE      SOURCE

                            1      $debug
                            2      $title (*** READ AND WRITE SFRs ***)
                            3      $pagewidth(98)
                            4      $nopaging
                            5      $nosymbols
                            6      ;*****************************************************************
                            7      ;                                                                 *
                            8      ; SFR.SRC - READ AND WRITE SPECIAL FUNCTION REGISTERS             *
                            9      ;                                                                 *
                           10      ;*****************************************************************
                           11
                           12             public rsfr, wsfr
                           13
                           14      eprom   segment code
----                       15             rseg    eprom
                           16
                           17      ;*****************************************************************
                           18      ;                                                                 *
                           19      ; RSFR - Read Special Function Register                           *
                           20      ;                                                                 *
```

```
                       21         ;       enter:  R0 contains address of SFR to read              *
                       22         ;       exit:   accumulator contains value and C = 0            *
                       23         ;               if invalid SFR, C = 1                           *
                       24         ;                                                               *
                       25         ;****************************************************************
                       26
                       27         ; Let's get lazy and define a macro to do all the work.
                       28
                       29         rsfr:
0000 B88003            30 +2              cjne    r0,#80h,SKIP00      ;p0
0003 E580              31 +2              mov     a,80h
0005 22                32 +1              ret
                       33 +2      SKIP00:
0006 B88103            34 +2              cjne    r0,#81h,SKIP01      ;sp
0009 E581              35 +2              mov     a,81h
000B 22                36 +1              ret
                       37 +2      SKIP01:
000C B88203            38 +2              cjne    r0,#82h,SKIP02      ;dpl
000F E582              39 +2              mov     a,82h
0011 22                40 +1              ret
                       41 +2      SKIP02:
0012 B88303            42 +2              cjne    r0,#83h,SKIP03      ;dph
0015 E583              43 +2              mov     a,83h
0017 22                44 +1              ret
                       45 +2      SKIP03:
0018 B88803            46 +2              cjne    r0,#88h,SKIP04      ;tcon
001B E588              47 +2              mov     a,88h
001D 22                48 +1              ret
                       49 +2      SKIP04:
001E B88903            50 +2              cjne    r0,#89h,SKIP05      ;tmod
0021 E589              51 +2              mov     a,89h
0023 22                52 +1              ret
                       53 +2      SKIP05:
0024 B88A03            54 +2              cjne    r0,#8ah,SKIP06      ;tl0
0027 E58A              55 +2              mov     a,8ah
0029 22                56 +1              ret
                       57 +2      SKIP06:
002A B88B03            58 +2              cjne    r0,#8bh,SKIP07      ;tl1
002D E58B              59 +2              mov     a,8bh
002F 22                60 +1              ret
                       61 +2      SKIP07:
0030 B88C03            62 +2              cjne    r0,#8ch,SKIP08      ;th0
0033 E58C              63 +2              mov     a,8ch
0035 22                64 +1              ret
                       65 +2      SKIP08:
0036 B88D03            66 +2              cjne    r0,#8dh,SKIP09      ;th1
0039 E58D              67 +2              mov     a,8dh
003B 22                68 +1              ret
                       69 +2      SKIP09:
003C B89003            70 +2              cjne    r0,#90h,SKIP0A      ;p1
003F E590              71 +2              mov     a,90h
0041 22                72 +1              ret
                       73 +2      SKIP0A:
0042 B89803            74 +2              cjne    r0,#98h,SKIP0B      ;scon
0045 E598              75 +2              mov     a,98h
0047 22                76 +1              ret
                       77 +2      SKIP0B:
0048 B89903            78 +2              cjne    r0,#99h,SKIP0C      ;sbuf
004B E599              79 +2              mov     a,99h
004D 22                80 +1              ret
                       81 +2      SKIP0C:
004E B8A003            82 +2              cjne    r0,#0a0h,SKIP0D     ;p2
0051 E5A0              83 +2              mov     a,0a0h
0053 22                84 +1              ret
```

```
                          85 +2  SKIP0D:
0054 B8A803               86 +2          cjne   r0,#0a8h,SKIP0E       ;ie
0057 E5A8                 87 +2          mov     a,0a8h
0059 22                   88 +1          ret
                          89 +2  SKIP0E:
005A B8B003               90 +2          cjne   r0,#0b0h,SKIP0F       ;p3
005D E5B0                 91 +2          mov     a,0b0h
005F 22                   92 +1          ret
                          93 +2  SKIP0F:
0060 B8B803               94 +2          cjne   r0,#0b8h,SKIP10       ;ip
0063 E5B8                 95 +2          mov     a,0b8h
0065 22                   96 +1          ret
                          97 +2  SKIP10:
0066 B8D003               98 +2          cjne   r0,#0d0h,SKIP11       ;psw
0069 E5D0                 99 +2          mov     a,0d0h
006B 22                  100 +1          ret
                         101 +2  SKIP11:
006C B8E003              102 +2          cjne   r0,#0e0h,SKIP12       ;acc
006F E5E0                103 +2          mov     a,0e0h
0071 22                  104 +1          ret
                         105 +2  SKIP12:
0072 B8F003              106 +2          cjne   r0,#0f0h,SKIP13       ;b
0075 E5F0                107 +2          mov     a,0f0h
0077 22                  108 +1          ret
                         109 +2  SKIP13:
0078 D3                  110             setb    c                  ;C = 1 if invalid SFR
0079 22                  111             ret
                         112
                         113     ;*******************************************************************
                         114     ;                                                                  *
                         115     ; WSFR - Write Special Function Register                           *
                         116     ;                                                                  *
                         117     ;        enter:  R0 contains address of SFR                        *
                         118     ;                accumulator contains value to write               *
                         119     ;        exit:   value written to SFR and C = 0                    *
                         120     ;                if invalid SFR, C = 1                             *
                         121     ;                                                                  *
                         122     ;*******************************************************************
                         123
                         124     wsfr:
007A B88003              125 +2          cjne   r0,#80h,SKIP14        ;p0
007D F580                126 +2          mov     80h,a
007F 22                  127 +1          ret
                         128 +2  SKIP14:
0080 B88103              129 +2          cjne   r0,#81h,SKIP15        ;sp
0083 F581                130 +2          mov     81h,a
0085 22                  131 +1          ret
                         132 +2  SKIP15:
0086 B88203              133 +2          cjne   r0,#82h,SKIP16        ;dpl
0089 F582                134 +2          mov     82h,a
008B 22                  135 +1          ret
                         136 +2  SKIP16:
008C B88303              137 +2          cjne   r0,#83h,SKIP17        ;dph
008F F583                138 +2          mov     83h,a
0091 22                  139 +1          ret
                         140 +2  SKIP17:
0092 B88803              141 +2          cjne   r0,#88h,SKIP18        ;tcon
0095 F588                142 +2          mov     88h,a
0097 22                  143 +1          ret
                         144 +2  SKIP18:
0098 B88903              145 +2          cjne   r0,#89h,SKIP19        ;tmod
009B F589                146 +2          mov     89h,a
009D 22                  147 +1          ret
                         148 +2  SKIP19:
```

```
009E B88A03      149 +2            cjne   r0,#8ah,SKIP1A      ;tl0
00A1 F58A        150 +2            mov     8ah,a
00A3 22          151 +1            ret
                 152 +2  SKIP1A:
00A4 B88B03      153 +2            cjne   r0,#8bh,SKIP1B      ;tl1
00A7 F58B        154 +2            mov     8bh,a
00A9 22          155 +1            ret
                 156 +2  SKIP1B:
00AA B88C03      157 +2            cjne   r0,#8ch,SKIP1C      ;th0
00AD F58C        158 +2            mov     8ch,a
00AF 22          159 +1            ret
                 160 +2  SKIP1C:
00B0 B88D03      161 +2            cjne   r0,#8dh,SKIP1D      ;th1
00B3 F58D        162 +2            mov     8dh,a
00B5 22          163 +1            ret
                 164 +2  SKIP1D:
00B6 B89003      165 +2            cjne   r0,#90h,SKIP1E      ;p1
00B9 F590        166 +2            mov     90h,a
00BB 22          167 +1            ret
                 168 +2  SKIP1E:
00BC B89803      169 +2            cjne   r0,#98h,SKIP1F      ;scon
00BF F598        170 +2            mov     98h,a
00C1 22          171 +1            ret
                 172 +2  SKIP1F:
00C2 B89903      173 +2            cjne   r0,#99h,SKIP20      ;sbuf
00C5 F599        174 +2            mov     99h,a
00C7 22          175 +1            ret
                 176 +2  SKIP20:
00C8 B8A003      177 +2            cjne   r0,#0a0h,SKIP21     ;p2
00CB F5A0        178 +2            mov     0a0h,a
00CD 22          179 +1            ret
                 180 +2  SKIP21:
00CE B8A803      181 +2            cjne   r0,#0a8h,SKIP22     ;ie
00D1 F5A8        182 +2            mov     0a8h,a
00D3 22          183 +1            ret
                 184 +2  SKIP22:
00D4 B8B003      185 +2            cjne   r0,#0b0h,SKIP23     ;p3
00D7 F5B0        186 +2            mov     0b0h,a
00D9 22          187 +1            ret
                 188 +2  SKIP23:
00DA B8B803      189 +2            cjne   r0,#0b8h,SKIP24     ;ip
00DD F5B8        190 +2            mov     0b8h,a
00DF 22          191 +1            ret
                 192 +2  SKIP24:
00E0 B8D003      193 +2            cjne   r0,#0d0h,SKIP25     ;psw
00E3 F5D0        194 +2            mov     0d0h,a
00E5 22          195 +1            ret
                 196 +2  SKIP25:
00E6 B8E003      197 +2            cjne   r0,#0e0h,SKIP26     ;acc
00E9 F5E0        198 +2            mov     0e0h,a
00EB 22          199 +1            ret
                 200 +2  SKIP26:
00EC B8F003      201 +2            cjne   r0,#0f0h,SKIP27     ;b
00EF F5F0        202 +2            mov     0f0h,a
00F1 22          203 +1            ret
                 204 +2  SKIP27:
00F2 D3          205               setb    c                ;if reached here, invalid
00F3 22          206               ret                      ;SFR, set carry flag
                 207               end


REGISTER BANK(S) USED: 0

ASSEMBLY COMPLETE, NO ERRORS FOUND
```

```
DOS 3.31 (038-N) MCS-51 MACRO ASSEMBLER, V2.2
OBJECT MODULE PLACED IN IS.OBJ
ASSEMBLER INVOKED BY:  C:\ASM51\ASM51.EXE IS.SRC EP


LOC  OBJ            LINE     SOURCE
                       1     $title (*** IS ROUTINES ***)
                       2     $debug
                       3     $pagewidth (98)
                       4     $nopaging
                       5     $nosymbols
                       6     $nolist                   ;next line contains $include(macros.src)
                      76     ;previous line contains $list
                      77     ;*******************************************************************
                      78     ;                                                                  *
                      79     ; IS.SRC - IS SUBROUTINES                                          *
                      80     ;                                                                  *
                      81     ; These subroutines test a byte to see if it matches a certain    *
                      82     ; condition.  If so, they return with the carry flag set; if not, *
                      83     ; they return with the carry flag clear.  These subroutines only  *
                      84     ; alter the carry flag; the accumulator is left intact.           *
                      85     ;                                                                  *
                      86     ;*******************************************************************
                      87
                      88             public  isgrph, ishex, isdig, isalph
                      89     eprom   segment code
----                  90             rseg    eprom
                      91
                      92     ;*******************************************************************
                      93     ;                                                                  *
                      94     ; ISGRPH - IS the byte an ascii GRaPHic code (i.e., in the range  *
                      95     ;          20H to 7EH)                                             *
                      96     ;                                                                  *
                      97     ;       enter:  ASCII code in accumulator                          *
                      98     ;       exit:   C = 1 if code is ASCII graphic character           *
                      99     ;               C = 0 if code is ASCII control character           *
                     100     ;                                                                  *
                     101     ;*******************************************************************
0000 B42000          102     isgrph: cjne    a,#20h,$+3      ;set C if < space
0003 10D703          103             jbc     cy,isgrp2       ;if set, clear and return
0006 B47F00          104             cjne    a,#7fh,$+3      ;set C, if graphic
0009 22              105     isgrp2: ret
                     106
                     107     ;*******************************************************************
                     108     ;                                                                  *
                     109     ; ISHEX - IS character ascii HEX?                                  *
                     110     ;                                                                  *
                     111     ;       enter:  ASCII code in ACC                                  *
                     112     ;       exit:   C = 1 if in range 0-9, a-f, or A-F                 *
                     113     ;               C = 0 otherwise                                    *
                     114     ;       uses:   isdigt                                             *
                     115     ;                                                                  *
                     116     ;*******************************************************************
000A C0E0            117     ishex:  push    acc
000C 120000     F    118             call    isdig
000F 400B            119             jc      skip            ;if digit, then ishex = true
0011 D2E5            120             setb    acc.5           ;convert to lowercase
0013 B46100          121             cjne    a,#'a',$+3      ;C = 1 if < 'a'
0016 10D703          122             jbc     cy,skip         ;if 1, clear and return
0019 B46700          123             cjne    a,#'f'+1,$+3    ;carry set if hex
001C D0E0            124     skip:   pop     acc
001E 22              125             ret
```

```
                       126
                       127       ;*************************************************************************
                       128       ;                                                                        *
                       129       ; ISDIG - IS ascii code a DIGit?                                         *
                       130       ;                                                                        *
                       131       ;        enter:  ASCII code in accumulator                               *
                       132       ;        exit:   C = 1 if code is character in range 0-9                 *
                       133       ;                C = 0 otherwise                                         *
                       134       ;                                                                        *
                       135       ;*************************************************************************
001F B43000            136       isdig:  cjne    a,#'0',$+3          ;carry set if < 0
0022 10D703            137               jbc     cy,skip2            ;if set, clear and return
0025 B43A00            138               cjne    a,#'9'+1,$+3        ;carry set if digit
0028 22                139       skip2:  ret
                       140
                       141       ;*************************************************************************
                       142       ;                                                                        *
                       143       ; ISALPH - IS ascii code an ALPHabetic character                         *
                       144       ;                                                                        *
                       145       ;        enter:  ASCII code in accumulator                               *
                       146       ;        exit:   C = 1 if code is character in range a-z or A-Z          *
                       147       ;                C = 0 otherwise                                         *
                       148       ;                                                                        *
                       149       ;*************************************************************************
0029 B46100            150 +2    isalph: cjne    a,#'a',$+3          ;JIR
002C 4005              151 +2            jc      SKIP00
002E B47B00            152 +2            cjne    a,#'z'+1,$+3
0031 400C              153 +2            jc      yes
                       154 +2    SKIP00:
0033 B44100            155 +2            cjne    a,#'A',$+3          ;JIR
0036 4005              156 +2            jc      SKIP01
0038 B45B00            157 +2            cjne    a,#'Z'+1,$+3
003B 4002              158 +2            jc      yes
                       159 +2    SKIP01:
003D C3                160               clr     c                   ;if reached here, can't be alpha
003E 22                161               ret
003F D3                162       yes:    setb    c                   ;must be alphabetic character
0040 22                163               ret
                       164               end

REGISTER BANK(S) USED: 0

ASSEMBLY COMPLETE, NO ERRORS FOUND
```

SET MEMORY TO VALUE

```
DOS 3.31 (038-N) MCS-51 MACRO ASSEMBLER, V2.2
OBJECT MODULE PLACED IN SET.OBJ
ASSEMBLER INVOKED BY:  C:\ASM51\ASM51.EXE SET.SRC EP


LOC  OBJ            LINE     SOURCE

                       1     $title (*** SET MEMORY TO VALUE ***)
                       2     $debug
                       3     $pagewidth (98)
                       4     $nopaging
                       5     $nosymbols
                       6     $nolist                   ;next line contains $include(macros.src)
                      76     ;previous line contains $list
```

```
                        77      ;************************************************************************
                        78      ;                                                                       *
                        79      ; SET.SRC - SET MEMORY TO VALUE                                         *
                        80      ;                                                                       *
                        81      ;       FORMAT: S<char><add><cr>                                        *
                        82      ;                                                                       *
                        83      ;               S       - set command                                   *
                        84      ;               <char>  - memory space selector                         *
                        85      ;                         B = bit address space                         *
                        86      ;                         R = SFRs                                      *
                        87      ;                         I = internal data RAM                         *
                        88      ;                         X = external data RAM                         *
                        89      ;                         C = code ROM                                  *
                        90      ;               <add>   - address to set                                *
                        91      ;               <cr>    - carriage return                               *
                        92      ;                                                                       *
                        93      ;       FOLLOWED BY:    <cr>    - examine next                          *
                        94      ;                       <value> - write data                            *
                        95      ;                       -       - examine previous                      *
                        96      ;                       Q       - quit                                  *
                        97      ;                       <sp>    - write same value into next            *
                        98      ;                                 address                               *
                        99      ;                                                                       *
                        100     ;************************************************************************
                        101             extrn   code (rsfr, wsfr, outadd, getpar, getcmd)
                        102             extrn   code (inline, outhex, outdat, outstr)
                        103             extrn   bit (bit_sp, cde_sp, int_sp, reg_sp, ext_sp)
                        104             extrn   data (parmtr, linbuf)
                        105             public  setcmd, getval
                        106
 0007                   107     bel     equ     07H
 000D                   108     cr      equ     0DH
                        109     eprom   segment code
----                    110             rseg    eprom
                        111
0000 850083     F       112     setcmd: mov     dph,parmtr      ;address to examine and/or set
0003 850082     F       113             mov     dpl,parmtr+1
0006 1100       F       114     set4:   acall   getval
0008 C0E0               115             push    acc
000A 1100       F       116             acall   outadd
000C D0E0               117             pop     acc
000E 1100       F       118             acall   outdat
0010 7800       F       119             mov     r0,#linbuf+2    ;start address of data
0012 1100       F       120             acall   inline
0014 E500       F       121             mov     a,linbuf+2      ;check response character
0016 B45103             122             cjne    a,#'Q',set6     ;if 'Q', quit set command
0019 020000     F       123             ljmp    getcmd
001C B42D0F             124     set6:   cjne    a,#'-',set7     ;if '-', decrement address pointer
001F C0E0               125 +1          push    acc             ;DEC_DPTR
0021 1582               126 +1          dec     dpl
0023 E582               127 +1          mov     a,dpl
0025 B4FF02             128 +2          cjne    a,#0FFH,SKIP00
0028 1583               129 +1          dec     dph
002A D0E0               130 +2  SKIP00: pop     acc
002C 80D8               131             sjmp    set4
002E B40D03             132     set7:   cjne    a,#0dh,set8     ;if <cr>, examine next location
0031 A3                 133             inc     dptr
0032 80D2               134             sjmp    set4
0034 1100       F       135     set8:   acall   getpar          ;otherwise a value has been entered
0036 E500       F       136             mov     a,parmtr+1      ;convert value and
0038 1100       F       137             acall   putval          ; put it into memory
003A A3                 138             inc     dptr            ;check out next location
003B 80C9               139             sjmp    set4
                        140
```

```
                     141      ;*********************************************************************
                     142      ; PUTVAL - PUT VALue into memory                                      *
                     143      ;                                                                     *
                     144      ;        enter:  DPTR contains address; ACC contains value; one       *
                     145      ;                 memory space selector bit set as follows:          *
                     146      ;                  bit_sp = 1 for bit address space                  *
                     147      ;                  cde_sp = 1 for code ROM                           *
                     148      ;                  int_sp = 1 for internal data RAM                  *
                     149      ;                  reg_sp = 1 for SFRs                               *
                     150      ;                  ext_sp = 1 for external data RAM                  *
                     151      ;                If bit address space selected then                  *
                     152      ;                  R4 = byte address                                 *
                     153      ;                  R5 = read byte value                              *
                     154      ;                (register bank 0 assumed)                           *
                     155      ;                Bit will be inserted into read byte                 *
                     156      ;                  value and byte value will be                      *
                     157      ;                  written back to byte address                      *
                     158      ;        exit:   value written into memory                           *
                     159      ;                                                                     *
                     160      ;*********************************************************************
003D A882            161      putval: mov     r0,dpl          ;put address in R0 also
003F 300002     F    162              jnb     int_sp,put2
0042 F6              163              mov     @r0,a           ;internal data RAM
0043 22              164              ret
0044 300002     F    165      put2:   jnb     ext_sp,put3
0047 F0              166              movx    @dptr,a         ;external data RAM
0048 22              167              ret
0049 300001     F    168      put3:   jnb     cde_sp,put4
004C 22              169              ret                     ;sorry, can't write into code space
004D 30001E     F    170      put4:   jnb     bit_sp,put5
0050 5401            171              anl     a,#1            ;make sure bits 1-7 = 0
0052 530007          172              anl     0,#7            ;reduce R0 to bit address
                     173                                      ;writing byte value
0055 7DFE            174              mov     r5,#0feh        ;build mask in R5
0057 08              175              inc     r0
0058 D802            176      put9:   djnz    r0,put7
005A 8006            177              sjmp    put8
005C 23              178      put7:   rl      a               ;adjust bit position
005D CD              179              xch     a,r5
005E 23              180              rl      a               ;adjust mask
005F CD              181              xch     a,r5
0060 80F6            182              sjmp    put9
0062 CC              183      put8:   xch     a,r4            ;byte value read earlier
0063 5D              184              anl     a,r5            ;turn off bit of interest
0064 4C              185              orl     a,r4            ;set if entered value = 1
                     186                                      ;byte value to write in accumulator
0065 A803            187              mov     r0,3            ;recover byte address
0067 B88000          188              cjne    r0,#80h,$+3     ;SFR address?
006A 5005            189              jnc     put10           ;yes if C = 1
006C F6              190              mov     @r0,a           ;Done!! Whew!
006D 22              191              ret
006E 300005     F    192      put5:   jnb     reg_sp,put6
0071 1100       F    193      put10:  acall   wsfr
0073 4001            194              jc      put6
0075 22              195              ret
0076 900000     F    196      put6:   mov     dptr,#put11     ;no memory space selected
0079 1100       F    197              acall   outstr          ;unrecoverable error, send message
007B 020000     F    198              ljmp    getcmd          ; and get another command
007E 07              199      put11:  db      bel,cr,'Error: no memory space selected',cr,0
007F 0D
0080 4572726F
0084 723A206E
0088 6F206D65
008C 6D6F7279
0090 20737061
```

```
0094 63652073
0098 656C6563
009C 746564
009F 0D
00A0 00
                          200
                          201    ;**********************************************************************
                          202    ; GETVAL - GET VALue from memory                                      *
                          203    ;                                                                     *
                          204    ;         enter:  DPTR contains address; One memory space selector    *
                          205    ;                   bit set (see PUTVAL)                              *
                          206    ;         exit:   Value in ACC, C = 0                                 *
                          207    ;                 If bit address space selected then                  *
                          208    ;                   R3 = byte address                                 *
                          209    ;                   R4 = read byte value                              *
                          210    ;                   (register bank 0 assumed)                         *
                          211    ;                 If invalid SRF address, advance                     *
                          212    ;                   DPTR until valid address found                    *
                          213    ;                                                                     *
                          214    ;**********************************************************************
00A1 A882                 215    getval: mov     r0,dpl          ;put addres in R0 also

00A3 300002     F         216            jnb     int_sp,get2
00A6 E6                   217            mov     a,@r0           ;internal data RAM
00A7 22                   218            ret                     ;C = 0
00A8 300002     F         219    get2:   jnb     ext_sp,get3
00AB E0                   220            movx    a,@dptr         ;external data RAM
00AC 22                   221            ret                     ;C = 0
00AD 300003     F         222    get3:   jnb     cde_sp,get4
00B0 E4                   223            clr     a
00B1 93                   224            movc    a,@a+dptr       ;code ROM
00B2 22                   225            ret                     ;C = 0
00B3 30002D     F         226    get4:   jnb     bit_sp,get5
                          227                                    ;So you want to read a bit
                          228                                    ;value, do you?
                          229                                    ;This gets tricky!!
00B6 5300F8               230            anl     0,#0f8h         ;reduce R0 to byte address
00B9 B88000               231            cjne    r0,#80h,$+3     ;if >= 80H, read SFR
00BC 400A                 232            jc      get6
00BE AB00                 233            mov     r3,0            ;save byte address in R3
00C0 1100       F         234            acall   rsfr            ;first get byte value
00C2 FC                   235            mov     r4,a            ;save byte value in R4
00C3 500D                 236            jnc     get8            ;now extract bit
00C5 A3                   237            inc     dptr            ;if C = 1, try next SFR
00C6 80D9                 238            sjmp    getval
00C8 E8                   239    get6:   mov     a,r0            ;internal RAM, therefore
00C9 03                   240            rr      a               ;build byte address in acc
00CA 03                   241            rr      a               ;by translating as shown
00CB 03                   242            rr      a
00CC D2E5                 243            setb    acc.5           ;eg: bit 35H is at byte address 26H
00CE F8                   244            mov     r0,a
00CF FB                   245            mov     r3,a            ;save byte address in R3
00D0 E6                   246            mov     a,@r0           ;read byte value
00D1 FC                   247            mov     r4,a            ;save byte value in R4
00D2 AF82                 248    get8:   mov     r7,dpl          ;extract bit by rotating
00D4 530707               249            anl     7,#7            ;right DPL mod 8 times
00D7 0F                   250            inc     r7
00D8 DF02                 251    get9:   djnz    r7,get10
00DA 8003                 252            sjmp    get11
00DC 03                   253    get10:  rr      a
00DD 80F9                 254            sjmp    get9
00DF 5401                 255    get11:  anl     a,#1            ;here's your bit
00E1 C3                   256            clr     c
00E2 22                   257            ret
```

```
00E3 300009    F       258      get5:   jnb     reg_sp,get7
00E6 1100      F       259      get13:  acall   rsfr
00E8 5004              260              jnc     get12           ;if invalid SFR address,
00EA A3                261              inc     dptr            ;increment DPTR & try again
00EB 08                262              inc     r0
00EC 80F8              263              sjmp    get13
00EE 22                264      get12:  ret
00EF 8085              265      get7:   jmp     put6            ;Error: no memory space selected
                       266              end

REGISTER BANK(S) USED: 0

ASSEMBLY COMPLETE, NO ERRORS FOUND
```

MACROS.SRC

```
;*********************************************************************
; JLT - Jump to "label" if accumulator Less Than "value"             *
;*********************************************************************
%*define(jlt(value,label))
        (cjne   a,#%value,$+3           ;JLT
        jc      %label)

;*********************************************************************
; JGT - Jump to "label" if accumulator Greater Than "value"          *
;*********************************************************************
%*define(jgt(value,label))
        (cjne   a,#%value+1,$+3         ;JGT
        jnc     %label)

;*********************************************************************
; JLE - Jump to "label" if accumulator Less than or Equal to "value"*
;*********************************************************************
%*define(jle(value,label))
        (cjne   a,#%value+1,$+3         ;JLE
        jc      %label)

;*********************************************************************
; JGE - Jump to "label" if accumulator Greater than or Equal to      *
;       "value"                                                      *
;*********************************************************************
%*define(jge(value,label))
        (cjne   a,#%value,$+3           ;JGE
        jnc     %label)

;*********************************************************************
; JOR - Jump to "label" if accumulator Out of Range of "lower_value"*
;       and "upper_value"                                            *
;*********************************************************************
%*define(jor(lower_value,upper_value,label))
        (cjne   a,#%lower_value,$+3     ;JOR
        jc      %label
        cjne    a,#%upper_value+1,$+3
        jnc     %label)
```

```
;*********************************************************************
; JIR - Jump to "label" if accumulator In Range of "lower_value" and*
;      "upper_value"                                                *
;*********************************************************************
%*define(jir(lower_value,upper_value,label)) local skip
        (cjne   a,#%lower_value,$+3         ;JIR
        jc      %skip
        cjne    a,#%upper_value+1,$+3
        jc      %label
%skip:  )

;*********************************************************************
; DECREMENT DPTR                                                    *
;*********************************************************************
%*define(dec    dptr) local skip
        (push   acc                 ;DEC_DPTR
        dec     dpl
        mov     a,dpl
        cjne    a,#0FFH,%skip
        dec     dph
%skip:  pop     acc)

;*********************************************************************
; PUSH DPTR ONTO STACK                                              *
;*********************************************************************
%*define(push   dptr)
        (push   dpl                 ;PUSH_DPTR
        push    dph)


;*********************************************************************
; POP DPTR FROM STACK                                               *
;*********************************************************************
%*define(pop    dptr)
        (pop    dph                 ;POP_DPTR
        pop     dpl)

;*********************************************************************
%*define (enable_register_bank(n))
        (%if (%n) then (setb   rs0) else (clr    rs0) fi
       %if (%n gt 1) then (setb   rs1) else (clr    rs1) fi)

;*********************************************************************
%*define (send_message(p))
        (push   dpl
        push    dph
        mov    dptr,#%p
        call   outstr
        pop    dph
        pop    dpl)
```

V12.M51

```
DATE : 04/21/91
DOS 3.31 (038-N) MCS-51 RELOCATOR AND LINKER V3.0, INVOKED BY:
C:\ASM51\RL51.EXE MAIN.OBJ,GETPAR.OBJ,IO.OBJ,CONVRT.OBJ,LOAD.OBJ,DUMP.OBJ,SFR.
>> OBJ,IS.OBJ,SET.OBJ TO V12 DATA(ONCHIP(30H))CODE(EPROM(0))


INPUT MODULES INCLUDED
  MAIN.OBJ(MAIN)
  GETPAR.OBJ(GETPAR)
  IO.OBJ(IO)
```

```
   CONVRT.OBJ(CONVRT)
   LOAD.OBJ(LOAD)
   DUMP.OBJ(DUMP)
   SFR.OBJ(SFR)
   IS.OBJ(IS)
   SET.OBJ(SET)


LINK MAP FOR V12(MAIN)


            TYPE     BASE       LENGTH     RELOCATION     SEGMENT NAME
            ----     ----       ------     ----------     ------------

            REG      0000H      0008H                     "REG BANK 0"
            DATA     0008H      0018H      ABSOLUTE
            BIT      0020H      0001H.5    UNIT           BITRAM
                     0021H.5    000EH.3                   *** GAP ***
            DATA     0030H      0033H      UNIT           ONCHIP

            CODE     0000H      0703H      UNIT           EPROM
```

Note: The following symbol table has been abreviated for this printout. Only symbols declared as "public" are shown.

```
SYMBOL TABLE FOR V12(MAIN)

VALUE          TYPE          NAME
-----          ----          ----

-------        MODULE        MAIN
N:0014H        PUBLIC        BUFLEN
D:0044H        PUBLIC        ENDBUF
C:0121H        PUBLIC        ERROR
C:00BCH        PUBLIC        GETCMD
D:0030H        PUBLIC        LINBUF
C:004CH        PUBLIC        NOTICE
B:0020H        PUBLIC        P_BIT
N:8000H        PUBLIC        RAM
X:0100H        PUBLIC        RIOT
C:2000H        PUBLIC        ROM
B:0020H.3      PUBLIC        X13_BIT
-------        ENDMOD        MAIN

-------        MODULE        GETPAR
B:0020H.5      PUBLIC        BIT_SP
B:0020H.6      PUBLIC        CDE_SP
B:0021H.1      PUBLIC        EXT_SP
C:0166H        PUBLIC        GETPAR
B:0020H.7      PUBLIC        INT_SP
D:0044H        PUBLIC        PARMTR
B:0021H        PUBLIC        REG_SP
-------        ENDMOD        GETPAR

-------        MODULE        IO
C:020EH        PUBLIC        INCHAR
C:022EH        PUBLIC        INLINE
C:028DH        PUBLIC        OUT2HX
```

```
C:01DEH         PUBLIC          OUTCHR
C:029AH         PUBLIC          OUTHEX
C:0282H         PUBLIC          OUTSTR
B:0021H.3       PUBLIC          R_FLAG
B:0021H.4       PUBLIC          R_IDLE
C:02CBH         PUBLIC          SERIAL_IO_USING_INTERRUPTS
B:0021H.2       PUBLIC          T_FLAG
D:0050H         PUBLIC          TB_COUNT
-------         ENDMOD          IO

-------         MODULE          CONVRT
C:0330H         PUBLIC          ATOH
C:033CH         PUBLIC          HTOA
C:0350H         PUBLIC          TOLOWR
C:0348H         PUBLIC          TOUPPR
-------         ENDMOD          CONVRT

-------         MODULE          LOAD
C:0358H         PUBLIC          LOAD
-------         ENDMOD          LOAD

-------         MODULE          DUMP
C:0422H         PUBLIC          DUMP
C:0453H         PUBLIC          OUTADD
C:047DH         PUBLIC          OUTDAT
-------         ENDMOD          DUMP

-------         MODULE          SFR
C:04DDH         PUBLIC          RSFR
C:0557H         PUBLIC          WSFR
-------         ENDMOD          SFR

-------         MODULE          IS
C:05FAH         PUBLIC          ISALPH
C:05F0H         PUBLIC          ISDIG
C:05D1H         PUBLIC          ISGRPH
C:05DBH         PUBLIC          ISHEX
-------         ENDMOD          IS

-------         MODULE          SET
C:06B3H         PUBLIC          GETVAL
C:0612H         PUBLIC          SETCMD
-------         ENDMOD          SET
```

V12.HEX

```
:10 0000 00 02 00 78 30 02 34 02 80 03 00 00 30 02 2F 02 80   A8
:10 0010 00 06 00 00 30 02 2A 02 80 09 00 00 30 02 25 02 00   9A
:10 0020 00 31 00 00 30 02 20 02 80 0F 00 00 30 02 1B 02 80   ED
:10 0030 00 12 20 03 03 02 02 CB 02 80 0C 02 20 03 02 20 06   DE
:10 0040 00 02 20 09 02 20 0C 02 20 0F 02 20 12 43 6F 70 79   57
:10 0050 00 72 69 67 68 74 20 28 63 29 20 49 2E 20 53 63 6F   D2
:10 0060 00 74 74 20 4D 61 63 4B 65 6E 7A 69 65 2C 20 31 39   5B
:10 0070 00 38 38 2C 20 31 39 39 31 A2 94 92 01 A2 93 92 02   5E
:10 0080 00 A2 95 92 03 20 01 03 02 20 00 20 03 17 C2 0B D2   85
:10 0090 00 0C D2 0A 75 50 0B 75 89 20 75 8D 98 D2 8E D2 AF   0F
:10 00A0 00 D2 AB 80 0B 75 98 52 75 8D F3 75 89 20 D2 8E C2   B4
:10 00B0 00 00 90 01 00 74 01 F0 90 01 3D 51 82 75 81 07 90   1C
:10 00C0 00 01 44 51 82 78 30 51 2E E5 30 B4 0D 03 02 06 12   FE
:10 00D0 00 B4 41 00 40 4C B4 5B 00 50 47 54 1F 14 23 F8 04   53
:10 00E0 00 90 00 ED 93 C0 E0 E8 93 C0 E0 31 66 22 01 21 01   69
```

```
:10 00F0 00 21 01 21 04 22 01 21 01 21 01 30 01 21 01 21 01   DD
:10 0100 00 21 01 21 03 58 01 21 01 21 01 21 01 21 01 21 01   A6
:10 0110 00 21 06 12 01 21 01 21 01 21 01 21 01 21 01 21 01   D9
:10 0120 00 21 90 01 4A 51 82 E5 30 31 DE E5 31 31 DE 80 8C   AB
:10 0130 00 74 BC C0 E0 74 00 C0 E0 C0 45 C0 44 22 0D 4D 4F   07
:10 0140 00 4E 35 31 00 0D 56 31 32 3E 00 07 0D 45 72 72 6F   4B
:10 0150 00 72 3A 20 49 6E 76 61 6C 69 64 20 43 6F 6D 6D 61   FF
:10 0160 00 6E 64 20 2D 20 00 78 32 79 44 7F 04 B6 2C 02 80   02
:10 0170 00 08 31 81 30 04 08 B6 2C 05 08 09 09 DF EE 31 B8   D2
:10 0180 00 22 C2 04 E6 B1 DB 50 13 D2 04 E4 F7 09 F7 19 E6   02
:10 0190 00 B1 DB 50 07 71 30 31 9C 08 80 F4 22 C0 07 C0 00   E9
:10 01A0 00 A8 01 08 7F 04 C4 33 C0 E0 E6 33 F6 E7 33 F7 D0   94
:10 01B0 00 E0 DF F3 D0 00 D0 07 22 C2 05 C2 06 C2 07 C2 08   A2
:10 01C0 00 C2 09 E5 31 B4 42 02 D2 05 B4 43 02 D2 06 B4 49   B1
:10 01D0 00 02 D2 07 B4 52 02 D2 08 B4 58 02 D2 09 22 C0 E0   B7
:10 01E0 00 A2 D0 B3 92 E7 20 03 0D 30 0A FD C2 AB C2 0A F5   DC
:10 01F0 00 4C D2 AB 80 07 30 99 FD C2 99 F5 99 C2 E7 B4 0D   96
:10 0200 00 04 74 0A 80 DB D0 E0 30 00 03 12 02 A7 22 20 03   2E
:10 0210 00 0D 30 0B FD C2 AB C2 0B E5 4D D2 AB 80 07 30 98   61
:10 0220 00 FD C2 98 E5 99 C2 E7 B4 03 03 02 00 BC 22 12 02   A2
:10 0230 00 0E 12 03 48 B4 10 04 B2 00 80 F3 12 01 DE F6 08   77
:10 0240 00 B8 44 08 90 02 67 51 82 02 00 BC B4 04 02 80 03   E3
:10 0250 00 B4 7F 0A 18 18 90 02 63 12 02 82 80 D1 B4 0D CE   C6
:10 0260 00 76 00 22 1B 5B 44 00 07 0D 45 72 72 6F 72 3A 20   C4
:10 0270 00 43 6F 6D 6D 61 6E 64 20 74 6F 6F 20 6C 6F 6E 67   7D
:10 0280 00 0D 00 E4 93 60 06 12 01 DE A3 80 F6 22 C0 E0 C4   F4
:10 0290 00 54 0F 12 03 3C 12 01 DE D0 E0 C0 E0 54 0F 12 03   F1
:10 02A0 00 3C 12 01 DE D0 E0 22 C0 E0 C0 83 C0 82 20 91 FD   7C
:10 02B0 00 30 90 FA 90 01 01 C2 E7 B4 0D 02 74 0A F0 D2 92   B4
:10 02C0 00 C2 92 D2 92 D0 82 D0 83 D0 E0 22 C0 E0 C0 D0 A2   2D
:10 02D0 00 0C 82 96 40 27 30 0C 0A C2 0C 75 4F 0C 75 51 09   E0
:10 02E0 00 80 1A D5 4F 17 75 4F 08 E5 4D A2 96 13 F5 4D D5   D9
:10 02F0 00 51 0A 75 51 09 33 F5 4D D2 0C D2 0B 20 0A 2C E5   69
:10 0300 00 50 B4 0B 09 C2 97 15 50 75 4E 08 80 1E D5 4E 1B   70
:10 0310 00 75 4E 08 E5 4C D3 13 92 97 F5 4C E5 50 B4 02 03   A3
:10 0320 00 75 4E 10 D5 50 05 75 50 0B D2 0A D0 D0 D0 E0 32   A2
:10 0330 00 C2 E7 B4 3A 00 40 02 24 09 54 0F 22 54 0F B4 0A   11
:10 0340 00 00 40 02 24 07 24 30 22 12 05 FA 50 02 C2 E5 22   9E
:10 0350 00 12 05 FA 50 02 D2 E5 22 90 03 BC 51 82 51 0E B4   2C
:10 0360 00 1A 02 80 31 B4 03 03 02 00 BC B4 3A F0 79 00 71   80
:10 0370 00 AC F5 F0 60 20 05 F0 71 AC F5 83 71 AC F5 82 71   DD
:10 0380 00 AC 71 AC F0 A3 15 F0 E5 F0 70 F6 E9 60 CF 90 04   25
:10 0390 00 0B 51 82 80 0F E5 31 71 48 B4 47 03 02 80 00 90   11
:10 03A0 00 03 F2 51 82 51 0E B4 1A FB 02 00 BC 51 0E 71 30   9F
:10 03B0 00 C4 F8 51 0E 71 30 48 FA 29 F9 EA 22 48 6F 73 74   73
:10 03C0 00 20 64 6F 77 6E 6C 6F 61 64 20 74 6F 20 53 42 43   BA
:10 03D0 00 2D 35 31 0D 5E 5A 20 3D 20 65 6E 64 2D 6F 66 2D   E2
:10 03E0 00 66 69 6C 65 0D 5E 43 20 3D 20 63 61 6E 63 65 6C   DC
:10 03F0 00 0D 00 65 6F 66 20 2D 20 66 69 6C 65 20 64 6F 77   3F
:10 0400 00 6E 6C 6F 61 64 65 64 20 4F 4B 00 07 0D 45 72 72   1E
:10 0410 00 6F 72 3A 20 5E 5A 20 54 65 72 6D 69 6E 61 74 65   20
:10 0420 00 73 00 85 44 83 85 45 82 75 52 00 79 52 20 03 05   07
:10 0430 00 30 0B 0C 80 03 30 98 07 51 0E B4 13 02 51 0E D1   CB
:10 0440 00 B3 C0 E0 91 53 D0 E0 91 BF 91 7D 40 DB D1 B3 C0   08
:10 0450 00 E0 80 F2 20 05 20 20 08 1D 20 07 1A E5 83 51 8D   39
:10 0460 00 E5 82 51 8D E5 30 B4 53 08 74 20 31 DE 74 3D 31   9E
:10 0470 00 DE 74 20 31 DE 22 E5 83 60 E6 02 00 BC C4 20 05   84
:10 0480 00 02 51 9A C4 51 9A E5 30 64 53 60 DD E5 83 65 46   B4
:10 0490 00 70 11 E5 82 65 47 70 0B 20 08 05 20 05 02 91 CE   9A
:10 04A0 00 02 00 BC A3 20 08 12 30 05 06 E5 82 54 07 60 09   4B
```

```
:10 04B0 00 E5 82 54 0F C3 70 07 91 CE 74 0D 31 DE D3 22 C0   94
:10 04C0 00 E0 B1 D1 40 02 74 2E F7 09 77 00 D0 E0 22 74 20   09
:10 04D0 00 31 DE 79 52 E7 70 01 22 31 DE 09 80 F7 B8 80 03   FE
:10 04E0 00 E5 80 22 B8 81 03 E5 81 22 B8 82 03 E5 82 22 B8   43
:10 04F0 00 83 03 E5 83 22 B8 88 03 E5 88 22 B8 89 03 E5 89   68
:10 0500 00 22 B8 8A 03 E5 8A 22 B8 8B 03 E5 8B 22 B8 8C 03   D4
:10 0510 00 E5 8C 22 B8 8D 03 E5 8D 22 B8 90 03 E5 90 22 B8   D2
:10 0520 00 98 03 E5 98 22 B8 99 03 E5 99 22 B8 A0 03 E5 A0   BD
:10 0530 00 22 B8 A8 03 E5 A8 22 B8 B0 03 E5 B0 22 B8 B8 03   F2
:10 0540 00 E5 B8 22 B8 D0 03 E5 D0 22 B8 E0 03 E5 E0 22 B8   50
:10 0550 00 F0 03 E5 F0 22 D3 22 B8 80 03 F5 80 22 B8 81 03   AE
:10 0560 00 F5 81 22 B8 82 03 F5 82 22 B8 83 03 F5 83 22 B8   8D
:10 0570 00 88 03 F5 88 22 B8 89 03 F5 89 22 B8 8A 03 F5 8A   A9
:10 0580 00 22 B8 8B 03 F5 8B 22 B8 8C 03 F5 8C 22 B8 8D 03   2F
:10 0590 00 F5 8D 22 B8 90 03 F5 90 22 B8 98 03 F5 98 22 B8   0B
:10 05A0 00 99 03 F5 99 22 B8 A0 03 F5 A0 22 B8 A8 03 F5 A8   ED
:10 05B0 00 22 B8 B0 03 F5 B0 22 B8 B8 03 F5 B8 22 B8 D0 03   1A
:10 05C0 00 F5 D0 22 B8 E0 03 F5 E0 22 B8 F0 03 F5 F0 22 D3   2D
:10 05D0 00 22 B4 20 00 10 D7 03 B4 7F 00 22 C0 E0 12 05 F0   3F
:10 05E0 00 40 0B D2 E5 B4 61 00 10 D7 03 B4 67 00 D0 E0 22   1D
:10 05F0 00 B4 30 00 10 D7 03 B4 3A 00 22 B4 61 00 40 05 B4   0F
:10 0600 00 7B 00 40 0C B4 41 00 40 05 B4 5B 00 40 02 C3 22   B3
:10 0610 00 D3 22 85 44 83 85 45 82 D1 B3 C0 E0 91 53 D0 E0   95
:10 0620 00 91 7D 78 32 51 2E E5 32 B4 51 03 02 00 BC B4 2D   D5
:10 0630 00 0F C0 E0 15 82 E5 82 B4 FF 02 15 83 D0 E0 80 D8   B8
:10 0640 00 B4 0D 03 A3 80 D2 31 66 E5 45 D1 4F A3 80 C9 A8   7C
:10 0650 00 82 30 07 02 F6 22 30 09 02 F0 22 30 06 01 22 30   F1
:10 0660 00 05 1E 54 01 53 00 07 7D FE 08 D8 02 80 06 23 CD   E5
:10 0670 00 23 CD 80 F6 CC 5D 4C A8 03 B8 80 00 50 05 F6 22   4F
:10 0680 00 30 08 05 B1 57 40 01 22 90 06 90 51 82 02 00 BC   0B
:10 0690 00 07 0D 45 72 72 6F 72 3A 20 6E 6F 20 6D 65 6D 6F   37
:10 06A0 00 72 79 20 73 70 61 63 65 20 73 65 6C 65 63 74 65   2E
:10 06B0 00 64 0D 00 A8 82 30 07 02 E6 22 30 09 02 E0 22 30   F1
:10 06C0 00 06 03 E4 93 22 30 05 2D 53 00 F8 B8 80 00 40 0A   59
:10 06D0 00 AB 00 91 DD FC 50 0D A3 80 D9 E8 03 03 03 D2 E5   04
:10 06E0 00 F8 FB E6 FC AF 82 53 07 07 0F DF 02 80 03 03 80   AD
:10 06F0 00 F9 54 01 C3 22 30 08 09 91 DD 50 04 A3 08 80 F8   A1
:03 0700 00 22 80 85                                           CF
:00 0000 01                                                    FF
```

APPENDIX H SOURCES OF 8051 DEVELOPMENT PRODUCTS

Allen Systems, 2151 Fairfax Road, Columbus, Ohio 43221

Product: FX-31 8052-BASIC SBC
Description: 8052 single-board computer with built-in BASIC interpreter
Product: CA-51
Description: 8051 cross assembler
Host Computer: IBM *PC* and compatibles
Product: DP-31/535
Description: Single-board computer based on Siemens 80535 CPU

Applied Microsystems Corp., 5020 148th Ave. N. E., P.O. Box 97002, Redmond, WA 98073-9702

Product: EC7000
Description: 8051 microcontroller emulator
Host Computer: IBM *PC* and compatibles

Aprotek, 1071-A Avenida Acaso, Camarillo, CA 93010

Product: PA8751
Description: 8751 programming adaptor
Features: Adapts 8751 EPROM microcontrollers to any EPROM programmer as a 2732

Avoset Systems, Inc., 804 South State St., Dover, Delaware 19901

Product: XASM51
Description: 8051 cross assembler
Host Computer: IBM *PC* and compatibles
Product: AVSIM51

Description: 8051 family simulator
Host Computer: IBM *PC* and compatibles

Binary Technology, Inc., P. O. Box 67, Meridan, NH 03770

Product: SIBEC-II
Description: 8052 single-board computer with built-in BASIC interpreter

Cybernetic Micro Systems, Inc., P.O. Box 3000, San Gregorio, CA 94074

Product: CYS8051
Description: 8051 cross assembler
Host Computer: IBM *PC* and compatibles
Features: Does not generate relocatable modules
Product: SIM8051
Description: 8051 simulator
Host Computer: IBM *PC* and compatibles
Product: CYP8051
Description: EPROM programmer for 8751
Interface: RS232 serial

Decmation, Inc., 3375 Scott Blvd., Suite #236, Santa Clara, CA 95054

Product: ASM51, PLM51, SIM51, etc.
Description: 8051 development software
Host Computer: VAX, PDP-11, IBM *PC* and compatibles

HiTech Equipment Corp., 9560 Black Mountain Road, San Diego, CA 92126

Product: 8051SIM
Description: 8051 simulator
Host Computer: IBM *PC/XT* or Z80 CP/M microcomputers

Huntsville Microsystems, Inc., P.O. Box 12415, 4040 South Memorial Parkway, Huntsville, AL 35802

Product: SBE-31, HMI-200-8051
Description: 8051 emulator
Interface: Serial
Host Computer: Various computers with CP/M or MS-DOS operating system

Logical Systems Corp., 6184 Teall Station, Syracuse, NY 13217

Product: UPA8751
Description: 8751 programming adaptor
Features: Adapts 8751 to 2732 socket for installation into any EPROM programmer

Product: SIM51
Description: 8051 simulator and debugger
Host Computer: IBM *PC* (MS-DOS, CP/M) and compatibles

Micromint, Inc., 4 Park Street, Vernon, CT 06066

Product: BCC52
Description: 8052 single-board computer with built-in interpreter

Nohau Corp., 51 East Campbell Ave., Suite 107E, Campbell, CA 95008

Product: EMV51-PC
Description: 8051 emulator
Host Computer: IBM *PC* and compatibles

Relational Memory Systems, Inc., P.O. Box 6719, San Jose, CA 95150

Product: ASM51, RLINK, RLOC, RLIB, OBJCON
Description: 8051 software development tools
Host Computer: IBM *PC* and compatibles

Scientific Engineering Laboratories, Inc., 104 Charles Street, Suite 143, Boston, MA 02114

Product: XPAS51
Description: 8051 PASCAL cross compiler
Host Computer: IBM *PC* and compatibles

Single Board Systems, P.O. Box 3788, Salem, OR 97306

Product: SBS-52
Description: 8052 single-board computer with built-in BASIC interpreter

Software Development Systems, Inc., 3110 Woodstock Drive, Downers Grove, Illinois 60515

Product: A51
Description: 8051 cross assembler
Host Computer: Various systems running MS-DOS, Xenix, or Unix

Universal Cross Assemblers, P.O. Box 384, Bedford, Nova Scotia B4A 2X3

Product: CROSS-16
Description: 8051 cross assembler
Host Computer: IBM *PC* and compatibles

URDA, Inc., 1811 Jancey St., Suite #200, Pittsburgh, PA 15206

Product: SBC-51
Description: PC board version of SBC-51 described in this text

Z-World, 2065 Martin Ave. #110, Santa Clara, CA 95050

Product: IBM *PC* co-processor

Description: Co-processor plugs into *PC* or *PC/AT;* runs Intel's ISIS operating system and software development tools

IC MANUFACTURERS

The following companies are manufacturers and/or developers of the 8051 and derivative ICs.

Intel Corp., 3065 Bowers Avenue, Santa Clara, CA 95051

Siemens Components, Inc., 2191 Laurelwood Road, Santa Clara, CA 95054

Signetics/Philips, 811 East Arques Ave., Sunnyvale, CA 94088-3409

Advanced Micro Devices, Inc., 901 Thompson Place, P.O. Box 3453, Sunnyvale, CA 94088-3453

Fujitsu Microelectronics, Inc., 3320 Scott Blvd., Santa Clara, CA 95054-3197

BIBLIOGRAPHY

Articles

Boyet, H., and R. Katz. The 8051 one-chip microcomputer. *Byte* (Dec. 1982): 288–311.

Ciarcia, S. Build the BASIC-52 computer/controller. *BYTE* (July 1986): 104–17.

———. Build an intelligent serial EPROM programmer. *BYTE* (Oct. 1986): 103–19.

———. Build a trainable infrared master controller. *BYTE* (Mar. 1987): 113–23.

———. Build a gray-scale video digitizer. Part I: Display/receiver. *BYTE* (May 1987): 95–106.

———. Build a gray-scale video digitizer. Part II: Digitizer/transmitter. *BYTE* (June 1987): 129–38.

———. Build the Circuit Cellar IC tester. Part I: Hardware. *BYTE* (Nov. 1987): 303–13.

Dinwiddle, G. An 8031 in-circuit emulator. *BYTE* (July 1986): 181–94.

Katausky, J. Built-in BASIC interpreter turns controller chip into versatile system core. *Electronic Design* (Dec. 13, 1984).

Koehler, B. Microcomputer doubles as Boolean processor. *Electronic Design,* vol. 28, no. 11 (1980).

Kornstein, H. MCS-51: A microcomputer optimized for control. *Electronic Product Design* (May 1981).

Messick, P., and J. Battle. Build a MIDI input for your Casio SK-1. *Keyboard* (Aug. 1987): 34–40.

Modares, J. Increased functions in chip result in lighter, less costly portable computer. *Design News* (Aug. 19, 1985).

Natarajan, K. S., and C. Eswarn. Design of a CCITT V.22 modem. *Microprocessors and Microsystems,* vol. 12, no. 9 (Nov. 1988): 532–35.

Schenker, J. Controller chip cuts keyboard redesign to weeks. *Electronics* (Jan. 21, 1988): 42E–42F.

Vaidya, D. M. Microsystem design with the 8052-BASIC microcontroller. *Microprocessors and Microsystems,* vol. 9, no. 8 (Oct. 1985): 405–11.

———. Software development for the 8052-BASIC microcontroller. *Microprocessors and Microsystems,* vol. 9, no. 10 (Dec. 1985): 481–85.

Williamson, T. Using the 8051 microcontroller with resonant transducers. *IEEE Transactions on Industrial Electronics,* vol. IE-32, no. 4 (Nov. 1985).

Books

Ayala, K. J. *The 8051 Microcontroller: Architecture, Programming, and Applications.* New York: West Publishing, 1991.

Boyet, H., and R. Katz. *The 8051–Programming, Interfacing, Applications.* New York: MTI Publications, 1981.

8-Bit Embedded Controllers (270645), Intel Corp., Santa Clara, CA, 1991.

MCS-51 Macro Assembler User's Guide (9800937-03), Intel Corp., Santa Clara, CA, 1983.

MCS-51 Utilities User's Guide for 8080/8085-Based Development Systems (121737-003), Intel Corp., Santa Clara, CA, 1983.

Stewart, J.W. *The 8051 Microntroller: Hardware, Software and Interfacing.* Englewood Cliffs, N.J.: Prentice Hall, 1993.

INDEX